PROBLEMS AND PROSPECTS OF ENVIRONMENT POLICY—INDIAN PERSPECTIVE

Edited by
M.S. Bhatt
Shahid Ashraf
Asheref Illiyan

AAKAR

PROBLEMS AND PROSPECTS OF ENVIRONMENT POLICY—
INDIAN PERSPECTIVE

Edited by

M.S. Bhatt
Shahid Ashraf
Asheref Illiyan

First Published 2008

ISBN 978-81-89833-59-6

Published by

AAKAR BOOKS

28 E Pocket IV, Mayur Vihar Phase I, Delhi-110 091
Phone : 011-2279 5505 Telefax : 011-2279 5641
aakarbooks@gmail.com; www.aakarbooks.com

Composed by
Limited Colors, Delhi-110 092

Printed at
Arpit Printographers
arpitprinto@yahoo.com

Acknowledgements

The present book is a collection of revised versions of the articles presented in two national seminars on 'National Environment Policy (NEP) 2006' organized by the Department of Economics, Jamia Millia Islamia, New Delhi during February and August 2007. Many people and organizations generously rendered their assistance and cooperation, first in the seminar arrangements and then in processing the proceedings in the form of a book. But for their help, it would have been difficult to translate promise into performance. We would like to place on record our gratitude to all such persons and institutions. In particular, the following deserve special mention.

Professor Mushirul Hasan, Vice-Chancellor, Jamia Millia Islamia, for his encouragement and excellent leadership at various stages of the seminar and publication of the book;

All participants in the aforementioned national seminar, particularly the contributors to the present volume, who took lot of pains to revise their papers as per our requirements;

All our colleagues in Department of Economics, Jamia Millia Islamia, for their unflagging support and conferring on us the honour of being the editors of the book.

Our students at the department for their valuable support and assistance.

The ministerial staff of Department of Economics - Abida Begum, Nazim, Shamsad Khan and Arif Ali for their back-up services;

Mr. K.K. Saxena of Aakar Books for his commendable work in getting the book printed in an excellent way;

Last but not least, Ministry of Environment and Forests (MoEF), Government of India; Nelson Mandela Centre for Peace and Conflict Resolution, Jamia Millia Islamia; Wetland International South Asia, for their financial help and collaboration.

M.S. Bhatt
Shahid Ashraf
Asheref Illiyan

Contributors

Ahmad, Furqan, Associate Research Professor, Indian Law Institute, New Delhi. Email: *furqan49ahmad@yahoo.com*

Ahmed, Farhan, Research Associate, India Infrastructure Publishing, B-17, Qutab Institutional Area, New Delhi-110016. Email: *ahmed_eco@rediffmail.com*

Ashraf, Shahid, Professor, Department of Economics, Jamia Millia Islamia, New Delhi-110025. Email: *sashraf2@rediffmail.com*

Azad, Naushad Ali, Professor, Department of Economics, Jamia Millia Islamia, New Delhi-110025. Email: *azad_jmi@yahoo.com*

Baig, M.A., Lecturer, Department of Economics, Jamia Millia Islamia, New Delhi-110025. Email: *mabaig76@yahoo.com*

Baig, T.A., Ravenshaw College, Cuttack, Orissa.

Bazlur Rahman Khan, Research Associate, Hamdard Study Circle, Delhi.

Bhatt, M.S., Professor and Head, Department of Economics, Jamia Millia Islamia, New Delhi-110025. Email: *ab1sul2@yahoo.com*

Biswas, Pradip Kumar, Reader, Department of Economics, College of Vocational Studies, Sheikh Sarai Phase II, New Delhi-110016. Email: *pkbiswas63@hotmail.com*

Chakraborty, Debashis, Assistant Professor, Indian Institute of Foreign Trade, IIFT Bhawan, B-21, Qutab Institutional Area, New Delhi-110016. Email: *debchakra@gmail.com*

Goel, Yugank, National Institute of Technology, Surat, Gujarat.

Govindrajalu, K., Reader in Economics, C.B.M. College, Kovaipudur, Coimbatore-641042. Email: *krishg2@yahoo.co.in*

Illiyan, Asheref, Lecturer, Department of Economics, Jamia Millia Islamia, New Delhi-110025. Email: *ashrafjamia@yahoo.com*

Karunakaran, C.E., Programme Coordinator (Environment), Centre for Ecology and Rural Development, Puducherry. Email: *cekarun@gmail.com*

Kaul, Sapna, Lecturer, Department of Economics, Kalindi College, University of Delhi.

Mukherjee, Sacchidananda, Research Scholar, Madras School of Economics, Gandhi Mandapam Road, Chennai-600 025, Tamil Nadu. Email: *sachs.mse@gmail.com*

Mukhija, Ketan, NALSAR University of Law, Hyderabad. Email: *ketanmukhija@gmail.com*

Nazrul Islam, A.K.M., ICCR Research Scholar, Department of Economics, Jamia Millia Islamia, New Delhi-110025. E-mail: *Nazrul2002@yahoo.com*

Prasad, Archana, Reader, Centre for Jawaharlal Nehru Studies, Jamia Millia Islamia, New Delhi-110025. Email: *archie.prasad@gmail.com*

Reddy, B. Sudhakara, Professor, IGIDR, Mumbai. Email: *sreddy@igidr.ac.in*

Sankar, U., Professor Emeritus, Madras School of Economics, Chennai. Email:*usankar@mse.ac.in*

Saravanan, Velayutham, Professor, Centre for Jawaharlal Nehru Studies, Jamia Millia Islamia, New Delhi-110 025. Email: *Saravanan.JMI@gmail.com*

Sodhi, Probhjit, Coordinator, UNDP, GEF, Small Grants Programme, New Delhi. Email: *prabhjot.sodhi@ceeindia.org*

Suhaib Ismail Mohammed, Bank of America, Gurgaon. Email: *shuhaibismail@rediffmail.com*

Sultan, Salma, PGT Economics, MAF Academy, Sector-62, Noida.

Tandon, Deepti, Lecturer in Economics, Delhi College of Arts and Commerce, University of Delhi. Email: *deep.tandon@gmail.com*

Verma, Sheetal, M. Phil Student, Centre for Economic Studies and Planning, Jawaharlal Nehru University, New Delhi-110067. Email: *sheetalverma1@gmail.com*

Contents

Macro Perspectives

Introduction

I

Economy-environment linkages have become very frail. The growing trade-off between the existing patterns of growth and ecological sustainability is considered the principal determinant of this frightening situation. The situation has become complex partly due to the very nature of the issues involved and partly due to the frame in which it has been debated. Its consequences have been widely documented. Among other things, growing concentration of wealth and marginalization of the poor, both within and across the countries, is bound to widen the prevailing conflict between growth and the quality of environment. The trade-off has to be minimized, if not eliminated. This is not possible without redefining the meaning and purpose of economic growth.

In spite of the growing number and frequency of international conferences/seminars, much needs to be done in terms of effective policies and their implementation. This does not mean that nothing positive has happened. Development vis-à-vis environment has become a serious research agenda during the last fifty years. This in turn has raised the levels of understanding of the issues involved. Experts from all the major disciplines of knowledge have analyzed the problems from their perspectives. Ways and means of integrating ecology and development in a symbiotic way have been discussed in a number of international conferences and organizations such as UN

Conference on Human Environment (Stockholm, 1972), World Conservation Strategy (1980), Our Common Future (1987), Caring for the Earth (1991), UN Conference on Environment and Development (1992), and The Rio Summit (1992). These have been associated with positive developments, like growth of environmental movements and exploration of new research paradigms. Cultural ecology, biodiversity and cultural diversity interface, and rediscovering the conventional ways of protecting the environment have emerged as new areas of research and teaching.

There is need for a paradigm shift in our approach to understand and analyze economy-environment relationship. This cannot be achieved without concerted efforts on all fronts – social, political, economic, educational, administrative, local, state, country and global levels. We have to evolve a framework in which economic challenges are solved in consonance with the environment and not at the cost of it. This is a challenging task, as "Economy–ecology bind us in ever tightening networks – economy and ecology must be completely integrated in decision making and law making processes' [Our Common Future, 1987][1]".

India is committed to make a tangible contribution to the global efforts to reduce trade-off between the existing pattern of growth and the quality of environment. Our Constitution imparts seminal importance to a clean environment [Articles 48 A and 51 A (g)]. A number of acts has been enacted and policies have been formulated in response to these Constitutional mandates. Appropriate institutional structures have been put in place. Many new policy instruments have been tried from time to time. Budgetary allocation has registered a marked increase. Civil society's involvement in protecting the quality of environment has also grown exponentially over the years.

Courts and other Constitutional bodies have pronounced some landmark judgments as a result of public interest litigations. Green movements have also become more organized and broad based. All these developments have indeed raised the level of awareness about the gravity of the situation and yielded positive results.

Be that as it may, the present state of environmental health in our country is far from satisfactory. We are facing rural as well as urban environmental problems, like loss of biodiversity and wildlife, land degradation, depletion of groundwater, over-exploitation of marine resources, degradation of ecosystems, air, water and noise pollution, over-increasing assimilative and stock wastes. Besides, other factors like overgrowing population, inadequate awareness about the importance of environmental capital and lack of effective enforcement mechanisms have led to this state of affairs. Moreover, globalization is purported to have an impact on the pace of deforestation, loss of biodiversity, water, air, noise pollution level, etc. The existing legal-cum policy framework has proved to be exceedingly inadequate and in many cases ineffective.

Against this background, the National Environmental Policy–2006 (henceforth NEP) assumes historical significance. It attempts to highlight the challenges posed by the degradation of the quality of environment at all levels and proposes policy frameworks and guidelines. It also imparts importance to the reduction of indoor air pollution; improved drinking water quality; sanitation and public health. Extensive awareness raising and education have been identified as the key strategies in the implementation of the policy. The specific objectives of the policy are: conservation of critical environmental resources, ensuring intra-generational and inter-generational equity, integration of environmental concerns into action, achieving efficiency in the use of environmental resources and environmental governance. Regardless of its novel and encouraging features, the NEP has several weaknesses in its policy guidelines as well as implementation strategies. For example, it has no clear long-term vision for the fast-growing Indian economy and does not propose an alternative imaginative framework to the present growth strategy. Traditional knowledge, cultural diversity, a holistic lifestyle, etc. have been globally accepted as very critical for the preservation of natural resources. Although the NEP has highlighted their importance in the preservation of environment, clear policy guidelines for their operationalization have not been formulated. Similarly, the role of the institutions, especially micro-level institutions, has not been adequately delineated. The policy has not given proper guidelines regarding the use of water resources, sharing of international and national river waters, saving the rivers from drying up, preservation of ground and rainwater for multiple uses. A clear policy and institutional arrangements for unforeseen natural disasters has also not been given enough priority in the policy.

NEP can still yield dividends, provided effective and appropriate implementation mechanisms are accorded timely and desired attention. These are critical inputs. Instruments of implementation, resource needs, an enabling institutional and legal framework, monitoring and evaluation mechanisms and back-up strategies are to be worked out with a lot of care and consideration. The role of the central, state, district/local governments, NGOs and civil society should be properly delineated and harmonized. Last but not east, the NEP has to be implemented in conjunction with other policy instruments, like Forest Policy, Industrial Policy, Trade Policy, Transport Policy, Population Policy, Energy Policy, Land Use Policy, Educational Policy, etc. Without this, there are bound to be overlaps and duplications causing loss of resources and even diluting the gains. This assumes added significance in the context of a decentralized development administration regime, which we have put in place in the country after the 73rd and 74th Amendment Acts.

NEP has already become an official policy document. Its litmus test lies in its implementation. Every country has to formulate its own enabling legal and administrative mechanism for the enforcement of environmental policy. In view of the enormity of the issues involved, agro-climatic conditions, biodiversity, cultural diversity, difference in the levels of development and the nature of the federal polity, this framework in the context of India should be formulated in a manner so that it works vertically within the economy and horizontally with the global economy. Implementing instruments range from Command and Control Instruments (CCI) to Market-Based Instruments (MIBs). Both have advantages and limitations. There are examples where both have worked. This should be kept in mind. The actual mix will ultimately depend upon the goals of economic development and socio-political framework. Generally, three management tools – technological restrictions, cooperative institutions and economic incentives – have been used by policy makers. Choosing any of these or a combination will depend upon their effectiveness, information requirements, type and nature of environmental quality. Any policy instrument should meet the requirements of efficiency, effectiveness, equity and flexibility. The required practical conditions are: an adequate information base, administrative capacity, strong institutional structures, regulatory mechanisms and political flexibility.

In creating an enabling environment for implementing the NEP, besides government machinery (at various levels) the civil society has to shoulder a critical role. This demands an implementation model wherein all stakeholders contribute in the successful implementation of the policy. Given the ground realities and challenges, the basic considerations in building this model should be as follows:

1. Economic and environmental systems are determined simultaneously. This requires a proper understanding of the working of the two systems. For example, we should keep in mind the mechanical underpinnings of the natural sciences and behavioural underpinnings of social sciences. Economy should be treated as a part of the earth's eco-system and it should be compatible with it. "The pre-eminent challenge is to design an eco-economy on the principles of ecology. A redesigned economy can be integrated into the eco-system, in a way that will stabilize the relationship between the two, enabling progress to continue."[2]
2. Environmental resources are fast becoming scarce and using them in a way involves an opportunity cost, which needs to be taken into consideration. This warrants great attention to the valuation of environmental costs and benefits. The techniques for achieving these are either not available or are inadequate. There is a need for using alternatives, like Contingent Valuation Method (CVM), Hedonic Pricing Method (HPM), Travel Cost Method (TCM), etc. on a much larger scale and on systematic basis. This cannot be done with the existing status and state of the database.
3. Both government and market interventions in the absence of required conditions can generate environmental bads, which can result in irreversible intra-generational and inter-generational damages. Therefore, the role of government and market should

be carefully examined. Neither should be taken for granted.

4. Environmental protection costs enormous resources – men, material and finances. These resources are very scarce and can hamper implementation. There is thus a need for a realistic assessment of resource needs and their availability.
5. Economic growth alone cannot solve all the problems (as suggested by some). In fact, it can add to and complicate the problems. Therefore, working of environment-economy linkages (both with reference to poverty and prosperity) should be the basis of any policy and its implementation.
6. Environmental problems are local, trans-local and global in nature, each having its own features and specificities. There is thus a need for context-specific instruments appropriate to each level.
7. Sustainability of economic development and possible sustainability rules/indicators must be based on both equity and efficiency considerations.
8. Implementation of rules and requirements should be worked out, keeping in consideration: eco-diversity, cultural diversity and differences in the levels of socio-economic development of the country.

This obviously speaks for informed debates on the various aspects of the NEP. We believe that these debates would help to understand its salient features and challenges ahead. Educational institutions at all levels have to shoulder great responsibilities in the optimal implementation of the NEP. This demands integration of environmental education with the mainstream educational curriculum at all levels.

Against this backdrop, Department of Economics, Jamia Millia Islamia, New Delhi organized two national seminars on: 'National Environment Policy-2006: Objectives, Strategies and Implementation' and 'National Environmental Policy-2006 – Missing Links' in 2007. The articles presented in these seminars and subsequently revised by the authors form the subject matter of the present book. The main objective of these seminars was to involve environmentalists, academicians, policy planners, researchers, NGOs and other stakeholders for free and informed answers to the above-mentioned issues. How far we have succeeded shall partly be reflected in the pages of the present volume. We are conscious that many crucial theoretical and policy issues could not find a place in this volume. With all humility, we do accept the primacy of these issues. But, due to obvious reasons, these could not be accommodated. Whatever we have presented is expected to prove a useful addition to the received literature on environment and environment policies. It will serve, in some way, as a fruitful guide to effective implementation of appropriate policies. The volume has been organized in three sections. **Section I** deals with the macro issues while **Sections II** and **III** present sectoral and regional/micro perspectives respectively.

Macro Perspective

This section presents a macro perspective of the economy-ecology bind. In a bunch of eleven chapters it is difficult to deal with all the relevant issues. Given this, we have tried to put together some macro themes, which in our considered opinion form the bedrock of policy formulation and implementation. Our economy is part of the earth's eco-system. Both are determined simultaneously. Any policy on environment has, therefore, to be premised on a proper understanding of the working of environment-economy linkages. Sheetal Verma's paper demonstrates the significance of this theme. She argues: *There seems to be a complete lack of comparable analysis that demonstrates whether any particular economy is consistent with the natural environments and in this regard the drawback in the entire economics-driven approach towards sustainability is the absence of any necessary binding constraint on the activities of the economic agents. The constant interactions between the economy and the environment gave the concept of sustainability, which requires a boundary condition to be satisfied so as to take care of the considerations of intergenerational equity in resource use. Various channels of prudent environmental management can bring forth the desired changes and it is hoped that all the stakeholders will make concerted efforts in this direction. While a number of steps have indeed been taken, a lot more needs to be done to exploit the synergies that a conservationist approach towards natural wealth and a sustainable path of development hold in their wake.*

Formulating a policy is always a challenging task but its implementation is doubly challenging. It needs a judicious mix of implementing instruments. These range from Command and Control Instruments (CCIs) to the Market-Based Instruments (MBIs). Both have advantages and limitations. There are examples where both have worked. Market-based instruments have mostly been tried in the developed countries. Given the complexity and enormity of the problem, no set of instruments can work in all situations. Professor U. Sankar in Chapter II deals with these issues with reference to the specificities of the Indian economy. He concludes: *As noted in the National Environment Policy (NEP) and as recommended by the Central Pollution Control Board (CPCB), the legal barriers to the introduction of Economic Instruments (EIs) should be removed by amending the environmental legislations. Civil liability law, civil sanctions, and processes should govern most situations of non-compliance. When pollution is measurable, EIs such as pollution charges and tradable permits should be considered. The design of the charge system and the permit system must be based on a participatory approach.*

In cases where measurement of pollution is difficult or/and transaction costs of implementing a charge system or a permit system are high, indirect EIs, such as taxes on polluting inputs and outputs, environmental user charges and bank guarantees be adopted. Regarding monitoring and enforcement of environmental laws and rules, the efforts of the Central Pollution Control Board (CPCB) and the State Pollution Control Boards (SPCBs) should be supplemented by Non-Governmental Organizations (NGOs), civil society and affected parties, serving as watchdogs. Market pressures, media coverage, green rating by independent and reputed agencies and public recognition of firms achieving compliance beyond the prescribed levels create an environment for sound environmental management.

In light of Hartwick's Rule[3] of sustainable development, the possibilities of replacing environmental capital with manmade capital have brought the role of technology to the centre-stage. Many believe that technology can provide a solution to daunting environmental problems, like developing non-conventional sources of energy. But Hardin's seminal paper on 'The Tragedy of Commons' 1968[4], and subsequent extensions of this theme suggest that technology or technical solutions may not work while dealing with certain types of environmental resources, like common property resources. Technological advancements have indeed created new opportunities and challenges. Professor B.S. Reddy in Chapter III outlines the promises, prospects and potential challenges of using technology for the improvement of the quality of environment. He writes: *The question that we must ask is: to which of our modern values can we attribute to the gross destruction of our environmental resource base? The answer is clear: we equate growth with development, our obsession with GDP as a measure of efficiency and affluence, to our passion for technology as a panacea for all our ills. Let us all hope that the people of the world will look back and say of our generation that we saw the challenge, that we answered the call and that we didn't flinch in the face of our responsibility to build a better world. It will not be easy. It has never been easy. But we will try.*

Many people argue that, "the only way to attain a decent environment in most countries is to become rich" (Beckerman 1994).[5] Studies premised on this formulation have attempted to test the validity of the Simon Kuznet's Inverted U-Hypothesis to the relationship between the quality of environment and growth. In contrast to this, Meadows et al. (1972)[6] argued strongly that the relationship between the two is negative. In view of this contrasting positioning, the interface among development the quality of environment and human development assumes added significance. Dr. Mukherjee and Dr. Chakraborty in Chapter IV re-examine this relationship with reference to Indian States. The authors argue: *The Environment Quality (EQ) ranks of the states exhibit variation over time, implying that environment has both spatial and temporal dimensions. Ranking of the states across different environmental criteria (groups) shows that different states possess different strengths and weaknesses in managing various aspects of EQ.* The paper concludes that individual states should adopt environmental management practices based on their local (at the most disaggregated level) environmental information.

The Green Revolution has brought both positive and negative consequences. These have been widely documented. Some of the serious negative consequences or the costs of achieving high yields (through the high intensity use of groundwater coupled with high doses of fertilizers, pesticides, insecticides and weedicides) are degradation of soil and bio-accumulation. These issues have not received proper weight and treatment in our policy prescriptions. The consequences are more than obvious. Foodgrain production has decelerated. Bio-accumulation is fast becoming a serious determinant of ill health. There is need for more informed assessment and analysis of these issues. Against this background, Chapter V by Pradip Kumar acquires relevance. He contends: *The mismanagement of soil nutrients originates in the divergence of their uptake and replenishment. In India, the cultivation of high-yielding varieties (HYV) crops using high intensity of groundwater and high doses of chemical fertilizers for long without actually studying the nutrient contents of the soil led to the degradation of soil quality. One of the likely consequences is the deceleration of the foodgrains production that started in the early 1990s. In brief, arbitrariness in the choice of crops, application of fertilizers, use of irrigation and the related state policies militated adversely on the quality of soil and groundwater reducing the sustainability of the cultivation. Further, this has raised the cost of cultivation and would reduce the competitiveness of the Indian farmers in the global market. The Environment Policy 2006 attributed these problems to "institutional failures resulting in lack of clarity or enforcement of rights of access and use of environmental resources, policies which provide disincentives for environmental conservation (and which may have origins in the fiscal regime), market failures (which may be linked to shortcomings in the regulatory regimes), and governance constraints"(p.4). In reality, institutions that are supposed to take care of these problems are yet to evolve and for that the National Environment Policy did not suggest any suitable measure. What is needed is to promote suitable institutions at various levels, which would shape foodgrain price policy and land utilization pattern, input price policy and input uses and consumer awareness and consumption pattern.*

One of the major and multi-dimensional environmental challenges facing the world today is the climate change and its all-embracing colossal ramifications. These have been variously described as the ultimate weapons of mass destruction or a threat worse than terrorism or a nuclear war. Experts and policy makers are in a real bind as how to meet these challenges and work out solutions. Professor Kurunakaran in Chapter VI deals with some of the key issues and offers very critical insights. He argues: *The collective response of the nations to the environmental challenge is halting and far from adequate, primarily because mitigation would affect corporate profits in the energy sector. But more important, the response fails to factor in the only principle that would help in establishing a truly sustainable method of sharing the burden of mitigation and adaptation-equity. It is now well established that time is past the stage where one could say that is it still possible to avoid major damage to life on earth. The only hope of the line drawn at limiting the warming to 2^0C over the pre-industrial level is to avoid totally unacceptable damage. The nations of the earth face the stark reality that the major responsibility for causing the problem lies with the developed world while the major impacts of the problem visit the least developed countries, who are also the least able to cope with it. The urgency of devising a method to share the adaptation responsibility on the 'polluter pays' principle and on equitable terms is not yet seen in international parleys. Equally, the mitigation responsibility sharing where the developing countries need to consume fossil energy at a faster rate than that of the developed countries in order to meet their developmental goals is not getting addressed in a manner that will be seen as fair and just by the former. Avoiding equity is easier said than understood. The different ways of approaching this in the form of solution in a patently unequal world need to be carefully addressed. Intra-country equity is also an issue that cannot be wished away, contrary to what many country elites would like to see happen.*

Of late, many novel and interesting aspects of environment-economy linkages have been explored by researchers across the sub-discipline of knowledge. One of these emerging areas is the causality between biodiversity and cultural diversity and its empirical testing. Some studies suggest that the causality between the two is very robust. Preservation of one can serve as a gateway to the preservation of the other. However, research in these directions is in its nascent stage. There is a need for replicating this research in a systematic and sustained manner. It should also capture context-specific dimensions. Professor Bhatt and Nazrul in Chapter VII present a synoptic description of various underlying issues of this emerging area of research, teaching and policy. The authors argue: *While the causality between the two is obvious, the direction of this causality is not conclusive. There can be two-way links between cultural diversity and biological diversity. The extinction of one will certainly have a negative effect on the other. India, being one of the twelve mega-diversity countries of the world in terms of cultural as well as biological diversity, too is facing similar threats from dominant global cultures and many indigenous and smaller cultures are under threat. The National Environment Policy-2006 has recognized only the significance of the preservation of endangered biodiversity. The success even on this front would certainly depend upon the effective implementation strategies. The preservation of cultural diversity can play a very crucial role in the implementation of such policy, as it doesn't require much scarce resource and elaborate formal legal and administrative/institutional mechanisms.*

The existing pattern of growth is further exponentially marginalizing the marginalized people across the globe. Billions of people are fighting losing battles for survival. Poverty accompanied by deterioration in the quality of environment due to high growth has resulted in loss of traditional sources of livelihood. Policy responses to meet the consequences of this grim situation have been far from satisfactory. In most of the cases these have been highly indifferent and insensitive, even to the ground realities. Dr. Archana Prasad provides a telling example of this grim scenario in the context of India and shows how NEP fails to provide even an adequate framework, which could help the marginalized people in India to meet the adverse consequences of environmental degradation. In assessing the framework of the NEP with reference to these challenges, Dr. Archana outlines and explains the internal contradictions within the policy. She argues: *All the egalitarian principles espoused by the policy negate the measures that it promotes. In this context the policy is matched by other policy measures taken by the government, like the restructuring of the environmental impact assessment process.* Dr. Archana is of the view that: *Implementation of the NEP may render the marginalised people even more helpless in dealing with current challenges.*

The Indian Penal Code (enacted in 1860) made pollution punishable. It took us another hundred and fourteen years to enact the first complete environmental legislation in the form of Water (prevention and control of pollution) Act (1974). Since then, a number of Acts, such as Air Act (1981), Environment Protection Act (1986), Forest Conversation Act (1988), etc. have been enacted for the direct and exclusive control of various forms of pollution (both old and new) and maintenance of ecological stability. Courts, of late, have also pronounced a few landmark judgments, which have brought the existing environmental laws, their implementation, inconsistencies and shortcomings to the centre-stage of the debates on environmental policy. It is being widely felt that there is a need for reforms and reviews in the existing environmental laws. Dr. Furquan Ahmad's paper provides an illustrative justification for such a rethink. He outlines and examines critical legal inputs necessary for the effective implementation of NEP. The author suggests: *The right to clean environment under Aricle 21 of the Constitution should be drafted in such a manner that it does not remain a paper work in view of enormous pressure of development but at the same time it cannot be made absolute so as to bring the country's progress to a standstill. Thus, there must be a balance between right and duty (i.e. duty to protect and improve environment). There are multiple controlling agencies under the general and environmental laws. They are at the central, state, regional and municipal levels. All those authorities must work in coordination with each other instead of creating watertight compartments. The law must envisage establishment of a new set of environment courts or tribunals. The long delays and procedural technicalities have greatly hampered environmental justice.*

Through the ages, the processes of evolution have driven innumerable species to extinction and by spontaneous mutation created new species. The process has often been gradual and secular in periodicity. Many biological scientists believe that the world is currently witnessing a wave of human-induced mass extinction of species, resulting in, among other things, loss of genetic material for food crops, keystone ecosystem functions and non-use benefits. Other reasons aside, preservation of biodiversity is critical to the very survival of human life on earth. Hence the need for appropriate policy responses. These have to be at three levels: international, national and local. As mentioned above, India being one of the mega diversity regions of the world obviously has to shoulder a major role in the preservation of biodiversity. In view of this, the Biological Diversity Act 2002 (BDA) assumes seminal importance Dr. Mukhija and Dr.Goel in Chapter X spell out the salient features of this Act and critically assess its legislative and regulatory framework. They identify three major problems of the act: *First, the legislation facilitates bio-piracy by only providing for regulation against persons falling under the class of alien nationality or extra-territoriality and not limiting local exploitative practices; Second, it has been argued that the many-*

layered structure of licenses/permissions only succeeds in burying biodiversity under a mountain of bureaucracy that can only serve to alienate ordinary farmers from their resources while making international bio-piracy easier; Third, the Act is quite toothless on the issue of intellectual property rights (IPR). Given that the only stipulation is that IPR applications will have to go through the NBA there exists no substantial restriction. The Act has also been criticized for not providing citizens with the power to approach courts when they detect violations. It has further been argued that the BDA is unnecessarily soft on Indian corporate and other entities, requiring only 'prior intimation' for the use of bio-resources, rather than permission as in the case of foreigners. The empowerment of local community bodies under the BDA has again been found to be somewhat incomplete.

The Small Grants Programme (SGP),[7] funded by the United Nations Development Programme, Global Environment Facility (GEF), seeks to support initiatives, which demonstrate community-based innovative, gender sensitive, participatory approaches and lessons learned from other development projects that lead to reduce threats to the local and global environment. The GEF/SGP was launched in 1991 by United Nations Development Programme (UNDP) to assist developing countries in fulfilling their commitment towards the protection of the global environment. The Programme is sourced with a belief that global environmental problems can only be addressed adequately, if local people are involved in planning, decision-making and sharing roles and responsibilities at all levels. Even with small amounts of funding, communities can undertake activities, which will make a "significant difference" in their livelihoods and environment. UNDP GEF/SGP is currently offered in 83 countries worldwide. The programme started in India since 1997. The UNDP and

Ministry of Environment and Forests (MoEF), Government of India (GoI) administers Small Grants Programme (SGP). It is being implemented by Centre for Environment Education (CEE) as the National Host Institution (NHI) since September 2000. CEE has its presence felt in all the states and Union Territories of India through a local network of 7 Regional Offices and 23 Field Offices across the country. CEE is currently implementing 89 different projects from the total of 240 projects (151 completed) with an outreach to more than 500 villages reaching a population of more than 550,000 people. The emphasis is more in establishing low-cost, low external input and easy-to-manage technologies, with greater emphasis on enhancing community capacities and social mechanisms for better livelihoods. In the India programme, the annual grants disbursement has increased from USD 300,000 in the year 2001-02 to USD 1,000,000 USD in the year 2005-06 and continues to deliver more than 1,000,000 USD each year. The Government of India has also put forward contributions of 100,000 USD every year since 2005-06. The communities of practice are also raising an equal co-financing in the programme. Given the increasing success of the SGP nation wide, CEE is increasingly looking at ways and means to mainstream the priorities of the state and central governments and private sector so as to create a better interface between the governments, private sector and the responsive NGOs. Learning from the past gives creditability to the SGP, in terms of how responsive it has been in addressing the issues effectively. Probjit Sodhi in Chapter XI presents an overview of some of these success stories. The author highlights the major activities of SGP across the country. He profiles projects completed in Maharashtra, Gujarat, Madurai, etc. where SGP has emerged out as an excellent vehicle for environmental preservation. Sodhi asserts that: *SGP emphasizes partnership and community participation in planning, implementation and monitoring of projects in the five GEF thematic areas of biodiversity conservation, climate change mitigation, protection of international waters, prevention of land degradation and elimination of persistent organic pollutants. SGP aims to develop community-based approaches, strategic partnerships and innovative technologies that reduce threats to the global environment through local initiatives.*

Sectoral and Micro Perspectives

Eleven chapters dealing with sectoral and micro-level issues are presented in Sections II and III. Economy-environment linkages operate at various levels of aggregation – international, national, meso, and micro level giving birth to environmental goods and bads with specific characteristics at each level. Macro analysis often either sidetracks these or is unable to capture their role and impact, thereby adversely affecting policy implementation. Macro indicators of environmental quality also are not enough as these cannot capture the variations due to the context-specific variables. Hence there is need for sectoral and micro-level studies. This explains the rationale behind Section III. Sequencing of chapters in no way is a reflection upon their relative importance.

The central theme of trade-environment debate has been meshing up economy and ecology linkages in an open economy framework. According to Charles S. Pearson,[8] the trade-environment nexus raises a number of interesting questions, such as: How do trade and foreign investment affect environmental quality? What are the pollution heavens? Do the rules of liberal trade regimes support or undermine environmental protection? Are environmental regulation used as covert trade barriers? Is comparative advantage determined by environmental resource endowments? Should poor countries specialize in producing "dirty" products? Answers to these questions have often been intractable and opinions are highly contrasting.

Professor Azad and Professor Saravanan through their papers provide very informed insights about certain crucial aspects of this debate. The former, dealing with Sustainable Development Aspects of the Doha Round: Implications for Developing Countries in Chapter XII and the latter, dealing with consequences of export-led industrialization on environment and the role of state in historical context. Professor Azad contends: *Trade and development are imperatives for any developing country. However, development should not be achieved at the cost of the poor living standards of the nation. People are not only poor because of low economic growth but also because of the deteriorating environmental conditions. Development should therefore also envisage the environmental issues that make us uncomfortable today. Clean technology should be used, which is less harmful to nature.*

Professor Saravanan attempts to analyze how the export-earning industries affect environmental sustainability. The role of the state, judiciary and civil society is evaluated in facilitating this process. He has analyzed these twin problem using historiography with special reference to Tirupur Knitwear Industries in Tamil Nadu during 1980-2005. Professor Saravanan raises an important question: will the export earnings compensate for the environmental damages caused to the environment, ecology, health and natural resources in the geographical region? He opines: *On the one hand, the state has provided water diverted from the other river basins for these industries which led to conflict among people living in these river basins. On the other hand, the diverted water is not only getting polluted but also the entire natural resources. The liberalization and globalization process has accentuated the threat to sustainable environment and state became an agent for the corporate companies and industrialists. The powerful industrialist-political nexus has been successful in availing the facilities from the government while evading its social responsibility. The existing legal mechanism is also ineffective. The government is extending all kinds of privileges to promote the export-oriented industries to have a comfortable foreign exchange reserve while it has not looked at the damages to the environment. Precisely, the State approach towards the export-oriented knitwear industries, particularly dyeing and bleaching industries will not ensure the sustainable natural resources but it will certainly sustain the polluted environment not only for the present generation but also for the several generations in future.*

Lester Brown has rightly observed that, like the rest of the society, corporations have a stake in building an eco-economy. Profits do not fare well when an economy is declining or threatening to collapse.[9] Many companies around the world have demonstrated that being 'green' pays. Empirical studies suggest that there is positive relationship between equity prices and environmental performance of the companies. Prof. Shahid Ashraf in Chapter XIII examines the prospects of this emerging area of corporate management and investment opportunity. He opines: *A rapidly evolving change in corporate management and strategy has been the recognition of and response to environmental concerns by industries. Often called greening, this response is a process by which human activity is made compatible with biospheric capacity. Industries have begun to develop and implement policies, programmes and tools to meet their environmental opportunities and constraints. This process has occurred partly in response to a desire to reduce potential liabilities. However, it has also been fuelled by opportunities in new markets and demand by diverse constituents including the public, shareholders, customers, and employees. Expectations are that companies with a good (bad) environmental record should be valued at a premium (discount) by the stock market. Such a relation results from the emergence of 'ethical' (or Green) investing and from an increased awareness by investors of the potential negative consequences of corporate environmental damages. Two factors will influence companies' stock market valuation. Companies with a poor pollution record will generate less cash flows because of investment in anti-pollution equipment. Second, they will face sanctions from governments. This will affect future market valuations from ethical investors.*

It is said that there is no 'free ride' in economics and everything comes at a cost. The information technology revolution has changed the world beyond belief. Unfortunately, the costs of this revolution seldom receive the desired space and attention. Policy responses do not even recognize the existence of some of these costs. For example, with the massive growth of information technology, electronic waste (e-waste) is fast emerging as a serious form of pollution. The stocks of 'e-waste' are bound to grow day in and day out. Policy makers are clueless about the potential solutions. Asheref Illiyan in Chapter XIV looks into the various dimensions of this problem and proposes solutions. He expostulates: *While most of the developed countries have passed national e-waste management policies and set up scientific recycling units, it is ironic that India, which has a well-developed software and hardware industry in the world does not have a scientific and well-developed recycling industry and a national policy on e-waste management. The existing law (Hazardous Waste Management and Handling Rules 2003) is not comprehensive enough. Even National Environment Policy 2006 does not speak much about e-waste. Therefore, there is a need for promulgation of a comprehensive national e -waste management and recycling policy.*

Way back in 1939, John Steinbeck set the scene of his seminal and widely acclaimed novel *The Grapes of Wrath* against the backdrop of droughts, poor soil conservation and water scarcity, which combined to ruin agricultural institutions, forcing people to leave Okhalahama to New

York in search of survival and jobs. These people in turn were uprooted, only to be caught up in a web of exploitation and despair. What happened in Okhalahama in 1939 is now happening across the world on a devasting scale with dire consequences. Water scarcity and water pollution are sentencing millions of people to hydrological poverty and vicious exposure to water pollution-related diseases. Major rivers are running dry. Water levels are falling fast. Around 70% of water around the world is used for irrigation, 20% for industrial use and 10% for residential purposes. Rising affluence is increasing demand from each source. This is bound to shift the use of water from agriculture to industry and bourgeoning services sector with serious implications for the world food security and poverty eradication. Ecological historians have explored water scarcity/water pollution as one of the major determinants which triggered the fall of early civilizations, like the Sumerian, Roman and Mayan civilizations. Though every major country of the world has a water policy in place, these are proving ineffective. The situation in India is far from satisfactory. Supply-demand gaps are mounting day in and day out and millions of people lack enough water to meet their basic needs of drinking, bathing and food. The supply of portable drinking water is skewed across states, households and classes of people. Major irrigation projects have brought prosperity to some people and places but wrought ruin and diseases to others. Industrialization and consequent industrial effluents have badly affected water bodies and aqua eco-systems, with catastrophic consequences on livelihoods, employment patterns and health status. Three micro studies (one from Tamil Nadu and two from Delhi) deal with the water scenario from different but related perspectives. These studies illustrate the location-specific consequences of both water scarcity and water pollution. Dr. Govindrajalu in Chapter XV describes the impact of polluted water (due to industrial effluents) on agricultural, health, livestock and fisheries by taking Noyyal River Basin as a case study. He avers: *The most serious environmental problems faced by the developing countries include water and air pollution. The problems of water pollution acquire greater relevance in the context of an agrarian economy like India. Industrial pollution has been a major factor responsible for degradation of the environment and polluting water resources. The industrial effluents generated by bleaching and dyeing industries in Tirupur (well known for hosiery and textile processing) are let into Noyyal River which originates from the Western Ghats and confluences with the river Cauvery. Consequently, the socio-economic background of the rural community, employment pattern, agricultural production, health status, source of drinking water, livestock population and fisheries production are considerably damaged. In the light of the National Environment Policy-2006, the issue of Noyyal River Basin has gained significance in two contexts: (a) Polluters pay the damage. (b) Ways for safe disposal of industrial effluent.*

Farhan and Sapana, in their respective chapters, deal with crucial aspects of pricing and determinants of demand for water. Using Contingent Valuation Method (CVM), Farhan works out the willingness to pay for the better quality and supply of water. He concludes: *Water wars are a reality now, right from micro level to global level. These conflicts are the result of cumulative effect of gross negligence and mismanagement of water resources over the years. The problem is not due to absolute shortage of water but due to absence of proper mechanism for conservation, distribution, and efficient use. Drinking water is the most crucial of all water-related problems. Presently, water and sanitation sector are highly subsidised. The challenge is not to eliminate the subsidy altogether but in a phased manner without mitigating the social objective in a cost-effective and economically efficient manner. The approach of cost recovery should be fostered with technical improvement in water delivery and administrative reforms. The National Water Policy document not only mentions covering the operation and maintenance (O&M) costs but also about the scarcity value of water. Pricing should be accompanied by efficiency improvement in the system. The consumer should not be penalized for the inefficiencies of the department in terms of distribution losses and unaccounted for water. The task, however, is to rationalize the pricing structure so that at least O&M costs are recovered fully in the phased manner.*

Sapna, while focusing on Delhi Jal Board's position as a major supplier of portable drinking water in Delhi, scans crucial and interesting facets of the water scene in Delhi. She argues: *The National Capital Territory of Delhi, where about fourteen million people reside at present, faces a huge divergence in the availability of water and its demand . Both qualitatively and quantitatively, the water situation in Delhi is in the vortex of severe crisis. Due to water scarcity, citizens are digging deeper into the water table.* Sapna concludes: *Factors like household income, literate females, expenditure on public water and family size are the most significant factors explaining variation in consumption of public water for domestic use at the aggregate level. However, at the disaggregated levels, some of the above-mentioned variables do not explain variation in consumption of water in all cases.*

The interface between poverty and declining environmental quality has been subjected to intense dogmatic academic debates, both in terms of the direction of the casualty between the two and possible solutions. As mentioned in the preceding pages, advocates of 'The Beyond the Limits'[10] view economic growth as the source of both poverty and deterioration in the quality of the environment. This view advocates a direct control over the growth process itself. Simon Kuznet,[11] on the other hand, suggests that growth provides the means to improve the welfare of future generations. Following Kuznets, many researchers have tried to extend this framework to study the growth and quality of environment relationship and suggest the existence of an Inverted-U Curve (popularly known as the Environmental Kuznets Curve, see Grossman and Kruzger 1993).[12] The empirical evidence, however, is far from conclusive. Environmental degradation and poverty are entwined in a complex way and it is very difficult to conclusively suggest which reinforces the other. The casualty appears to be interactive. There is a need to specify and test it with reference to homogenous reference frames and then match micro level results with macro evidences. Deepti Tandon examines the relevance of this suggested line of approach with reference to Delhi. She argues: *Many instances and empirical results go on to show that there is a strong interplay between poverty and environmental resource degradation. These trends, which mark an increasingly stressed relationship between the economy and the earth's ecosystem, are taking a growing economic toll. At some point, this could overwhelm the worldwide forces of progress, leading to economic decline. The present study establishes that poverty was a highly significant variable in explaining the environment resource degradation.*

Migration and its effect on socio-economic, demographic and cultural factors has recently attracted increasing attention from administrators, planners, social scientists and researchers seeking a thorough understanding of the process of population movement. At any particular point in time, the population size of any country or region is governed by three components: fertility, mortality and migration. Fertility and mortality largely operate within the biological framework through social, economic and cultural factors. Migration, on the other hand, is a purely socio-economic phenomenon involving social, psychological, economic, political, institutional and other determinants. It is usually classified as either international migration or internal migration. The former has greater political significance and formidable geographical and cultural barriers as well. On the other hand, the latter results in a change of usual place of residence within the political boundaries of a nation. It may consist of crossing of a village or town boundary as a minimum condition for qualifying the movement as internal migration. This type of migration tends to occur broadly within the same cultural and political bounds. Though internal migration does not change size of the population of the country as a whole, it does influence the social, economic and cultural characteristics of the inhabitants at both the places of origin and destination. When an individual or a family changes residence, the move is made and the destination is selected for a number of specific reasons. Sometimes these reasons may not be obvious, but are the result of social, economic, political and other factors occurring in different combinations. The factors associated with the process of migration may be broadly classified as: 'push factors', and 'pull factors'. Dr. M.A. Baig and T.A. Baig in Chapter XIX enumerate and explain socio-economic and environmental impact of migration with special reference to Cuttack (Orissa). Baigs argue that: *Lack of opportunities to earn bread and butter in the native places is the main reason for migration. Government should provide alternative job opportunities at rural places. Though government is doing well in launching and funding different programmes in this direction, it should closely monitor the implementation of such programmes. In addition, the government should try to provide vocational training to enhance their productivity. The New Environment Policy, 2006 (NEP) over-emphasizes the trade-off between growth and industrialization. Micro aspects of macro problems of environmental degradation are overlooked in the policy documents. The NEP lays down very broad and vague reforms and actions plans. Hence, there is a need for specific plans to deal with specific problems.*

The earth's forested area declined from five billion hectares to 2.9 billion hectares during the last one hundred years. Loss of forests on this scale has intensified global warming, decreased biodiversity, diminished agricultural productivity, increased soil erosion and desertification, precipitated the extinction of traditional cultures, triggering of water scarcity and massive loss of high altitude wetlands, livelihood/habitats/sources of food for wild life. Rudel and Roper (1997)[13] evaluated two competing explanations of deforestation, namely the 'Frontier Model' and the 'Immiserization Model'. The first model assumes capital investment (through modern patterns of growth) as the main determinant of deforestation with growth of timber, pharmaceutical, paper and agro-industries playing the villain. Massive forest areas were deliberately cleared in the Western countries using modern technology to fuel industrialization and development of the modern mass and rapid means of transportation, like roads, railways and shipbuilding since the industrial revolution. This model was subsequently followed by the developing countries. The Immiserization Model sees deforestation as a consequence of demographic transitions, mass poverty and low level of income. Cross-country evidences are not uniform and choice between the two models is not clear-cut. Local regional, national and global causes of deforestation (like the benefits of forests) should first be analyzed separately and then randomized. This obviously is very difficult but highly desirable. Nazrul et al. in Chapter XVIII justify the rational of such an approach. The authors argue: *Once green Patharia Range in Karimganj district of Assam became a victim of free riding. Many traditional inhabitants of this range either got extinct or displaced which sometimes causes more human suffering directly or indirectly. For example, thousands of monkeys, once living in the jungle, have no other option but to survive by sharing human interests. Crops, fruits, flowers from fruit-bearing trees or plants or even kitchen foods become their easy victims. This 'animal revenge' is not only causing insecurity, frustration or many social problems, it is also causing much economic loss. This effort, although a micro-level and focusing narrowly on the links between displaced monkeys and economic loss, proves that there is an overall institutional failure, which made the situation worse. Proper policy guidelines and careful implementations, involving all stakeholders, can ensure a better management of endangered natural resources and thus more economic and other benefits.*

Suhaib, at the end of this volume adds an interesting dimension to the fable of environmental degradation in India. India has vast and rich coastal eco-systems. These have been home to rich biodiversity and sources of livelihood to the people living in the costal areas. Of late, these eco-systems have been subjected to mindless exploitation with serious environmental consequences. Sand mining in Kerala is one such example. He contends: *The issue of sand mining in the state of Kerala has been a matter of contention in the recent years. Sand is one of the most important items needed in any kind of construction work. Owing to the exorbitant utilization of the river sand, the availability has dwindled, leading shortage. This has in turn damaged the local water bodies. A number of laws exists in the state against these activities. However, powerful vested interests have lobbied over the years to water these down so that today the protective spirit of the original notifications is all but lost.*

Like sand mining in Kerala, mining around the world has demonstrated the prophetic significance of Anton Chekhow's (1896) remarks in *Uncle Variya*, Act I. He lamented, "Man is endowed with reason and creative powers to increase and multiply his inheritance; yet up to now he has created nothing, only destroyed. The forests grow ever fewer, the rivers parch, the wildlife is gone, the climate is ruined, and with every passing day the earth becomes uglier and poorer."[14]

Notes and References

1. World Commission on Environment and Development (1987) *Our Common Future*, OUP, London.
2. For details see Brown, L. R., (2001), *Eco-economy* p. 21, Earthscan Publications Ltd, UK.
3. Hartwick, J.M., (1977), Intergenerational Equity and Investing of Rents from Exhaustible Resources, *American Economic Review*, 67 (5), pp. 972-74.
4. Hardin, G., (1968), The Tragedy of Commons *Science*, Vol. 16, pp. 1243-8.
5. Beckerman, W., (1994), Sustainable Development: Is it a Useful Concept?, *Environmental Values*, 3, pp. 191-209.
6. Meadows, D., et al., (1972), *Limits to Growth*, Universe Books, New York.
7. Information about SGP has been provided by Probjit Sodhi, Regional Director SGP in India. We acknowledge this help with usual caveats.
8. Pearson, S.C., (2000), *Economics and Global Environment*, Cambridge University Press, Cambridge, p. 1.
9. Brown, L. R., (2001), *Eco-Economy-Building an Economy for Earth*, Earthscan Publications Ltd., UK. pp. 261.
10. Meadows, D., et al., (1972), Limits to Growth, Universe Books, New York.
11. Kuznets, S., (1955), Economic Growth and Income Inequality, *American Economic Review*, Vol. 45 (1), pp.1-28.
12. Grossman and Kruger (1993), quoted in *Environmental Economics* by Nike Hanely, Ben White and Jason. F. Shogren, pp.129, Para-II, OUP, 2004 (Indian Edition).
13. Rudel, T., Roper, J., (1997), The Paths to Rainforest Destruction-Cross National Patterns of Tropical Deforestation, *World Development*, Vol.25, pp. 53-65.
14. Quoted in *Environmental and Natural Resource Economics*, by Tietenberg, T., (2004), pp. 183, Pearson Education Inc.,

Macro Perspectives

Chapter 1

Economy–Environment Linkages: Economic Growth, Quality of Environment and Sustainability

Sheetal Verma

I

"Nature is not a stable backdrop against which humans can orchestrate their affairs. That natural processes have their own dynamism and integrity must be borne in mind."

Mike Davis

As the limits to economic growth become increasingly realised, the above quote assumes great relevance for policymakers across the world for it seeks to re-emphasise that solely concentrating on economic growth may not always be the panacea for all problems plaguing their societies. From time immemorial, environment–economy linkages have been a major concern of all societies. As an economy proceeds along the path of economic development, it influences the nature, content and quality of the environment with which it interacts on a continuous basis, both for resource extraction as well as for waste disposal. The primitive economy was localised and constrained by the prevailing environmental conditions on several fronts. With the passage of time, the nature–economy interactions became increasingly complex and economic motivations began to override environmental concerns. Resultantly, as the development experience of most of the economies shows, environmental stresses and strains are no longer the unique attribute of developed economies but are now a ubiquitous phenomenon appearing in all economic systems. Most industrialised economies face acute water and air pollution threats—from the Thames in London to the river Yarra in Melbourne, from the pea soup fog of England to the Asian Brown Haze over the subcontinent. Even the relatively underdeveloped, poverty-stricken regions of Africa are reeling under the impact of overgrazing, soil erosion and desertification. Different kinds of pollution now threaten even the most precious of bio-spherical assets such as the Aral Sea, the sacred Ganges, and the Danube. Besides, several species are becoming extinct with each passing day, leading to a loss of biodiversity as well as raising serious concerns on grounds of inter-generational equity in resource use. The threats loom large, and in response to the same, the global community is trying to stem the tide by making concerted efforts in the form of international treaties and protocols.[1]

Given the topic of this paper, the primary question that comes to mind before initiating a discussion on sustainability is how economic growth and quality of environment are related. There is a huge body of literature relating to this subject. The views are divergent indeed, ranging from the resource-exploitative, cornucopian ideology to the extreme preservationist; and an equally bewildering variety of mechanisms are suggested to grapple with the problem of environmental degradation (price rationing, quantity rationing and liability rules). The paper shall delve into the finer nuances of the diverse streams of thought and their degree of relevance in the current Indian context. Finally, we shall conclude by summarising our findings and examining how far the growth momentum will be sustained in the coming decades.

II

Even in the classical era, Ricardo, Malthus, Mill, etc. had discussed the possible impact on environment due to an increasingly evident land constraint and rising population. Malthus's famous theory talks of the *preventive checks* imposed by nature in the form of droughts, pestilence, etc. if population breaches the optimum limit. If one were to take a closer look at Marx's analysis of capitalist generalised commodity production, one would realise that a major reason why capitalism as a mode of production is not able to replicate itself could be rooted in the environmental degradation over time. The rising environmental costs not only undermine the sustainability of the capitalist system but also reinforce the *Law of the Tendency of Declining Rate of Profit*. To elaborate, as newer technologies are introduced, the impact on environment is often devastating because of the toxic and non-degradable nature of the residuals of such a production process. Computers and other electronic industries using such cutting-edge technology are a case in point. Up to a great extent, the international trade theorists have begun to analyze the patterns of trade between the developed North and the underdeveloped South in terms of the lower level of awareness in poorer countries, which brings down the huge compliance costs resulting from stringent environmental regulations in the West. The case of the ship *Clemenceau* and more recently *Norway*, containing high levels of asbestos and therefore being returned by the Indian authorities is a brilliant example of such distortions in the market for environmental goods.

As the neo-classical era was ushered in the late 19th century, environmental goods came to be regarded like any other marketable commodity whose economic value is largely determined by the amount of personal utility it yields (i.e. through the interaction of the forces of demand

and supply). Market failures were regarded as an exception to the rule, which in turn necessitated government intervention. Environmental organisations established from the late 19th to the mid-20th century were primarily middle-class lobbying groups concerned with nature conservation, wildlife protection, and the pollution that arose from industrial development and urbanisation.

It wasn't until the post-war re-constructionist era that the sustainability issue came to the fore. While the post-war boom in the global economy and the enormous aid channelisation under the aegis of the Marshall Plan worked wonders to help the ailing economies of Europe in gaining a sound economic footing, it also had some indirect effects on the world climate in the form of oil spills, ozone hole, and global warming. It was finally realised that unless sincere efforts are made to abate the loss of natural endowments in the near future, the very basis of economic development might be undermined. After all, no amount of wealth could buy the pleasure of breathing in clean air and leading a disease-free life. Environmental awareness became the buzzword of the era and a vast array of ideologies was put forth to take on the issue. A number of these ideologies was basically anti growth, which emphasised the huge social and environmental costs that 'living in a growth' economy entails. *Easterlin's paradox* (i.e. survey data including that material affluence and human happiness were not closely correlated), *Hirsch's* 'positional goods' concept (i.e. that the enjoyment of a range of commodities is necessarily restricted to a small group of high-income earners, despite the illusion that all sections of society might one day participate in such consumption), and *Scitovsky's* 'joyless economy' analysis (again emphasising human need for more than mere material affluence) is representative of 'social limits' thinking.

The furore caused by the reports brought forth by *Club of Rome* and Meadow's *Limits to Growth* (Meadows et al, 1972), with a discernible Malthusian strain, led to the emergence of a new genre of economists who continued to argue that economic growth remained both feasible (a growing economy need not run out of natural resources) and desirable (economic growth need not reduce the overall quality of life). What was required, however, was an efficiently functioning price system which would accommodate higher levels of economic activity while still preserving an ambient environmental quality. The 'depletion effect' of resource exhaustion would be countered by technical change (including recycling) and substitutions which would augment the quality of labour and capital, and allow for, among other things, the continued extraction of lower quality non-renewable resources (Pearce and Turner 1990).

In the less-industrialised or developing world, environmentalism has been more closely involved in 'emancipatory' politics and grass roots activism on issues such as poverty, democratisation, and political and human rights, including the rights of women and indigenous peoples. Examples of such movements include the Chipko movement in India, which linked forest protection with the rights of women, and the Assembly of the Poor in Thailand, a coalition of movements fighting for the right to participate in environmental and development policies.[2]

Although a small number of bilateral and multilateral international environmental agreements were in force before the 1960s, since the 1972 United Nations Conference on the Human Environment in Stockholm, the variety of multilateral environmental agreements has increased to cover most aspects of environmental protection as well as many practices with environmental consequences, such as trade in endangered species, the management of hazardous waste, especially nuclear waste, and armed conflict. The changing nature of public debate on the environment was reflected also in the organisation of the 1992 United Nations Conference on Environment and Development (the Earth Summit) in Rio de Janeiro, Brazil, which was attended by some 180 countries and various business groups, non-governmental organisations, and the media. In the 21st century, the environmental movement has combined the traditional concerns of conservation, preservation, and pollution with more contemporary concerns with the environmental consequences of economic practices as diverse as tourism, trade, financial investment, and the conduct of war. Environmentalists are likely to intensify the trends of the late 20th century, during which some environmental groups increasingly worked in coalition not just with other emancipatory organisations, such as human rights and indigenous people's groups, but also with corporations and other businesses.

Table 1.1 gives a panoramic view of the different environmental ideologies that have occupied the centre-stage of the discussion on economic and environmental linkages. Apart from these, the newly emergent branch of *institutional economics* and the seminal contribution of Ronald Coase suggested a bargaining solution to the environmental problems, even in the absence of State intervention, if the property rights are well defined and transferable. The approach recognises market failures due to asymmetries of information, but at the same time maintains that State action may not always hold the key, given that there are no inbuilt incentives for politicians and bureaucrats to resist pressures from special interest groups, which results in a massive loss of environmental wealth. The *material balance approach*, on the other hand, attempts to find an (economic) optimum level of pollution, given certain simplifying assumptions. However, the data deficiencies and the limitations of the static approach make optima an impracticable policy option.

Table 1.1

TECHNOCENTRIC		ECOCENTRIC	
Extreme Cornucopian	Accommodating	Communalist	Deep Ecology
Resource exploitative, growth oriented position	Resource conservationist and "managerial" position	Resource preservationist position	Extreme preservationist position
Economic growth ethic in material value terms Maximise Gross National Product It is taken as axiomatic that unfettered market mechanisms or central planning (depending on the ruling political ideology) in conjunction with tech innovation will ensure infinite substitution possi-bilities capable of mitigating long-run physical resource scarcity	Infinite substitution is not considered realistic but sustainable growth is a practicable option as long as certain resource management rules (e.g. for renewable resource sustainable yield management) are followed	Pre-emptive environmental constraints on economic growth are required, because of physical and social limits Decentralised socio-economic system is necessary for sustainability	Minimum 'resource-take' socio economic system (e.g. based on organic agriculture and de-industrialisation) Acceptance of bio-ethics (i.e. non-conventional ethical thinking which confers moral rights or interests on non-human species)
Instrumental value in nature	Instrumental value (i.e. of recognised value to the humans) in nature	Instrumental and intrinsic value in nature (i.e. valu-able in its own right regardless of human experience)	Intrinsic value in nature

Environmental Ideologies (Source: adapted from O'Riordan and Turner, 1983).

Figure 1.1

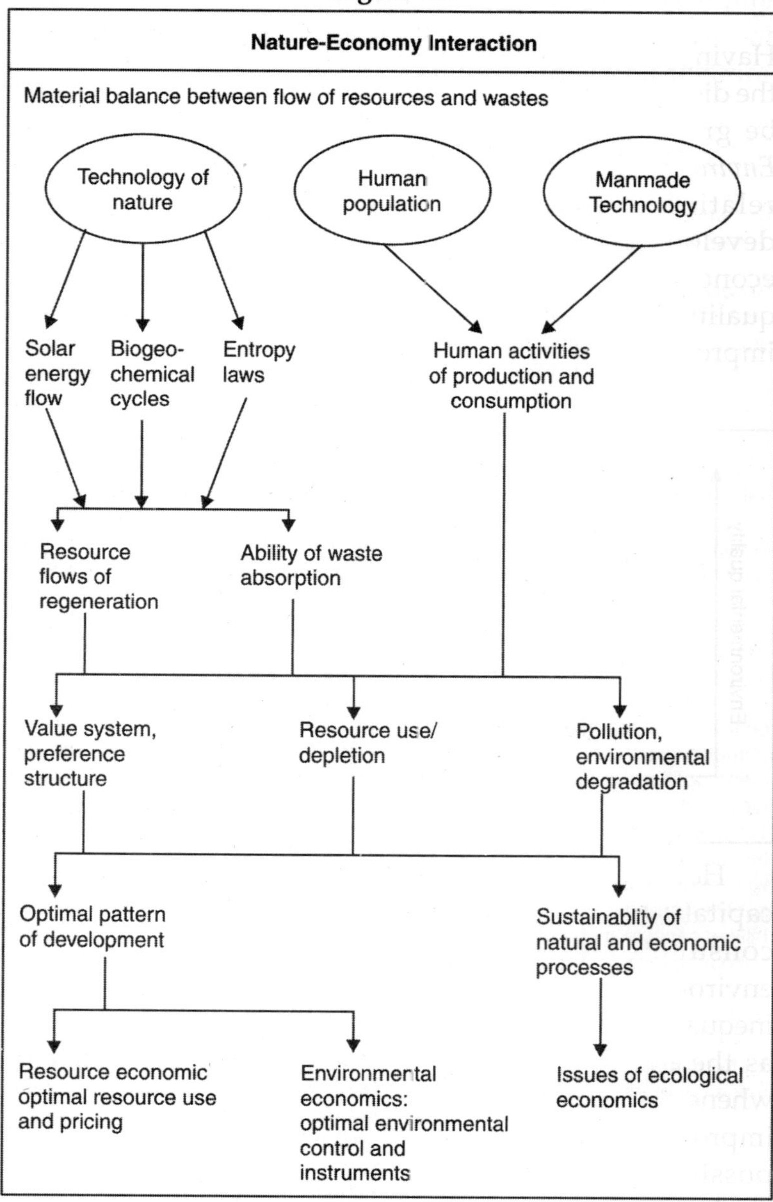

Adapted from Sengupta, Ramprasad: Class notes, Environmental Economics, Centre for Economic Studies and Planning, JNU, 2005.

III

Having analysed the key ideologies that form the crux of the discussion on environment–economic linkages, it would be grossly unfair if one were to overlook the famous *Environmental Kuznet's curve hypothesis* on the most likely relation between the environment and economic development. It states that, during the initial phase of economic development, degradation of environment in quality and quantity can take place, but subsequently it improves for the better as depicted in Figure 1.2.

Figure 1.2

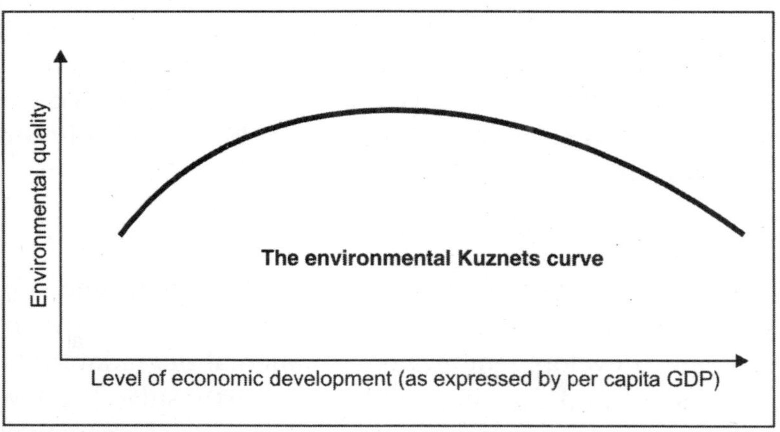

Heavy emphasis on economic growth, as viewed from capital formation addressed to enhance production and consumption, neglect (or unconcern) by people on environmental matters, high poverty levels; high income inequality and lack of community-level institutions are cited as the main factors. The improvement in environment, whenever it occurs, is attributable to technical change and improvement in efficiency in the use of energy resources, possibility of substitution by resources not linked to environmental resources, increased awareness among people about hazards of environmental degradation, and better implemented environmental regulations. This also means that a faster growth of market of eco-friendly products and technology would require a faster development process, so that a country crosses the threshold level quicker and quickly expands the demand for environmental services. New paths of development for the newly industrialising countries need to be identified which may enable them to avoid the dirty phases of development, by adopting cleaner technology and growth strategy from the beginning.

Experience of countries across the world seems to indicate the converse, in that environment and development seem to be complements only in the early stages of development. Once take-off is achieved, they become substitutes but with the caveat that they can be traded off against each other only up to a certain limit, and only for certain environmental functions.

IV

The study conducted by Kadekodi, 1995 (Tables 1.2 and 1.3) shows that in the Indian context, the relationship between economic development and environmental degradation is quite mixed and does not yield any robust results. Yet, the cells in Table 1.3 which are highlighted in yellow, show that there is a not-so insignificant degree of correlation between the key economic variables (e.g. per capita income, poverty, population density etc.) and their environmental counterparts (proxied by deforestation and area under wastelands).

In India too, the policymakers, the State and the people being equal stakeholders in development, need to collaborate in their efforts in conserving resources. Legislations and policies to this end seem to abound and some of them, like the Joint Forest Management initiative in 1988, the Ganga Action Plan, the National Environment

TABLE 1.2: ENVIRONMENT AND ECONOMY LINKAGES IN INDIA

1 **States**	2 **Per Capital income (1993–94)**	3 **Mean income of the poor per yr (1993–94)**	4 **Gini Ratio per capita income**	5 **Sen poverty index**	6 **Deforestation between 1987–97 (percentage)**	7 **Total Wasteland as % of area (1988–90)**	8 **Head count ratio as%, NCAER, 1994**	9 **Population density (1991)**
Orissa	3028	1319	0.42	0.3	–11.7	13.31	55	203
Bihar	3691	1587	0.39	0.21	–7.36	1.85	42	497
West Bengal	3157	1745	0.35	0.24	–5.24	5.69	51	767
Madhya Pradesh	4166	1605	0.41	0.15	2.69	20.71	4	149
Maharashtra	5525	1595	0.45	0.13	4.77	25.78	34	257
Tamil Nadu	5122	1591	0.43	0.16	–7.16	17.67	34	429
Assam	5070	1976	0.34	0.12	–9.71	20.29	33	286
Karnataka	4769	1357	0.49	0.18	0.43	14.1	33	235
Uttar Pradesh	4185	1535	0.42	0.22	8.11	22.27	40	473
Kerala	5778	1999	0.4	0.13	–0.65	4.25	30	749
Gujarat	5288	1495	0.49	0.19	–7.31	23.36	39	211
Rajasthan	4229	1672	0.41	0.2	7.01	32.19	40	129
Andhra Pradesh	5046	1396	0.42	0.08	–13.75	21.63	21	242
Haryana	6368	1922	0.37	0.11	–6.21	8.46	27	372
Punjab	6380	1771	0.39	0.15	81.07	7.51	32	403

TABLE 1.3: CORRELATION COEFFICIENTS AMONG ECONOMIC AND ENVIRONMENTAL VARIBALES

1	2 Per capita income (1993–94)	3 Mean income of the poor per yr (1993–94)	4 Gini ratio per capita income	5 Sen poverty index	6 Deforestation between 1987–97 (percentage)	7 Total wasteland as % of area (1988–90)	8 Head count ratio as %, NCAER, 1994	9 Population density –1991
2	1	0.4669	0.0778.	–0.7814	0.412	–0.1359	–0.8199	–0.017
3		1	–0.6999	–0.425	0.1878	–0.3718	–0.2789	0.4844
4			1	0.0917	–0.0705	0.3844	–0.071	–0.4488
5				1	–0.0784	–0.0525	0.9939	0.0817
6					1	–0.0269	–0.1252	0.0471
7						1	–0.0339	–0.7088
8							1	0.08338
9								1

Policy are praiseworthy indeed; but as is often noticed and also affirmed by a study conducted by Menon Ajit 2006, developments with regard to different environmental legislations and policies in particular sectors are often contradictory in nature, sending confused signals as to the future of environmental policy itself.

The environment assumed a central role in India, to a large extent, as a result of the first major international conference on the environment, namely, the United Nations Conference on the Human Environment (UNCHE) held in Stockholm in 1972. In preparation for this meeting, each member state was asked to prepare a report on the state of the environment. India set up a committee on human environment under the chairmanship of Pitambar Pant, a member of the Planning Commission. The outcome was three reports: one on the state of the environment, a second on the problems of human settlement, and a third on the possible strategies to manage resources. Environmental goals were subsequently incorporated in the Fifth Five-Year Plan onwards. Legislations such as the Wildlife Protection Act, 1972 and the Water (Prevention and Control of Pollution) Act, 1974 were passed soon after as well [Divan and Rosencranz 2001]. [3]

While the discursive thrust of much of environmental policy making in the late 1970s and early 1980s was on incorporating environmental principles in sectoral planning, something that was matched with legislative intervention, the latter part of the 1980s saw the focus shift towards sustainable development. The importance of this shift was that the link between social and environmental concerns was more forcefully articulated. Again, this was at least partly due to international developments. The World Commission on Environment and Development (WCED) published 'Our Common Future' (the Brundtland Commission Report) in 1987, a report that highlighted the importance of both inter-generational and intra-generational equity with regard to environmental management. Equally important was the fact that the Brundtland Commission Report explicitly recognised the linkages between the rights of communities and the management of the environment Lafferty, 1998.

The 1988 National Forest Policy (NFP) was the first 'environmental' policy document in India that explicitly recognised the linkages between environmental and social concerns in terms of community rights to natural resources Ghate 1992. While the NEP 2006 seems to be promising in this regard as it explicitly recognises the community rights over natural resources, the market-based approach towards environmental management is fraught with risk because, given the underdeveloped nature of our economy, it is often the case that the poor and the underprivileged may be marginalised by the workings of market mechanism. The hard reality is that these very people form the core group that has the maximum impact on their immediate environment (e.g. tribals may not be included in the market process but are entirely dependent on forest produce for their survival). It has also been seen that even conservation efforts have certain undesirable effects sometimes. For instance, by imposing restrictions on fishing and mangrove use, the official conservation strategies have led to instability of the local economy in some of the villages, which may pose a risk of further denudation of forests, making the goal of a holistic management approach even further unrealisable.

> Menon Ajit (2006) suggests, *At one level that however well-intended policies and laws are, they get diluted in course of time – largely because of other developmental priorities. These priorities themselves have changed in the last 15 years or so with a movement towards a neo-liberal regime. Policies and law, as a result, are envisaged much*

> *more as ways to create the conditions for better management than as a means through which to uphold particular claims to resources or particular normative goals of development. Moreover, talk about decentralised natural resource management and community-based management initiatives need to be taken with a pinch of salt as these are loaded terms that often lack substance in practice. As a result, while no doubt necessary to engage with and fight for the recognition of community and individual rights to land and other resources, one should be aware of the obstacles that are likely to be in the way given the priorities of the time.*

India too had to bear the brunt of unmitigated environmental destruction in the pursuit of impressive growth rates, which were considered the sine qua non of the take-off to high growth trajectory. While a number of steps has indeed been taken, a lot more needs to be done to exploit the synergies that a conservationist approach towards natural wealth and a sustainable path of development hold in their wake.

V

The foregoing analysis shows that economic research into monetary valuation of environmental commodities is still in a state of flux, although considerable progress has been made. The three economic functions, resource supply, waste assimilation and aesthetic commodity, can be regarded as a general function of natural environments, i.e. the function of life support. But, in the absence of a demand curve and market price for many environmental commodities, a number of non-market methods for estimating their value have been devised. For value data collection techniques—travel cost method, participation/unit day value method, hedonic pricing and contingent valuation—have been extensively tested. A growing literature has suggested that non-use values (bequest and existence values) and option value should be counted as part of the total economic value of the natural resource. It is only by substantially improving our understanding of economy–environment interactions that we will gain a better grasp of the economic issues.

Economic development can therefore be viewed as a process of adaptation to a changing environment, while itself being a source of environmental change. Economic production systems become more roundabout and complex as development proceeds. Development can be regarded as a process of moving through a succession of ecological niches. Niche occupancy is variable and a niche may sometimes be destroyed by means external to a society's own development process. The co-evolutionary perspective has been designed to study the ongoing feedback process between two evolving systems. If the interactions prove favourable to society, the development process continues. Since learning, knowledge and evolution are interrelated, additional co-evolutionary potential remains untapped. However, the magnitude and the extent of this development potential, which will determine how tolerable survival will be, remain uncertain.

After a brief glimpse of the literature dealing with environment–economy interlinkages, there seems to be a complete lack of a comparable analysis that demonstrates whether any particular economy is consistent with the natural environments which are necessarily linked to that economy. But the fact remains that, if we are interested in sustaining an economy, it becomes important to establish some conditions for the compatibility of economies and their environments. The drawback in the entire economics driven approach towards sustainability is the absence of any necessary binding constraint on the activities of the economic agents to make them consistent with their natural environments, more so when they are vastly dispersed over time and space. This is the very reason why we face a 'tragedy of commons' and sometimes even a 'tragedy of anti-commons' in the use of environmental resources which are often collectively owned by the society, or some community as a whole. In the absence of a commandeering 'Walrasian auctioneer', it is practically impossible to orchestrate the activities of millions of 'rational'[4] economic agents who try to maximise their respective welfares, either as producers or as consumers.

> As Pearce and Turner (1990)[5] Illustrate, *"Uncertainty still surrounds the exact nature and extent of the global interdependencies between economic growth and the supporting environmental systems. It is difficult to fully quantify the risks to future human well being posed by acid rain, ozone depletion and the greenhouse effect. Future necessary global economic growth will further diminish the sector of nature in which self regulating natural systems can regenerate free of human intervention."*

VI

The two-way relationship between environment and economic growth gave the concept of environmental sustainability of the development process by pointing out the biophysical limits of use of nature as a source of resources and a sink for waste disposal.

'Sustainable development', as defined by the World Commission on Environment and Economic Development, 1987 requires a boundary condition to be satisfied so as to take care of the considerations of inter-generational equity in resource use. Emphasis has been laid on such use of natural resources that would enable the future generations to experience the present generation's level of well being.

Munasinghe (1994), an expert in this field, describes 'sustainable development' as *'a process for improving the range of opportunities that will enable individual human beings and communities to achieve the aspirations and full potential over a sustained period of time, while maintaining the resilience of economic, social and environmental systems*[1].

In other words, sustainable development requires increases both in adaptive capacity and in opportunities for improvement of economic, social and ecological systems. Improving adaptive capacity will increase resilience and sustainability. Expanding the set of opportunities for improvement will give rise to development. Adapting this general concept, a more focused and practical approach towards making development more sustainable would seek continuing improvements in the present quality of life at a lower intensity of resource use, thereby leaving for future generations an undiminished stock of productive assets (i.e. manufactured, natural and social capital) that will enhance opportunities for improving their quality of life. (Gunderson and Holling, 2001)

Munasinghe even introduces a whole new concept of sustainable economics called 'sustainomics'[6] which can be considered as a '*trans-disciplinary, integrative, balanced, heuristic and practical meta-framework for making development more sustainable*'. Going by the above definition, the environmental, social and economic criteria have an important role to play in the sustainomics framework. Environmental sustainability comes into play because of the concerns relating to resource degradation, pollution and loss of biodiversity, and hence overall viability and health of ecological systems (defined in terms of a comprehensive, multi-scale, dynamic, hierarchical measure of resilience). Social sustainability deals with reduction of the vulnerability and maintenance of the health of social and cultural systems. Enhancing human capital through education, strengthening social values and institutions are important in this framework such that a social system develops within it a dynamic ability to adapt to change across a range of spatial and temporal scale, rather than the conservation of some ideal 'static' state. Economic sustainability refers to maximisation of the flow of income that can be generated while at least maintaining the stock of assets (or capital) which yields this income flow.

Figure 1.3

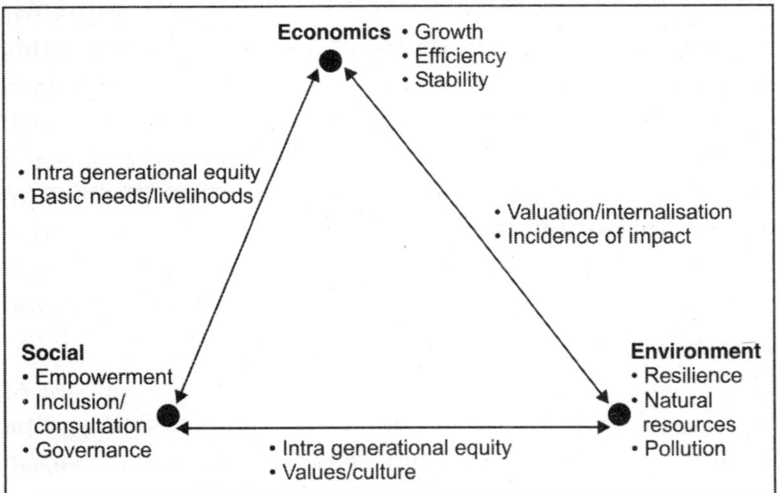

Elements of Sustainable Development
Adapted from Munasinghe (1993, 1994)

Gunderson and Holling (2001) use the term 'panarchy' to denote such a hierarchy of systems and their adaptive cycles across scales. A system at a given level is able to operate it in stable (sustainable) mode, because it is protected by the slower and more conservative changes in the super system above it, while being simultaneously invigorated and energised by the faster cycles taking place in the sub-systems below it. In other words, both conservation and continuity from above, and innovation and change from below, are integral to the panarchy-based approach.

The very mention of the word 'sustainability' is enough to give us a picture of what this entire discussion is all about and why more and more people are gradually shifting their emphasis from economic growth to larger issues such as natural endowments, community participation, and the role of market-based valuation in resource conservation, etc. To put it simply, the term is self explanatory and indispensable in any analysis of nature–economy interactions. But what often escapes notice is the holistic coverage of this approach, with its cultural, social, economic and political dimensions. Figure 1.3 makes an attempt to capture the essence of this approach. Mere focus on the environment–economy nexus would be like treating a general equilibrium problem in partial equilibrium framework; for, no matter how efficient the functioning of markets for environmental goods, how is one to assign a cost to the multifunctionality of natural goods? The following section elaborates on some of these issues.

The contribution of various economists cited in this paper is valuable indeed but even now we clearly lack a framework or some clearly defined guidelines which can be adhered to, introduce an element of sustainability in the growth process. The elaborate mathematical models with constraints on resource use and endless number of equations might seem impressive on paper and may indeed be extend the horizons of our understanding of the issues at hand, but a foolproof mechanism to bring about the same is yet to be devised.

VII

While the entire discussion on sustainability might lead us to believe that the stock of natural resources be held constant over time, it does not at all mean that we use none of the exhaustible resources at all. Rather, by following the techniques of prudent environmental management, one can actually improve the sustained yield and the waste assimilative capacity of the natural resources, e.g. by allowing the rivers to recharge the underground aquifers, by adopting energy plantation[7] to cut down on the dependence on fossil fuels). One way to ensure sustainability of natural resources is by increasing the efficiency in resource use. However, the primary channel through which such efficiency gains can be brought about is technological progress. But there is nothing to ensure that the new technology would be necessarily less polluting. (e.g. the shift from burning fuel wood to fossil fuels has lead to massive air pollution). Moreover, as development requires a constantly growing stock of resources to draw upon, there is no guarantee that technological progress will continue for a very long time. Once again, quoting Pearce and Turner, '*there may indeed be backstop technologies that will free us from natural resources, but they cannot be brought into existence simply by assuming that they are there*'. Moreover, while there is considerable scope for substituting natural capital by manmade capital, it needs to be borne in mind that manmade capital is not independent of natural capital and it is important to remember the multifunctionality of natural resources, a feature which is not shared by manmade capital.

The point of the entire discussion is that it is important to exercise restraint in using natural resources rather than simply squandering them away in a blind quest for economic growth. Despite the advancements in the frontiers of scientific knowledge, there is no certainty about the ways in which environments function (e.g. the knowledge of complex bio-geochemical cycles, etc. is rather limited), whether internally or in terms of their interactions with the economy. Moreover, if we decide to surrender natural capital, there is often a sting in the tail: irreversibility. Given these limitations, though economic wisdom would suggest maintenance of an optimal stock of natural capital in place of the existing stock of resources, it would be wiser to follow the prudent approach of resource conservation, more so when there is a rationale in terms of uncertainty and irreversibility for conserving the existing stock, at least until we have a clearer understanding of what the optimal stock is and how it might be identified. A few points that are put forth by the economists in this regard are:

(a) Dematerialisation of economic process that would reduce their net aggregate impact on nature

(b) Decarbonisation of energy to circumvent the problems of pollution and exhaustibility of fossil fuels

(c) Increasing substitution of non-renewable resources by renewables

(d) Recycling of waste by converting it into a manmade resource

(e) Non-recyclable waste treatment before disposal, especially the harmful effluents

(f) Enhancement of primary productivity of bio-spheric space in ecosystems

(g) Facilitation of the redistribution process of income by increasing the productivity of wage goods and creating more employment opportunities

Finally, while it has been attempted to expose the various angles of the entire discussion, there is immense scope for advanced treatment of several issues which might have been raised thus. Especially in a developing country like ours, the need for a sustainable harvest of resources may not be perceived in the correct terms, given the vast proportion of poor for whom the immediate concern is to raise their standard of living and escape from the miserable conditions which affect their being. But, it is indeed commendable that a beginning in this direction has been made, and it is hoped that all the stakeholders will ensure that the momentum in this direction will be '*sustained*'.

Bibliography

- Divan, S. and A. Rosencranz (2001), *"Environmental Law and Policy in India."* (2nd edition), Oxford University Press, New Delhi.
- Encyclopaedia Britannica, Book of the Year 2002.
- Ghate, R. S. (1992), *"Forest Policy and Tribal Development: A Study of Maharashtra"*, Concept Publishing House, New Delhi.
- Gunderson, L. and Holling, C.S (2001), *"Panarchy: Understanding Transformations In Human And Natural Systems"*, Island Press, New York.
- Kadekodi, G.K. (1995), *"Operationalising Sustainable Development, Ecology-Environment Interactions at Regional Level"*. IVM Internal Publication, Institute For Environmental Studies, Amsterdam, The Netherlands.
- Lafferty, W M (1998), *"The Politics of Sustainable Development: Global Norms for National Implementation"* in J S Dyzek and D Scholsberg (eds), Debating the Earth: The Environmental Politics Reader, Oxford University Press, London.
- Menon, Ajit (2006), *"Environmental Policy, Legislation and Construction of Social Natur"*, Economic and Political Weekly, January 21, 2006
- Munasinghe, Mohan (1994), *"The Sustainomics Transdisciplinary Meta-Framework For Making Development More Sustainable: Applications To Energy Issues"* in Economics Of Environment And Development ed. by Pushpam Kumar
- Pearce, D.W. and Turner, Kerry (1990), *"Economics Of Natural Resources And The Environment"*, Chapter 1-4, Harvester Wheatsheaf, London UK
- Sengupta, Ramprasad: Class notes, Environmental Economics, Centre for Economic studies and Planning, JNU, 2005
- Srinivas, K.R. (2000), *"Biodiversity Bill: Nice Words, No Vision"*, Economic and Political Weekly, November 4, pp 3916-19

Endnotes

1. Transnational coalition building was and remains another important strategy for environmental organisations and for grass roots movements in developing countries, primarily because it facilitates the exchange of information and expertise but also because it strengthens lobbying and direct-action campaigns at the international level.
2. Encyclopaedia Britannica, Book of the Year 2002
3. Although the National Biodiversity Act, 2002, does recognise the need for local biodiversity committees (BMCs), the actual powers given to these committees again are mostly managerial in nature. Under Section 41, these committees are constituted for "promoting conservation, sustainable use and documentation of biological diversity including preservation of habitats, conservation of land races, folk varieties and cultivars, domesticated stocks and breeds of animals and microorganisms and chronicling of knowledge relating to biological diversity". What remains missing is any serious consideration of what the entitlements and rights of these committees are.
4. Humanists often put forth the notion of extended self rationality according to which individuals are capable of truly altruistic acts. This extended rationality also generates a strong obligation to abide by particular laws which are seen by the individual as promoting his/her meta-preferences, despite a potential tension between law and narrow self-interest.
5. W. Pearce and Kerry Turner, *Economy and Environment*, Chapter 1.
6. *Sustainomics* attempts to use two approaches to yield consistent and complementary results, i.e. to provide an integrated and a balanced treatment of economic, social and environmental viewpoints. First, when material growth is the main objective and uncertainty is not a serious problem, then the focus may be on optimising economic output subject to constraints that ensure social and environmental sustainability. Second, an alternative objective could be the sustainability of the environment. In this case emphasis tends to be on paths which are economically, socially, and environmentally durable or resilient, but not necessarily growth optimising.
7. 'Energy plantation' might seem to be an over-used term in the discussion on environmental economics, but the fact is that by planting trees like Jatropha, Pongamia, etc. on the vast areas of wastelands, we can not only get an assured supply of bio-fuels but also provide a sustainable means of livelihood to several communities which are dependent on natural resources.

Chapter 2

The Role of Economic Instruments in Prevention and Control of Industrial Pollution in India: Scope, Design and Implementation

U. Sankar

1. Introduction

Industrial pollution is a by-product of industrial activity. It becomes a negative externality when a firm fails to internalise the environmental costs in its production decisions and passes the costs on to the rest of society by emitting pollutants into air, water and land. Pigou (1920) suggested government intervention in the form of a levy of per unit tax on the output equal to the difference between the marginal social cost and the marginal private cost, corresponding to the social optimal output. Coase (1960) advocated a role for government only in the assignment of property rights for environmental resources and reduction in transaction costs to facilitate voluntary bargaining between polluters and pollutees.

The United Nations Conference on Environment and Development (UNCED, 1992) states the following objectives of environmental policy:

1. To incorporate environmental costs in the decisions of producers and consumers, to reverse the tendency to treat environment as a 'free good' and to pass

these costs on to other parts of society, other countries or to future generations,

2. To move more fully towards the integration of social and environmental costs into economic activities, so that prices will appropriately reflect the relative scarcity and total value of resources and contribute towards the prevention of environmental degradation, and
3. To include, wherever appropriate, the use of market principles in the framing of economic instruments and policies to pursue sustainable development (UNCED, 1992, Agenda 21, Chapter 8, p.85).

It endorses application of the 'polluter pays' principle and the precautionary principle to environmental policy. While the 'polluter pays' principle makes the polluter bear the costs of pollution, the precautionary principle recognises the existence of uncertainty and seeks to avoid irreversible damages via the imposition of a safety margin into policy.

In India, the important legislations dealing with industrial pollution are: the Water (Prevention and Control of Pollution) Act 1974, the Air (Prevention and Control of Pollution) Act 1981, and the Environment (Protection) Act 1986. Divan and Rosencranz (2001) characterise India's pollution control regime based on the environmental legislations and rules emanating from the legislations as 'prohibit and punish' regime.

Government of India (1992) Policy Statement for Abatement of Pollution favoured 'a mix of instruments in the form of legislation and regulation, fiscal incentives, voluntary agreements, educational programmes and information campaigns'. It recommended the 'polluter pays' principle, involvement of the public in decision making and new approaches for considering market choices 'to give industries and consumers clear signals about the cost of using environmental and natural resources'.

In India, economic instruments (EIs) remain the exception rather than the rule. The major factor inhibiting the introduction of EIs is that all the three environmental legislations come under criminal law. Government of India (2006) National Environmental Policy (NEP) recognises the problem. It states that 'the present environmental mechanism is predominantly based on doctrines of criminal liability, which have not proved sufficiently effective and need to be supplemented'. {4.(viii)} It notes that civil law 'offers flexibility, and its sanctions can be more effectively tailored to particular situations. The evidentiary burdens of civil proceedings are less daunting than those of criminal law. It also allows for preventive policing through orders and injunctions' {5.1.2.(ii)}.

The NEP recommends a review of the existing legislation to arrive at a judicious mix of civil and criminal processes and sanctions. It states that civil liability law, civil sanctions, and processes would govern most situations of non compliance. Criminal processes and sanctions would be available for serious, and potentially provable, infringements of environmental law, and their initiation would be vested in responsible authorities. Recourse may also be had to the relevant provisions in the Indian Penal Code and the Criminal Procedure Code. Both civil and criminal penalties would be graded according to the severity of the infraction {5.1.2(ii)}.

The NEP notes that EIs 'work by aligning the interest of economic actors with environmental compliance, primarily through the application of the 'polluter pays' principle and recommends an action plan for preparation and implementation on the use of EIs for environmental regulation in specified contexts' {5.1.(vi)}.

This paper considers the scope for application of EIs, choice among EIs, and problems in design and implementation of each EI in a given context. It reviews the efforts already made by the Ministry of Environment and Forests (MoEF) towards the introduction of EIs and indicates what needs to be done in developing and applying EIs for the prevention and control of industrial pollution.

The plan of the paper is as follows. Section 2 outlines the options for the design of an environmental policy regime based on predetermined environmental standards. Of the three possible regimes, India has been relying largely on the command and control regime. Within this regime, there is scope for introducing only indirect EIs, such as fiscal measures, bank guarantees and public opinion to influence polluters' behaviour. Section 3 deals with design and implementation of pollution charges, with focus on water pollution. Section 4 covers problems in developing markets for tradeable permits. Section 5 considers the need for reassessing the environmental policy in the context of globalisation and planned annual GDP growth of 10 per cent. Section 6 contains concluding remarks.

2. Command and Control Environmental Policy Regime, Policy Options and the Next Steps

Policy options:

Environmental policy regimes for pollution control in most countries are based on a two-stage approach. The first stage involves determination of environmental quality standards based on baseline pollution loads, stage of economic development, societal preferences and assessment of the feasibility of achieving the standards. In the second stage, there are three possible policy approaches to achieve the predetermined environmental targets, namely: (i) regulation or command and control policies, (ii) implementation of pollution charges, and (iii) development of markets for pollution permits. A mix of these three options is also feasible. Based on a general equilibrium framework and using the Pareto-efficiency criterion, Baumol and Oates (1987) show that an environmental policy regime based either on the standards and charges approach, or on the standards and permit approach minimises the aggregate compliance costs to society. These two approaches give signals to the producers to search for and adopt cost-effective methods of complying with the standards.

Command and control regime:

As mentioned earlier, the present policy regime for prevention and control of industrial pollution in India is the command and control type of regulation. Polluters are permitted to pollute up to the prescribed levels, generally based on pollution concentrations in emissions, and any excesses beyond the prescribed limits are prohibited. The enforcement is only on whether or not a firm complies with the standards, and not on the extent of compliance (Sankar, 2001a).

The punishments for violations of the standards are unrelated to (often lower than) the compliance costs. As a result, even when the enforcement is strict, it is cheaper for firms with high pollution abatement costs to pay the fines rather than to carry out the abatement. When political will for enforcement is weak, the regime creates an opportunity for 'rent seeking'. When court judgement results in an order for closure of polluting units, other issues like loss of output, loss of employment or/and loss of exports create pressures on governments to search for remedial measures.

Another problem with the command and control policy regime in India is that the standards are uniform for most regions and firms of varying sizes and vintages. Even though the legislations permit the State Pollution Control

Boards (SPCBs) to fix standards lighter than those prescribed by the Central Pollution Control Board (CPCB) in pollution hotspots, most SPCBs do not prescribe tighter standards for the hotspots.[1] It is well known that average and marginal abatement costs are higher for small and medium firms than for large firms. Small and medium firms also face structural adjustment problems and credit crunch.

Supportive measures:

The government has taken a few supportive measures with the intention of reducing the pollution compliance costs for firms. These measures include fiscal incentives, assistance in erecting common effluent treatment plants, creation of industrial estates with environmental support measures and technological modernisation programmes.

The fiscal incentives are: rebates on customs and excise duties on capital equipments and accelerated depreciation allowances for investments in pollution-control equipments. These policies reduce the annualised capital costs for complying with the regulations. If the net operating costs of the effluent treatment plants are positive, then the firms have no incentive to operate the plants, when the enforcement regime is weak. The Task Force appointed by the MoEF, (see NIPFP, 1997), recommended abolition of tax concession on installation of pollution-control equipment. Such fiscal incentives can be justified only when the firms set up plants for recovery and reuse of materials and water in such a way that the value of the recovered materials and water exceeds the operating cost of the units; then the firms have an incentive to operate the plants without any need for monitoring.

The government has been assisting small units in industrial clusters to erect and operate Common Effluent Treatment Plants (CETPs). Government help comes in the form of capital subsidy. Sankar (2001b, 2004) urges the need for creating incentive structures for self management of the CETPs and full internalisation of the environmental costs in the decisions of CETPs and polluting firms.

The government has initiated a few technological upgradation schemes for small and medium industries in cooperation with agencies such as United Nation Industrial Development Organisation (UNIDO).

Bank guarantees:

The West Bengal Pollution Control Board took the initiative of introducing a bank guarantee scheme as an alternative to closure/regulation. The possible applications are: (i) compliance with the conditions in consent to establish and consent to operate certificates, (ii) enforcing compliance with the standards, and (iii) restoration of land or the area a firm has degraded by discharging effluents. Under this scheme, the polluting firm and the State Pollution Control Board (SPCB) reaches a time-bound action plan for upgradation of technology/remediation of land/compliance with regulations. The firm executes a bank guarantee and if it fails to execute the job in time, it forfeits the amount. The SPCB may use the amount for environmental purposes. There is no legal barrier for introducing the scheme. The Kolkata High Court in W.B. 5938(W) of 2000 endorsed the scheme.

The problem with the scheme is that it is not standardised and is applicable only to some firm-specific compliance problems. It is not suitable for monitoring emissions when it is continuous. For environmental damages resulting from random accidents/breakdowns of equipments, performance fund is a better instrument than a bank guarantee.

Taxes on polluting inputs and outputs:

The National Institute of Public Finance and Policy (NIPFP,

1997) recommended taxes on complementary inputs and or outputs in cases where emissions are difficult to measure. An ecotax on a polluting output would raise the output price, thereby causing a reduction in its demand. An ecotax on a polluting input would increase the cost of production and hence the output price, thereby resulting in reduction in its demand. The advantages of ecotax are: (a) there is no legal barrier, i.e. it can be introduced in government budget, (b) the tax will generate additional revenue to the government which can be earmarked for environmental purposes and (c) the transaction costs of introducing ecotaxes are much lower than taxes on pollutants because the tax administration has reliable measures of inputs and outputs and the incremental costs of administering the ecotaxes will be small. If there is a proportional relation between output/input and pollution load, ecotax can achieve the environmental effectiveness criterion. If there is no proportional relationship between output/input and pollution load because of differences among firms in per unit emissions, then rebates from the ecotaxes should be given to firms who have better compliance records.

Chelliah, Appasamy, Pandey and Sankar (2007) have recommended ecotaxes on coal, automobiles, chlorine use in pulp and paper and viscose rayon industries, phosphate-based detergents, chemical pesticides, chemical fertilisers, lead acid batteries and plastics. They have suggested differential tax rates for chemical and organic fertilizers, and chemical and biopesticides.[2]

Legal reforms for introduction of direct EIs:

The Central Pollution Control Board (CPCB) set up a Legal Discussion Group in 2002 to examine legal measures to be taken for the introduction of pollution charges in India.[3] The Legal Discussion Group submitted its report to the MoEF in February 2003 (see CPCB, 2003). The Report noted that the Water (Prevention and Control of Pollution) Act, 1974 and the Environment (Protection) Act, 1986 empower the SPCBs only to impose physical barriers through regulations by way of prescribing standards for effluent disposal, directing polluters for provisions of pollution control equipment/structures, etc. These legislations do not permit imposition of monetary/fiscal barriers such as pollution charges/environment user fees. Violations of the provisions of the existing environmental legislation are tried under the criminal procedure code. It does not provide for partial compliance/violations of the provisions of Criminal Acts.

It suggested amendments to some sections in the Water Act 1974 and the Environment Act 1986 to enable imposition of pollution charges for violation of prescribed effluent standards. The suggested amendments are Sections 24,25,43 and 44 of the Water Act and Sections 3(2), 3(3) and 6 of the Environment Act. It also noted a few precedents with regard to imposing fiscal regulatory measures under the Environment Act. Some examples are (i) Hazardous Waste (Management & Handling) Rules, 1989-Section 16(3) providing for levying fine by SPCBs, and (ii) Aquaculture Authority and Loss of Ecology Authority constituted under Section 3(3) of the Environment (Protection) Act giving powers to assess and impose monetary barriers for past acts of pollution by industries.

The next steps:

It is necessary to identify cases when standards and regulation of the command and control type are preferable to EIs such as pollution charges and pollution permits. The precautionary approach is relevant when the carrying capacity/assimilative capacity of a region is near the threshold level or new emissions will result in irreversible damages, or there are serious uncertainties about the consequences of an economic policy on the environment.

In the context of industrial pollution, there is a case for prohibition of discharge of effluents containing hazardous wastes.

Even when a command and control policy is relevant, there is room for giving some choices to the polluters to prevent and control pollution. The regulator should set only the permissible levels of pollutants/pollution load and not the technologies, processes or input usages. The firm can use its private information to find a cost-effective solution for compliance with the regulations. For violations in excess of the permissible limits, the penalties should increase with the concentrations in the emissions so that the fines can act as deterrents.

With legal amendments to the three environmental legislations, it will be feasible to introduce the pollution charges or pollution permits. The choice between these two options depends on: (i) the Weitzman condition, (ii) political preference, and (iii) the transaction costs. Weitzman (1974) has shown that if the abatement cost function is expected to be flat relative to the damage function, then the quantity should be fixed. If, however, the cost function is expected to be steeper than the damage function, then pollution charges are preferred. Political preference also matters. The United States has preferred the permit approach whereas many western European Countries have preferred the charge system. Some feel that the right to permit is not consistent with the idea that the State is a trustee of natural resources. Finally, the relative transaction costs of designing and implementing the two systems also influence the choice. In general, administrators and scientists seem to prefer quantity targets because they are easily observable and measurable and meet the environmental effectiveness criterion.

3. Design and Implementation of Pollution Charges

There is a vast theoretical and empirical literature on design and implementation of pollution charges. See for example World Bank (2000), Sankar (2002) and World Bank, Ministry of Environment and Forests, and Confederation of Indian Industry (2002).

Baumol and Oates (1987) suggest that the pollution charges be equal to the marginal abatement costs corresponding to the prescribed standards. Econometric exercises on estimation of pollution abatement cost functions in India by Mehta, Mundle and Sankar (1997), Murty, James and Misra (1998) and Goldar, Misra and Mukherjee (2001) consider issues in the specification and estimation of the abatement cost functions, focusing on choice of pollutants, functional form, database and estimation techniques.

Conceptual and measurement problems:

There are some conceptual and policy issues in using the estimates of marginal abatement costs in the design of a pollution charge system. These are: (i) The estimated cost function is an expected cost function, applicable to an average firm. Mehta, Mundle and Sankar (1997) argue that the marginal costs of relatively high-cost producers should serve as the basis for setting charges and taxes. This would ensure that most producers find it cheaper to abate rather than pollute. But a higher charge may be difficult to implement because of 'political risk aversion' (see Helms, 2005). (ii) Even in the case of one pollutant, say biological oxygen demand (BOD), two charges are needed—one on the volume of waste water and another on BOD concentration in excess of the permissible limit. If there is a charge only on the volume of waste water, pollution concentration will increase, and if there is a charge only on BOD, water use will increase because of dilution. (iv) The Water Act prescribes limits on many parameters. Estimation of abatement cost function with multiple pollutants with flexible functional form results in econometric problems such as multicollinearity and loss of degrees of freedom. For this reason, one has to be content with one or two parameters which are important for pollution control. (v) The results are sensitive to the model and the database employed.

An alternative to the econometric cost-function approach is to find the actual abatement expenditure incurred by a firm and then estimate, on a normative basis, what would be the additional cost of complying with the standards. In case of water pollution from tanneries, the additional cost can be estimated for abatement requirement up to the prescribed level for each pollutant, e.g BOD, totally dissolved solids, sludge disposal and so on (see Sankar, 2001b). The advantages of the approach are that (a) it avoids problems in specification and estimation of abatement cost functions, and (b) the information requirements are less severe. The disadvantage is that the results cannot be generalised; we need many case studies.

Measurement of pollution:

Apart from determination of the charge base and the charge rate, some problems arise at the implementation stage. To administer the charge system effectively, reliable measures of pollution concentration/pollution loads are essential. If meters can be installed and operated effectively, then the implementation is easier. If metering is not possible because of the high installation and operating costs, one has to estimate the charge bases either by sample surveys or/and proxies for pollution, such as water usage , process adopted, installed capacity. In order to encourage the firms to adopt waste minimisation techniques, it is desirable to revise the charges periodically, based on the firm's past performance records. Metering may be feasible for large firms. For small and medium units, sample surveys or proxies may be the only alternative. For these units, supportive measures such as dissemination of information about pollution prevention and control, technological upgradation, access to environmentally sound technologies, and subsidy for treatment plants are needed to enable them to reduce the burden of high abatement costs.

Another issue relates to the subsidiarity principle. As water pollution is a 'local bad', there is a case for design and implementation of pollution charges for each location e.g. watershed or industrial cluster. The standards can be determined based on current pollution loads, assimilative capacity of area and the preferences of various stakeholders. A public–private partnership involving local bodies, polluting units, civil society, and technical experts is necessary for the design and implementation of pollution charge systems.

Experiences of developing countries in the design and implementation of pollution charges reveal that initially they choose a few pollutants (one or two) and low rates (see World Bank, Ministry of Environment and Forests and Confederation of Indian Industry, 2002). The charge is set at a low level, far below the average and marginal abatement costs. This is to ensure participation and compliance by polluters. Helms (2001) gives two reasons for starting with low charges. First, introduction of the charge system has a learning effect. The process of introducing the instrument educates the affected parties, and is often accompanied by media information on ways of substituting to mitigate the impact. Second, it gives adjustment time for the polluters as well as for the management to learn about the shapes of the cost and damage functions.

Monitoring and enforcement:

Any pollution control regime would require monitoring and

enforcement regardless of policy choice . Mehta, Mundle and Sankar (1997) discuss a menu of different pollution control incentive regimes. The first option is abatement charges with government clean-up. The second option is abatement charges with third party clean-up. The third option is levy of a tax on all firms that fail to achieve the source-specific standards and a per-unit subsidy to firms that carry on abatement beyond the prescribed limits. The fourth option is tradeable private permit systems. The degree of government intervention is the maximum with the first option and lowest with the fourth option.

Other options for monitoring and enforcement suggested in the literature include: (i) incentives for regulators for better enforcement, (ii) external audit, (iii) monitoring by industry associations, (iv) environmental rating of firms, (v) actions by local community or/and association of affected people.

4. Tradeable Permits

Compared with the charge system, a tradeable permit system has some attractive features, such as environmental effectiveness and greater reliance on market forces for achieving the quantity targets. However, introduction of the system raises ethical issues, like 'right to pollute' and capacity to create and operate perfect markets.

If pollution is a public bad and its impact is the same everywhere, the market can be the whole country. However, it is necessary to restrict the coverage to point sources and perhaps to large and medium units in order to reduce the transaction costs. With an estimate of baseline pollution load and desirable quantity target, permits can be allocated among the units either on the basis of 'grandfathering' or auctioning.[4] The regulator's main problems are certification of emission credits and overseeing market operations. In this case, with a large number of participants, one can expect competitive outcomes.

When the pollutant is uniformly distributed but its impact varies from region to region because of differences in their assimilative capacities, then one unit of emission must get different weight, depending on the assimilative capacity of a region. It is the job of the market regulators to fix the exchange ratio between any two regions. When the pollutant is non-uniformly distributed and its impact varies from region to region, then we need separate markets and the market sizes would be small. Perfect competition may not be feasible because of the thin markets and high transaction costs in the certification of permits and monitoring the market transactions.

In the Indian context, it is desirable to carry out pilot studies in the following areas—intra-firm permit system, tradeable permits in a watershed and tradeable permits among members of CETPs. At present, the standards are monitored at each emission point in a plant. As the control costs vary among emission points, and in order to minimise overall control costs, it is desirable to have a monitoring system for the plant as a whole and even for the firm as a whole in the form of a permissible pollution load. Such a system will result in equalisation of marginal abatements costs at the different points and hence minimisation of aggregate compliance cost.

Pandey and Bharadwaj (2004) develop a model to examine the compliance costs under an intra-plant emission trading system for a non-uniformly mixed assimilative plant and apply it to an integrated steel plant (Bokaro Steel Plant) in India. They find that intra-plant trades would result in significant savings to the steel plant system.

Sankar (2004) argues a case for experimenting a tradeable permit system among members of a CETP.

Tanners' abilities in reducing pollution loads via adoption of pollution-prevention measures or/and primary treatment and their marginal costs of reducing/abating pollution differ, because they are heterogeneous in terms of size, vintage, raw material processed, process adopted and management skills. A tradeable permit scheme based on the volume of effluent and concentrations of pollutants in the effluent discharged will provide an incentive to each tanner to search for the least cost options for reduction, of pollution loads. When a firm achieves pollution reduction it can use the surplus permit for capacity expansion or sell it to another member. Hence, with the same CETP capacity, tannery production can increase.

It is also desirable to experiment with a system of pollution charge/environmental user-fee system in a river basin or watershed. This requires a partnership among polluting units, local bodies and affected parties. This participatory approach provides an opportunity for learning by doing and mid-term corrections.

5. Policy Reform for Sustainable Development

One problem with the current environmental policy regime is the absence of link between the ambient and source standards, i.e. the ambient and source-specific standards are determined independent (see Mehta, Mundle and Sankar, 1997). Hence, it is quite possible that the quality of the environment could continue to deteriorate despite a high degree of compliance among individual polluters. This problem assumes importance in the context of a high annual GDP growth rate of 10 per cent contemplated during the Eleventh Five-Year Plan. To achieve a sustainable growth, the source-specific standards have to be tightened periodically to contain the pollution loads at the macro level.

India's external trade as a per cent age of her GDP has been increasing in recent years. India is also a member of the World Trade Organisation (WTO) and many multilateral environmental agreements. It has to comply with the Technical Barriers to Trade and Sanitary and Phyto-Sanitary Agreements of the WTO and other environmental requirements of the importing countries to retain/gain market access in the importing countries (see Sankar, 2006). India has also signed many multilateral environmental agreements. The Doha Round of negotiations includes liberalisation of trade in environmental goods and services. Environmental issues also now enter in to regional and bilateral trade agreements.

In this context, India's environmental policy must lay stress more on pollution prevention via adoption of environmentally sound technologies and cleaner processes, and recovery/reuse of materials from waste water, solid waste and emissions into air. Large industrial units can adopt environmentally sound technologies and adopt waste minimisation techniques. They have an incentive to do so to gain dynamic comparative advantage. Small and medium units, especially the export-oriented units, need government support in technological upgradation, gaining access to environmentally sound technologies, and establishing and operating CETPs and industrial complexes with environmental infrastructure.

6. Concluding Remarks

This paper considered the scope of EIs and issues in their design and implementation in India. Regulation in the form of command and control should be confined to situations where the precautionary principle is applicable. These situations include cases where environmental damage function is very steep, the carrying capacity of the receiving region is at the threshold level, irreversible damages are likely and pollution impacting 'incommensurable' ('incomparable') values.

As noted in the NEP and as recommended by the CPCB, the legal barriers to the introduction of EIs should be removed by amending the environmental legislations. Civil liability law, civil sanctions, and processes should govern most situations of non compliance. When pollution is measurable, EIs such as pollution charges and tradeable permits should be considered. The design of the charge system and the permit system must be based on the participatory approach. This approach provides an opportunity to understand the constraints and adjustment problems polluting firms would face at the implementation stage and consider their behavioural reactions to the proposed policy changes. This approach is necessary to provide flexibility in the design and implementation of the policies and also to decide on the choice of supportive measures to ensure a smooth and effective transition to the new policy regime.

In cases where measurement of pollution is difficult or/and transaction costs of implementing a charge system or a permit system are high, indirect EIs such as taxes on polluting inputs and outputs, environmental user charges and bank guarantees be adopted.

Regarding monitoring and enforcement of environmental laws and rules, the efforts of the CPCB and the SPCBs should be supplemented by NGOs , civil society and affected parties, serving as watchdogs. Market pressures, media coverage, green rating by independent and reputed agencies and public recognition of firms achieving compliance beyond the prescribed levels create an environment for sound environmental management. The NEP recognises that maintaining a healthy environment is not the state's responsibility alone, but that of every citizen.

References

Baumol, W.J. and W.E. Oates (1987), *The Theory of Environmental Policy*, 2nd edition, Cambridge University Press, Cambridge.

Central Pollution Control Board (2003), Legal Supports in Environmental Legislations to Introduce Pollution Charges/Environment User Fees for Water Pollution Control: An Assessment, Delhi.

Chelliah, R.J., P.P. Appasamy, Rita Pandey and U. Sankar (2007), *Ecotaxes on Polluting Inputs and Outputs*, Academic Foundation, New Delhi.

Coase, R.H. (1960), The Problem of Social Cost, *Journal of Law and Economics*, Vol.3, pp. 1–44.

Divan, S. and A. Rosencranz (2001), *Environmental Law and Policy in India*, Oxford University Press, New Delhi.

Goldar, Bishwanath, Smita Misra and Badal Mukherjee (2001), Water Pollution Abatement Cost Function. Methodological Issues and an Application to Small Scale Factories in an Industrial Estate in India, *Environment and Development Economics*, Vol. 6, Part 103–122.

Government of India (Ministry of Environment and Forests) (1992), Policy Statement for Abatement of Pollution, New Delhi.

____________ (2006), National Environmental Policy, New Delhi.

Helms, Dieter (2005), Economic Instruments and Environmental Policy, *The Economic and Social Review*, Vol. 36, No. 3.

Mehta, S., S. Mundle and U. Sankar (1997), *Controlling Pollution Incentives and Regulations*, Sage Publications, New Delhi.

Murty, M.N., A.J. James and Smita Misra (1998), *Economics of Water Pollution—The Indian Experience*, Oxford University Press, New Delhi.

National Institute of Public Finance and Policy (1997), *Report of the Task Force to Evaluate Market-based Instruments for Industrial Pollution Abatement*, New Delhi.

Pandey, Rita and Geetish Bharadwaj (2004), Comparing the Cost Effectiveness of Market-Based Policy Instruments versus Regulation: the Case of Emission Trading in an Integrated Steel Plant in India, *Environment and Development Economics*, Vol. 9, Part 1.

Pigou, A.C. (1920) *Economics of Welfare*, Macmillan, London.

Sankar, U. (2001a), Laws and Institutions Relating to Environmental Protection in India, presented at the conference on 'The Role of Law and Legal Institutions in Asian Economic Development' held at the

Erasmus University, Rotterdam 1–4 November 1998. MSE Occasional Paper No. 3.

____________ (2001b), *Economic Analysis of Environmental Problems in Tanneries and Textile Bleaching and Dyeing Units and Suggestions for Policy Action*, Allied Publishers, Chennai.

____________ (2002), On the Design and Enforcement of Fiscal Instruments for Pollution Control in India in M. Govinda Rao (ed.), *Development, Poverty and Public Policy*, Oxford University Press, New Delhi.

____________ (2004), Pollution Control in Tanneries, in Gopal K. Kadekodi (ed.) *Environmental Economics in Practice*, Oxford University Press, New Delhi.

____________ (2006), *Trade and Environment; A Study of India's Leather Exports*, Oxford University Press, New Delhi.

United Nations Conference on Environment and Development (1992), *Agenda* 21, Rio de Janeiro.

Weitzman, M.L., (1974), Prices Vs Quantities, *Review of Economic Studies*,Vol.14,No4, 477-91.

World Bank (2000), *Greening Industry New Roles for Communities, Markets and Governments*, Oxford University Press, New York.

World Bank, Ministry of Environment and Forests and Confederation of Indian Industry (2002), *International Workshop on Economic Instruments for Industrial Pollution Prevention and Control in India*, Confederation of Indian Industry, New Delhi.

Endnotes

1. When the SPCBs issue 'consent to establish' certificates, sometimes they disallow new units in highly polluted areas.
2. The project was funded by the Ministry of Environment and Forests (MoEF) to the Madras School of Economics under its Centre of Excellence in the Environmental Economics Programme.
3. The Legal Discussion Group was constituted by the CPCB to provide inputs for the Task Force on Economic Instruments of the MoEF.
4. Grandfathering, i.e. allocation of pollution permits proportional to baseline pollution levels to firms free of cost, is popular in the US. Political feasibility and industry acceptance favour this approach. The advantage of auctioning is that it generates revenue to government.

Chapter 3

Environment, Development and the Role of Technology

B. Sudhakara Reddy

You are probably well aware that our collective ecological footprint is on the rise due to the way we use our resources. I understand that when the British were leaving India, one of its gentlemen asked Mohan Das Karamchand Gandhi, rather sarcastically, 'Mr. Gandhi, how much time do you think it will take India to reach the level of the standard of living of the Briton?' Quick came the reply from the Mahatma, 'Sir, it took more than half of the earth's resources for Briton to reach that level. For a size and population of India, how many planets are required to reach that standard of living?' That is indeed what we should worry about—the unlimited appetite for resource utilisation.

The question that we must ask is: to which of our modern values can we attribute the gross destruction of our environmental resource base? Forests replaced by farmlands, and of course farmlands by Special Economic Zones (SEZs); wetlands drained; rivers channelled and dammed; erosion of the soil cover; silting of water courses; pollution of air, soil, seas and freshwater resources; and the near extinction of species. The answer is clear: we equate growth with development, our obsession with GDP as a measure of efficiency and affluence, to our passion for technology as a panacea for all our ills.

How should we view development? As a comprehensive and global process, embracing all aspects of the social system and its inter-relationship with the natural environment. In this dynamic inter-relationship, technology is the fundamental link between the social and the natural systems; at the same time, it is an essential instrument for the achievement of sustainable and environmentally sound development in the long run. The role of technology is to serve development goals, because development itself is a unified process, albeit with economic, social and environmental facets. In fact, each technological pattern implies specific approaches to the management of resources and is associated with a given value system and lifestyle.

What is the scenario now? Technological application has on the one hand, created new opportunities and fostered development and, on the other, created new problems particularly for environment.

The various criticisms of the utilization of modern technology can be classified into three broad categories: (i) environmental; (ii) economic; and (iii) social; but the overlap between these categories prevents an unambiguous classification.

Environmental Criticisms

The prolific advances in modern technology have led to spectacular increases in affluence, but this affluence has not resulted in an environment more conducive to the physical and mental well-being of man. Indeed, with the increasing deployment of modern technology, man's welfare has been threatened by escalating levels of pollution—pollution of the air that he breathes, the water that he drinks, the food that he eats, the quietness that he needs and the beauty of nature that he enjoys. At the same time, the nature of these technologies has a determining influence on the structure and functioning of human settlements. In particular, urban gigantism has become increasingly predominant; and with it, has followed the aggravation of psychological stresses and social tensions. Simultaneously, these giant cities have had major environmental impacts arising from their exorbitant demands for water, energy, sanitation, transportation and housing.

All this hyper-activity of production and consumption has involved a scale of 'exploitation of natural resources'. The word 'exploitation', which accurately describes the essence of the man–nature relationship implicit in modern technology, connotes the very opposite of efficient resource management. The effects of this irresponsibility are already evident in the disturbance of the finely adjusted ecological balances of nature. The gravity of the risks vary from relatively trivial ones, like automobile accidents to potentially catastrophic ones such as all-out nuclear warfare or destruction of the life-sustaining properties of the biosphere.

There is a perception that environmental problems may be potential threats to development and we must not and will not allow ourselves to be distracted from the imperatives of the economic development and growth by the illusory dream of an atmosphere free from smoke. Further, the powerful anti-environmental lobbies permit many modern technologies based on the plant or mineral resources of the region to use these resources irrationally and wastefully. Serious environmental effects follow, e.g. rayon factories denuding a whole region of its bamboo forests. This results, first, in setting up the unending exodus to cities that then cannot cope with the resulting environmental problems and, second, in upsetting the traditionally sound ways of managing rural land. If social participation and control on the resources cannot assume peaceful forms, it can only lead to explosions of violence (as we are witnessing in West Bengal).

What should we do? The answer is the development and diffusion of environmentally sound technologies, which fall into three broad categories:

(a) economic goals predominate;

(b) environmental concerns are crucial; and

(c) social goals are emphasised.

This view of development is fundamentally different from the one in which development is equated with growth. It is focused on human beings, rather than only on goods and services. It is principally concerned with the quality of life, and not merely with the quantity of goods and services. Not merely growth (and the magnitude of the GNP), but also the structure and benefits of growth (and the composition and distribution of the GNP) are of central importance. The environmental dimension must be concerned with the rational and sustained use, rather than indiscriminate rapid devastation, of the resource-bestowing and life-supporting bio-geophysical environment. Hence, this dimension must involve the exercise of several preferences in the choice of technologies, for example:

1. A preference for energy-production technologies based on renewable rather than depletable, energy sources (e.g. sun, wind and biogas, rather than oil or coal);
2. A preference for resource- and energy-saving, rather than resource- and energy-intensive, technologies;
3. A preference for technologies which produce goods that can be recycled and re-used, and that are designed for durability, rather than obsolescence;
4. A preference for technologies which blend into natural ecosystems by causing them minimal disturbance.

Criteria for selection of technology[1]

1. *Resource Development*

   (a) Does it make optimal use of local factors (manpower, capital, natural resources, etc.) by

   (i) Sustaining/generating employment

   (ii) Saving/generating capital, raw materials, including energy

   (iii) Does it increase the capacity to produce on a sustained, cumulative basis?

2. *Societal Development*

   (a) Does it reduce dependence and promote self-reliance based on mass participation at the local/national/regional levels, enabling the society to follow its own path of development?.

   (b) Does it reduce inequalities among occupational, ethnic, sex and age groups, between rural and urban communities, and between (groups of) countries?

3. *Environmental Development*

   (a) Does it minimise depletion and pollution by using renewable resources, through built-in waste minimisation, recycling and/or re-use and blending better with existing eco-cycles?

   (b) Does it improve the natural and manmade environment by providing for a higher level of complexity and diversity of the ecosystems, thereby reducing their vulnerability?

   (c) Does it accentuate the positive and eliminate the negative environmental impacts?

How can we achieve these objectives? The answer lies in the 'stake-holder oriented approach'. We need to identify specific (groups of) actors, and study their specific situations. We must learn to understand why things are as they are and what constraints and opportunities each actor experiences. Then to see how they could improve their situation, and what is needed to help them attain this. That leads to ideas for policies, programmes or projects.

Putting it differently, the environmental situation is a *system*. If we want to change the functioning of a system, we first need to understand how it works now. We can then try to devise improvements that would convert it into a system that is more beneficial. Then we can see what external actions could be devised to bring about such changes, and how these actions could be set up.

In an 'actor-approach', decisions are made by actors. An actor has a measure of information and knowledge, perceives a number of options, has a set of preferences, has economic means, has a measure of freedom of choice, and then makes the choice. That is the 'causality' we need to know if we are thinking of policy recommendations. After all, a 'policy' is just one more input into the decision process of the actor. There is a famous saying: 'You can take the horse to the water, but you cannot make him drink'. Policies can try to influence any of the factors. You can inform or educate (information, knowledge); add options and make them known (options perceived); try to modify attitudes (preferences); influence prices or incomes (economic means); restrict or empower (freedom of choice).

Once we have understood the behaviour of the system, and of the other actors in the system, we can try to see if we can design policies that would give improvement. Subsidise or tax particular activities or products; make new technical designs, set up projects, and much more. Only then can we have 'environmentally sound development'.

I am very wary of all good-willing 'policies' which help to 'provide a better a environment'. Many of the present day 'policies' are notoriously incapable helping the environment, and often do more harm than good. I am not a pessimist, and just as you do, I believe that good initiatives are possible. But these can only be designed in a detailed analysis of exactly what one wants to achieve, and of the ways of ensuring that the policies really works to that goal.

As polices do not function in isolation from the wider spheres of power and decision making, much of what can be achieved through these actions can only be sustained, institutionalised and scaled up by removing the obstacles at the local, regional or national levels.

As we look back at the true thinkers of our times as those who challenged the status quo, and think about the good that came from the questions they posed, a hundred years from now, let us all hope that the people of the world will look back and say of our generation that we saw the challenge, that we answered the call and that we did not flinch in the face of our responsibility to build a better world. It will not be easy. It has never been easy. But we will try.

Notes

1. Lecture notes on 'Technology and Development', Prof. A.K.N. Reddy, Department of Management Studies, Indian Institute of Science, Bangalore.

Chapter 4

How are Environment, Human Development and Economic Growth Related in India? A Cross-State Analysis

Sacchidananda Mukherjee and Debashis Chakraborty

1 Introduction

Since the initiation of economic reform in 1991, Indian States with better infrastructural conditions and enterprising governments have grown at a much higher rate as compared to the natural resource-rich economies (Bhandari and Khare, 2002); leading to higher growth of the economy as a whole. The enhanced growth is likely to improve the human development (HD) level and subsequently the public awareness on environmental sustainability and governance in a State (World Bank, 2006). However, given the higher level of Foreign Direct Investment (FDI) inflow in relatively more polluting sectors in India during the post-liberalisation period (Gamper-Rabindran and Jha, 2004); ensuring efficiency of environmental governance has rightly been identified as a key development question (Costantini and Salvatore, 2006; Kathuria and Sterner, 2005; Kathuria, 2004; Parikh, 2004; Murty et al., 2003; Sankar, 1998). The Supreme Court interventions are worth mention in this regard

(Antony, 2001; World Bank, 2006), although their limitations have also been highlighted (Venkatachalam, 2005).[1]

Apart from domestic reform measures, external factors have influenced the environmental scenario, as Indian firms (especially those in sectors like textile, marine products, leather, chemicals, etc.) have often complained that the environmental compliance requirements in the European Union (EU) and US are too stringent.[2] Nonetheless, owing to sanctions and regular factory visits by importing country officials, the compliance level has considerably improved (Tewari and Pillai, 2005; Sankar, 2006), with obvious positive implications on the domestic environment.

In this background, through a secondary data analysis, the current paper attempts to examine the relationship of environmental quality (EQ) with HD and economic growth (EG) separately for 14 major Indian States. For understanding the evolving performance of the States towards EQ, we consider two periods: Period A (1990–1996) and Period B (1997–2004). A brief literature survey on EQ, HD and EG is provided first, followed by a discussion on the methodology, results and the policy observations respectively.

2 Literature Review

2.1 Relationships between Environmental Quality and Economic Growth

For examining the relationship between Per Capita Income (PCI) or the Per Capita Net State Domestic Product (PCNSDP) in the case of States within a country and environmental degradation or pollution, generally the existence of an inverted U-shaped curve in the PCI vs. pollution plane ('Environmental Kuznets Curve') is verified. The relationship implies that, with the rise in PCI, environmental degradation continues up to a certain level of PCI, but improves afterwards, as with prosperity, the environment-consciousness increases (Esty and Porter, 2001–02). The literature shows that, while several local pollutants like Sulphur dioxide (SO_2), Suspended Particulate Matter (SPM), Carbon monoxide (CO), etc. support the EKC hypothesis; other pollutants reveal either monotonicity or an N-shaped relationship (Dinda, 2004). Studies based on both ambient concentration of pollutants (Baldwin, 1995; Grossman and Krueger, 1995; Selden and Song, 1994; Shafik and Bandyopadhyay, 1992) and the actual emissions of pollutants (Bruvoll and Medin, 2003; Carson et al., 1997) also support the EKC hypothesis.

It is observed that a composite environmental index represents the environment condition of a region, country or State in a better way than individual indicators in determination of the EKC relationship (Mukherjee and Kathuria, 2006).[3] The Environmental Degradation Index (EDI) created by Jha and Bhanu Murthy (2001) covering 174 countries observed an inverse link between EDI and HDI, indicating the existence of an inverted N-shaped global EKC rather than an inverted U-shaped one.

The EKC relationship in the Indian framework has been tested by Mukherjee and Kathuria (2006) for 14 major States over 1990–2001 by using 63 environmental variables, arranged under eight broad environmental groups. It is observed that the relationship between EQ and PCNSDP is slanting S-shaped, indicating that the economic growth has occurred in Indian States mostly at the cost of EQ. Kadekodi and Venkatachalam (2005) noted evidence of a strong linkage between various natural resources and environment with income and the status of livelihood, and concluded that the causal relationship between poverty and environment works in both directions.

2.2 Relationship between Environment and Human Well-being

The literature on the environmental impacts of the structural adjustment programme indicate that if the victims of depletion and degradation of natural environment are not identified and compensated for by the beneficiaries, the vulnerable sections face additional economic hardship, which may further fuel inequality (Dasgupta, 2001), and this socio-economic inequality leads to environmental inequality, ultimately affecting EQ (Boyce, 2003). It is agreed that achieving environmental sustainability requires a careful balance between HD activities and maintaining a stable environment that predictably and regularly provides resources and protects people from natural calamities (Melnick et al., 2005). The belief of considering environmental problems as the side effects of development process is now giving way to a new approach focusing on promotion of their integration instead (Ginkel et al., 2001). This objective has been incorporated in Target 9 of the United Nations' Millennium Development Goals (MDGs), which demands that environmental conservation and conservation of natural resources from quantitative depletion and qualitative degradation, should be an integral part of any economic and development policy.

The framework used by the UNDP in Human Development Reports (HDR) is considered to be a standard exercise, although it is often argued that there is enough room for including important socio-economic variables in the list of HD indicators (Nagar and Basu, 2001; Guha and Chakraborty, 2003). In India, the National Human Development Report 2001, brought out by the Planning Commission, also followed the UNDP methodology.

3. Methodology and Data

3.1 Environmental Quality Index (EQI)

The composite EQI for the States is postulated to be linearly dependent on the 63 selected variables, placed under eight broad categories and has been determined by adopting the HDI method. We assume X_{ij} to be the value of the ith indicator for j^{th} State of India with respect to X (or EQ), where X consists of a large number of indicators varying from 6 to 12 (see Appendix 3). Here the X's are AIRPOL, INDOOR, GHGS, ENERGY, FOREST, WATER, NPSP and LAND respectively (see Table 2).

In line with the HDI method, we transform the indicators into their standardised form, by which the adjusted values of X_{ij} (i.e., EX_{ij}'s) to be used for the analysis become:

$$EX_{ij} = (X_{ij} - X_i^*)/(X_i^{**} - X_i^*) \text{ or } EX_{ij} = (X_i^{**} - X_{ij})/(X_i^{**} - X_i^*)$$

where, X_i^* and X_i^{**} are the minimum and maximum values for the ith indicator of environmental quality X respectively.[4] Now, $EQIX_j$, i.e., the environmental quality index score for the jth State with respect to each individual environmental quality X (which constitutes of n number of indicators, n varies from 6 to 12), is arrived at by summing the EX_{ij}s over *i* by using the following formula:

$$EQIX_j = \frac{1}{n}\sum_{i=1}^{n} EX_{ij}$$

In a similar manner, EQI_j, i.e., the overall environmental quality index score for the jth State, is arrived at by summing the EX_{ij}s for all X over *i* by using the following formula:

$$EQI_j = \frac{1}{N}\sum_{i=1}^{N=63} EX_{ij} \forall X$$

The obtained EQIs measure the environmental well-being of the States, i.e. the States with higher score are characterised by a cleaner environment. The EQI_js (where j=1 to 14), thus arrived at is therefore used to obtain the $REQI_j$s (the rank of the jth State), where the States having higher EQI_j are assigned higher rank.

In order to obtain State-level secondary information on environment and natural resources from published government reports and other databases for both the time periods selected in our analysis, i.e. Period A (1990–96) and Period B (1997–2004), the sample is restricted only to 14 major Indian States, namely Andhra Pradesh (AP), Bihar (BH), Gujarat (GJ), Haryana (HR), Karnataka (KR), Kerala (KL), Madhya Pradesh (MP), Maharashtra (MH), Orissa (OR), Punjab (PB), Rajasthan (RJ), Tamil Nadu (TN), Uttar Pradesh (UP) and West Bengal (WB). The indicators with at least two observations are chosen, while one of them is located within the boundary of the two sample periods. The selected indicators have then been normalised using appropriate measures of size/scale of the States—geographical area, population and GSDP at current prices.

Here we need to distinguish between two key concepts, namely—endowment effect and efficiency in natural resource management effect. The depletion and degradation of natural resources and occurrence of environmental pollution is chiefly concerned with environmental management. On the other hand, the initial endowments of natural resources (forests, land and water) are determined by geographical, climatic and ecological factors. Quite understandably, the former is comparatively more influenced by human activities. By calculating the change in the natural resource position with respect to a base year we can isolate the two effects.[5] The current study focuses on the environmental management efficiency effect as well as the size effect of the States.

Table 4.1: Description of the Environmental Groups

Groups	Description	Number of variables
AIRPOL	Air Pollution	6
INDOOR	Indoor Air Pollution Potential	9
GHGS	Green House Gases (GHGs) Emissions	6
ENERGY	Pollution from Energy Generation and Consumption	6
FOREST	Depletion and Degradation of Forest Resources	8
WATER	Depletion and Degradation of Water Resources	12
NPSP	Nonpoint Source Water Pollution Potential	10
LAND	Pressure and Degradation of Land Resources	6
	Total	**63**

3.2 Human Development Index (HDI)

Following the principle of the NHDR 2001 (released in March, 2002) methodology, for calculation of the HDI, we consider three variables, namely—inflation and inequality adjusted per capita consumption expenditure (X_1); and composite indicator of educational attainment (X_2) and composite indicator on health attainment (X_3). With this formulation, following the HDI method, the HDI score for the jth State is given by:

$$HDI_j = \frac{1}{3}\sum_{i=1}^{3} X_i$$

where, X_i represents the normalised values of the three indicators selected for construction of the HDI score, obtained by using the following formula:

$$X_i = (X_{ij} - X_i^*)/(X_i^{**} - X_i^*)$$

where X_{ij} refers to attainment of the ith indicator by the jth State and X_i^{**} and X_i^* are the scaling maximum and minimum values of the indicators respectively (i = 1 to 3).

Although UNDP considers real GDP per capita in Purchasing Power Purity (PPP) USD for generating the HDI, the NHDR 2001 (2002) has preferred total inflation and inequality adjusted per capita consumption expenditure of a State (i.e. rural and urban combined) over that for the analysis. Here the monthly per capita consumption expenditure data obtained from National Sample Survey Organisation (NSSO) for two periods (1993–94 and 1999–2000), adjusted for inequality using estimated *Gini* ratios, and further adjusted for inflation to bring them to 1983 prices by using deflators derived from a State-specific poverty line (Raju, undated). We follow the NHDR methodology in our analysis and consider total inflation and inequality adjusted per capita consumption expenditure of a State as an explanatory variable.

The composite indicator on educational attainment (X_2) is constructed with two variables, namely literacy rate for the age group of seven years and above (e_1) and adjusted intensity of formal education (e_2). The weightages assigned to the two variables are 0.35 and 0.65 respectively. The idea is that literacy rate being an overall ratio alone may not indicate the actual scenario, and the drop-out rate needs to be incorporated. By considering the data on literacy rate (1991, 2001) and the adjusted intensity of formal education data (1993, 2002) for two periods, we arrive at X_2. The data is obtained from the '*7th All India Educational Survey (AIES): All India School Education Survey (AISES)*', published by NCERT (2002).

The composite indicator on health attainment (X_3) is arrived at by considering two variables, namely Life Expectancy (LE) at age one (h_1) and the reciprocal of Infant Mortality Rate (IMR) as the second variable (h_2). The weightages assigned to the two variables are 0.65 and 0.35 respectively. For h_1, which measures the life expectancy at age 1 (rural and urban combined), the two data points considered for the two periods are 1990–94 and 1998–2002 respectively. On the other hand, the IMR (per thousand) data is considered for two periods, namely 1992 and 2000. The IMR data is taken from Sample Registration System (SRS) Bulletins; Registrar General of India, New Delhi and LE data is taken from Indiastat database website (www.indiastat.com).

3.3 Economic Growth (EG)

Here, economic growth is measured by the PCNSDP of the States at constant (1993-94) prices. The PCNSDP for Periods A and B is the average PCNSDP for the years 1993-94 to 1995-96 and 1997–98 to 1999–2000 respectively. To understand the size of the economy and growth pattern of each of the 14 States, we have classified them into three categories with respect to their Gross State Domestic Product (GSDP) at constant 1993–94 prices, e.g. high-income States (having GSDP: greater than 3rd Quartile), medium-income States (GSDP: 1st to 3rd Quartile), and low-income States (GSDP: less than 1st Quartile). The data on EG of the States is obtained from EPW Research Foundation Database Software and RBI's Database on Indian Economy.

4. The Results

4.1 EQI

Table 4.2 shows the EQ scores and rankings of the States for Period A. Kerala, Karnataka and Maharashtra emerge as the toppers, while UP, Punjab and Haryana are the laggards. While topper Kerala scores handsomely in several indicators, it is among the laggards in the case of INDOOR, LAND and FOREST. Karnataka had a mixed performance—good for AIRPOL and GHGS, and moderate show elsewhere. Maharashtra, on the other hand, performed noticeably in INDOOR, GHGS, LAND and NPSP, but poorly in terms of ENERGY, WATER, FOREST and AIRPOL. The

TABLE 4.2: ENVIRONMENTAL QUALITY SCORES AND RANKS OF THE STATES: 1990-1996

	Airpol (1)	Indoor (2)	Ghgs (3)	Energy (4)	Land (5)	Water (6)	Forest (7)	Npsp (8)	Eqi Score (9)
Andhra Pradesh	0.876 (4)	0.320 (11)	0.685 (5)	0.524 (8)	0.592 (5)	0.535 (6)	0.506 (13)	0.489 (8)	0.544 (7)
Bihar	0.647 (8)	0.129 **(14)**	0.467 (11)	0.555 (7)	0.502 (10)	0.604 (5)	0.515 (12)	0.467 (11)	0.480 (11)
Gujarat	0.432 (12)	0.643 (3)	0.617 (7)	0.282 (13)	0.616 (4)	0.482 (9)	0.594 (6)	0.630 (2)	0.545 (6)
Haryana	0.783 (6)	0.574 (4)	0.494 (9)	0.450 (12)	0.183 (13)	0.381 (13)	0.671 (2)	0.333 (13)	0.475 (12)
Karnataka	0.912 **(1)**	0.435 (5)	0.901 **(1)**	0.673 (5)	0.535 (7)	0.516 (7)	0.573 (9)	0.541 (6)	0.607 (2)
Kerala	0.874 (5)	0.327 (10)	0.870 (3)	0.709 (3)	0.520 (8)	0.696 (3)	0.517 (11)	0.559 (5)	0.617 **(1)**
Madhya Pradesh	0.483 (11)	0.367 (8)	0.355 (13)	0.468 (10)	0.719 **(1)**	0.695 **(4)**	0.158 **(14)**	0.597 (4)	0.493 (10)
Maharashtra	0.653 (7)	0.715 (2)	0.697 (4)	0.473 (9)	0.652 (2)	0.514 (8)	0.581 (8)	0.599 (3)	0.605 (3)
Orissa	0.909 (2)	0.228 (12)	0.350 **(14)**	0.771 **(1)**	0.543 (6)	0.760 (1)	0.584 (7)	0.516 (7)	0.578 (4)
Punjab	0.644 (9)	0.803 **(1)**	0.427 (12)	0.274 **(14)**	0.181 **(14)**	0.244 **(14)**	0.840 **(1)**	0.267 **(14)**	0.456 (13)
Rajasthan	0.631 (10)	0.397 (7)	0.881 (2)	0.622 (6)	0.637 (3)	0.465 (10)	0.520 (10)	0.642 **(1)**	0.577 (5)
Tamil Nadu	0.896 (3)	0.412 (6)	0.656 (6)	0.458 (11)	0.513 (9)	0.381 (12)	0.612 (4)	0.478 (10)	0.525 (8)
Uttar Pradesh	0.152 **(14)**	0.222 (13)	0.600 (8)	0.682 (4)	0.363 (12)	0.447 (11)	0.620 (3)	0.478 (9)	0.443 **(14)**
West Bengal	0.248 (13)	0.357 (9)	0.473 (10)	0.758 (2)	0.369 (11)	0.699 (2)	0.611 (5)	0.442 (12)	0.508 (9)

laggards, like Haryana and UP suffered owing to poor show in the case of LAND, NPSP, etc. Interestingly, Orissa, moderate performer in terms of PCNSDP, obtained fourth rank on the EQI scale, due to a good show in the case of AIRPOL, ENERGY and WATER.

Table 4.3 shows the EQ scenario of the States for Period B. Kerala, Karnataka and Maharashtra are still the top performers, while Haryana, Bihar and Punjab are the laggards in this period. The toppers bettered their position in several sub-categories (Kerala in AIRPOL, INDOOR, GHGS, FOREST; Karnataka in ENERGY, FOREST, etc.) vis-à-vis Period A, although degradation in certain areas is also witnessed. For instance, the lower ranking of Karnataka in AIRPOL in Period B can be explained by rapid urbanisation, industrialisation and vehicular pollution.[6] Its relative performance on WATER also raises concern. On the other extreme, the laggards witnessed little improvement in several sub-categories (e.g. Punjab—AIRPOL, GHGS, ENERGY, LAND, WATER and NPSP; Bihar—AIRPOL, INDOOR, GHGS, LAND, FOREST and NPSP; Haryana—ENERGY, LAND, WATER and NPSP).

The comparison of the relative performance of the States on the EQ scale during Periods A and B reveals interesting results. Although the overall position of the top performers remained unchanged, some interesting movements of their ranking within the sub-categories is noticed. For instance, while Maharashtra's ranking has declined in LAND and FOREST,[7] its relative performance has improved in WATER. Karnataka's ranking, on the other hand, changed significantly across the categories: it improved in ENERGY and FOREST, but declined for AIRPOL, INDOOR and WATER. Although Kerala improved its relative performance in several sub-categories (e.g. FOREST[8], it suffered due to decline in others (e.g. WATER).[9] The ranking of Punjab and UP experienced a sharp decline in the case of FOREST, indicating degradation on that front.

Table 4.3: Environmental Quality Scores and Ranks of the States: 1997-2004

	Airpol (1)	Indoor (2)	Ghgs (3)	Energy (4)	Land (5)	Water (6)	Forest (7)	Npsp (8)	Eqi Score (9)
Andhra Pradesh	0.802 (3)	0.498 (7)	0.553 (5)	0.585 (5)	0.617 (3)	0.479 (9)	0.781 (9)	0.474 (9)	0.580 (6)
Bihar	0.433 (12)	0.141 **(14)**	0.428 (10)	0.574 (7)	0.436 (10)	0.675 (2)	0.480 (13)	0.422 (12)	0.455 (13)
Gujarat	0.310 (13)	0.718 (3)	0.547 (6)	0.231 (13)	0.653 **(1)**	0.539 (8)	0.769 (10)	0.599 (4)	0.564 (7)
Haryana	0.715 (5)	0.714 (4)	0.542 (7)	0.362 (10)	0.088 **(14)**	0.332 (13)	0.790 (7)	0.278 **(14)**	0.472 (12)
Karnataka	0.684 (6)	0.610 (6)	0.885 **(1)**	0.679 (3)	0.568 (7)	0.465 (11)	0.807 (5)	0.563 (6)	0.636 (3)
Kerala	0.791 (4)	0.467 (8)	0.882 (2)	0.644 (4)	0.534 (9)	0.541 (7)	0.942 **(1)**	0.598 (5)	0.656 **(1)**
Madhya Pradesh	0.510 (10)	0.453 (10)	0.302 **(14)**	0.349 (12)	0.647 (2)	0.689 **(1)**	0.230 **(14)**	0.627 **(1)**	0.497 (10)
Maharashtra	0.676 (7)	0.771 (2)	0.682 (4)	0.428 (9)	0.578 (6)	0.606 (5)	0.731 (11)	0.615 (2)	0.641 (2)
Orissa	0.823 (2)	0.189 (13)	0.381 (11)	0.745 (2)	0.612 (4)	0.673 (3)	0.864 (2)	0.527 (7)	0.593 (5)
Punjab	0.600 (9)	0.812 **(1)**	0.349 (13)	0.211 **(14)**	0.118 (13)	0.273 **(14)**	0.789 (8)	0.279 (13)	0.434 **(14)**
Rajasthan	0.670 (8)	0.459 (9)	0.807 (3)	0.564 (8)	0.603 (5)	0.470 (10)	0.804 (6)	0.614 (3)	0.606 (4)
Tamil Nadu	0.949 **(1)**	0.624 (5)	0.376 (12)	0.361 (11)	0.567 (8)	0.356 (12)	0.842 (3)	0.483 (8)	0.555 (8)
Uttar Pradesh	0.471 (11)	0.305 (12)	0.518 (8)	0.584 (6)	0.366 (11)	0.566 (6)	0.507 (12)	0.458 (10)	0.473 (11)
West Bengal	0.212 **(14)**	0.417 (11)	0.476 (9)	0.794 **(1)**	0.347 (12)	0.610 (4)	0.825 (4)	0.422 (11)	0.522 (9)

4.2 HDI

Table 4.4 compares the HDI scores and rankings of the States in the three sub-categories and the composite index for Periods A and B. Kerala, Punjab and Maharashtra were the toppers during Period A, and Haryana replaced Maharashtra at the top three in Period B. Kerala performed well in all categories in both periods. Although Punjab is placed comfortably in terms of consumption and health, its educational achievements look moderate. On the other hand, Bihar, UP and MP remained at the bottom for most of the categories, which is responsible for their poor overall HDI ranking. It is further observed that no major change has occurred in the overall HDI score of the States. The changes are usually noticed in terms of consumption ranking. The aggregate picture often hides the dynamics at the component level: while for MP the aggregate HDI score had gone up from 0.072 to 0.132, its consumption score had gone down from 0.052 to 0.000. The improvement in education actually enhances its HDI score. Interestingly, consumption expenditure score goes down both for poor States like Orissa (insignificant poverty reduction over NSSO 50th (1993–94) and 55th (1999–2000) round) and middle-income States like West Bengal (9 per cent poverty reduction over NSSO 50th and 55th round), which can perhaps be explained by income inequality in these two States.

4.3 Cross-Period Analysis between HDI and EQI

The comparison of the EQI score of the States over the two periods through Figure 4.1 shows that while the toppers are located in the north-eastern corner (Karnataka, Kerala, Maharashtra), the laggards are placed in the other extreme (Bihar, Haryana, Punjab, UP). Punjab, Bihar and Haryana witnessed a decline in the EQI score during Period B (1997–2004) vis-à-vis Period A. Punjab, Haryana, UP, Bihar,

Table 4.4: HDI Scores and Ranks of the States over the Sample Period

States	Consumption		Health		Education		Hdi Score	
	Period A (1)	Period B (2)	Period A (3)	Period B (4)	Period A (5)	Period B (6)	Period A (7)	Period B (8)
Andhra Pradesh	0.338 (8)	0.196 (10)	0.300 (8)	0.410 (7)	0.136 (11)	0.344 (11)	0.258 (9)	0.317 (9)
Bihar	0.000 **(14)**	0.025 (13)	0.125 (11)	0.143 (11)	0.000 **(14)**	0.000 **(14)**	0.042 **(14)**	0.056 **(14)**
Gujarat	0.575 (5)	0.636 (4)	0.275 (9)	0.374 (9)	0.484 (4)	0.531 (4)	0.445 (6)	0.514 (6)
Haryana	0.610 (3)	0.792 (3)	0.499 (4)	0.614 (3)	0.366 (8)	0.497 (6)	0.492 (4)	0.635 (3)
Karnataka	0.295 (9)	0.402 (8)	0.466 (5)	0.481 (5)	0.371 (7)	0.478 (8)	0.377 (8)	0.454 (7)
Kerala	0.831 (2)	1.000 **(1)**	1.000 **(1)**	1.000 **(1)**	1.000 **(1)**	1.000 **(1)**	0.944 **(1)**	1.000 **(1)**
Madhya Pradesh	0.052 (12)	0.000 **(14)**	0.000 **(14)**	0.000 **(14)**	0.165 (10)	0.396 (9)	0.072 (12)	0.132 (13)
Maharashtra	0.459 (7)	0.490 (6)	0.549 (3)	0.570 (4)	0.541 (2)	0.710 (2)	0.516 (3)	0.590 (4)
Orissa	0.258 (11)	0.069 (11)	0.083 (12)	0.089 (13)	0.235 (9)	0.377 (10)	0.192 (10)	0.178 (11)
Punjab	1.000 **(1)**	0.907 (2)	0.765 (2)	0.837 (2)	0.414 (5)	0.505 (5)	0.726 (2)	0.750 (2)
Rajasthan	0.294 (10)	0.307 (9)	0.241 (10)	0.312 (10)	0.038 (13)	0.317 (12)	0.191 (11)	0.312 (10)
Tamil Nadu	0.489 (6)	0.583 (5)	0.366 (6)	0.454 (6)	0.517 (3)	0.658 (3)	0.457 (5)	0.565 (5)
Uttar Pradesh	0.039 (13)	0.054 (12)	0.050 (13)	0.134 (12)	0.073 (12)	0.238 (13)	0.054 (13)	0.142 (12)
West Bengal	0.583 (4)	0.441 (7)	0.358 (7)	0.383 (8)	0.378 (6)	0.486 (7)	0.440 (7)	0.437 (8)

MP and West Bengal need to concentrate on the problem areas, given their lower ranking in both periods.

Figure 4.1: Comparison of EQI Scores over periods A and B

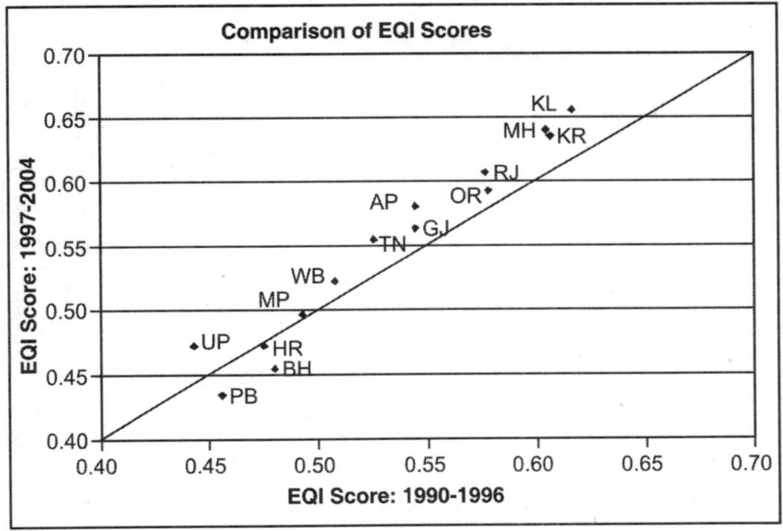

Figure 4.2: Comparison of HDI scores over periods A and B

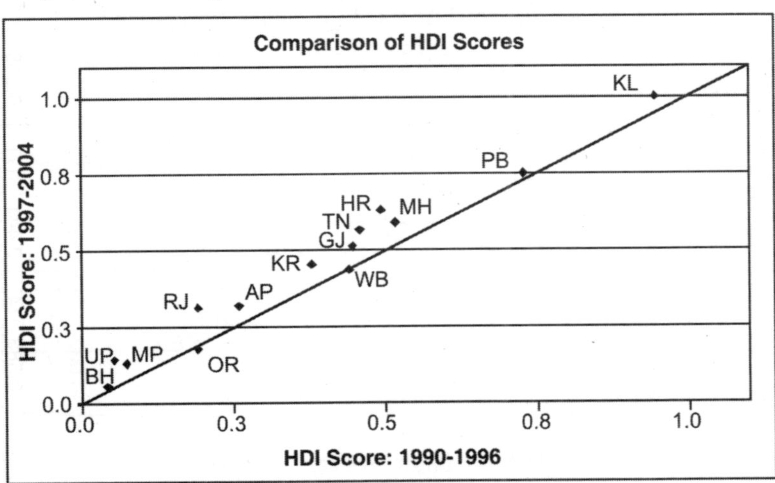

Figure 4.2 compares the HDI scores of the States over the two periods. Orissa and West Bengal witness decline in

HDI score during Period B. While the high-HD States are located in the north-eastern corner of the diagram (Punjab and Kerala), the laggards are placed at the diagonally opposite end (Bihar, UP, MP, Orissa and Rajasthan).

4.4 EQI and PCNSDP

The relationship between the EQI scores and PCNSDP of the States during Period A seems to be a convex one, as evident in Figure 4.3. While Maharashtra performed appreciably in both categories, UP put up a poor show on both counts. Orissa's poor performance in EG is partially balanced by its moderate show on the EQI front. Punjab and Haryana, on the other hand, have ensured their growth (located above the third quartile), at the cost of environmental sustainability. The environmentally sustainable States Kerala and Karnataka, however, have performed moderately in terms of EG. On the other hand, West Bengal and Tamil Nadu were mediocre in both respects.

Figure 4.3: PCNSDP Vs. EQI Score (Period A)

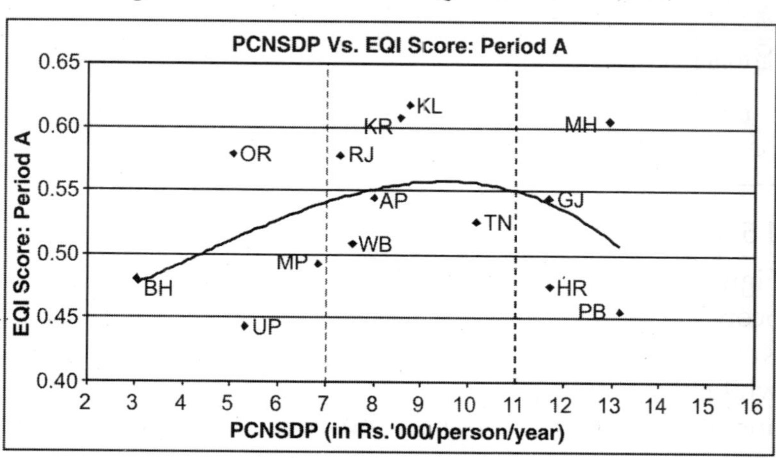

Figure 4.4 indicates a convex relationship between the EQI scores and PCNSDP of the States during Period B.

Except Orissa, the performance of the low-income States generally had deteriorated. While Maharashtra performs well on both counts, Bihar, MP and UP are the laggards. Despite improvement in the EQI Score, Orissa still remained at the bottom in terms of EG (i.e. below the first quartile line). Punjab and Haryana continue to perform well on PCNSDP but poorly on the EQI front. The middle-income States— Karnataka, Kerala and Andhra Pradesh—secure higher EQI scores. Interestingly, Rajasthan improves its position both in terms of EQI and EG vis-à-vis Period A. West Bengal and Tamil Nadu turn out to be the laggards among middle- income States.[10]

Figure 4.4: PCNSDP Vs. EQI Score (Period B)

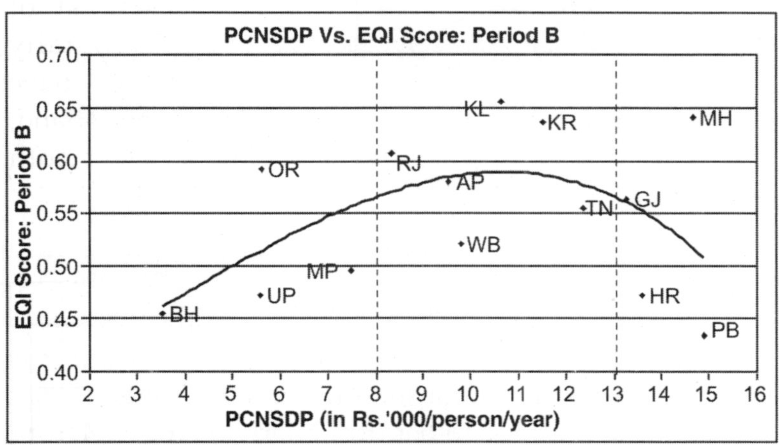

4.5 EQI and HDI

Figure 4.5 compares the relationship between EQI and HDI Scores of the 14 States during Period A. It is observed that Kerala and Maharashtra are the toppers on both counts while Bihar and UP are placed at the other extreme. Orissa and Rajasthan, on the other hand, had performed poorly in terms of HDI (placed below the first quartile), despite their good show on the EQ front. While Andhra Pradesh had performed moderately well in terms of EQI, its accomplishment in terms of HDI was rather limited. Gujarat and Tamil Nadu were moderately placed on both counts. Karnataka, on the other hand despite being a top performer in terms of EQI, fared moderately on the HDI front (placed below the second quartile line). The HDI had been quite high for Punjab and Haryana, but they performed poorly on the EQI front, primarily owing to over-exploitation of natural resources.[11]

Figure 4.5: HDI Score Vs. EQI Score (Period A)

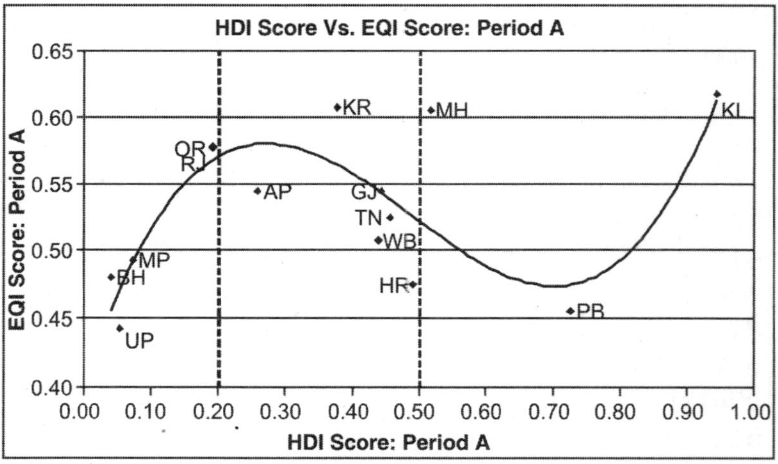

Figure 4.6 plots the EQI and HDI scores of the States in Period B. Kerala is still at the top on both fronts. Punjab and Haryana continue to perform poorly on the EQI front, despite the good HDI score. On the other hand, Bihar witnesses a decline in the EQI score, while the same increased for MP and UP. The EQI for Orissa is still high. All the middle HDI category States improved their performance in EQI during Period B, especially AP and Rajasthan. Broadly, the relationship between the EQI score and HDI score is found to be slanting N-shaped owing to the divergence in performance of toppers like Punjab, Haryana and Kerala.

Figure 4.6: HDI Score Vs. EQI Score (Period B)

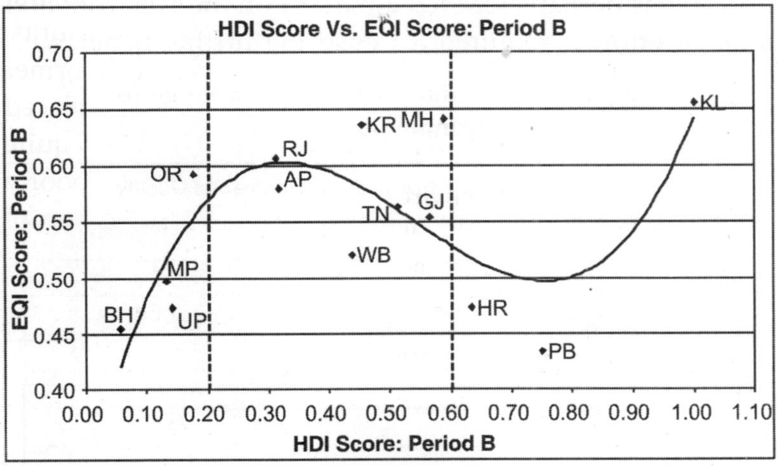

4.6 The Changing Perspective

Figure 4.7 shows the evolving economic growth-EQ-HDI scenario of the States. It is observed that, apart from Bihar, Orissa and UP, all States are growing a rate higher than 6 per cent. Bihar, Punjab and Madhya Pradesh have achieved their economic growth at the cost of their EQ. On the other hand, Karnataka and Rajasthan have grown fast, but maintained their EQ simultaneously. While Punjab, Kerala and AP grow at a similar momentum, only for Punjab change in EQ is negative. Interestingly, the change in EQ is comparable for UP, Tamil Nadu, Rajasthan and Karnataka, although their growth rates vary widely.

With the help of Figure 4.8, we compare the economic growth-HDI scenario for the States. For Orissa and West Bengal, economic growth is associated with negative impacts on incremental benefits of human development. On the contrary, high economic growth of Haryana, Tamil Nadu and Rajasthan is associated with significant change in human well-being. The impact of economic growth on the incremental benefits of HD is minimum for Bihar and

Punjab, while a vicious cycle can explain the result for Bihar, Punjab has perhaps reached a plateau on the HD front and a major effort is required for ensuring further benefits.

Figure 4.7: Economic Growth Vs. Change in Environmental Quality

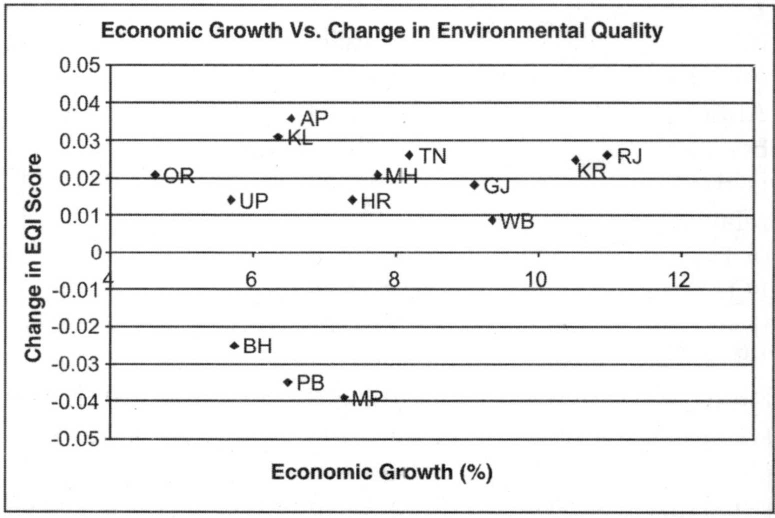

Figure 4.8: Economic Growth Vs. Change in HDI Score

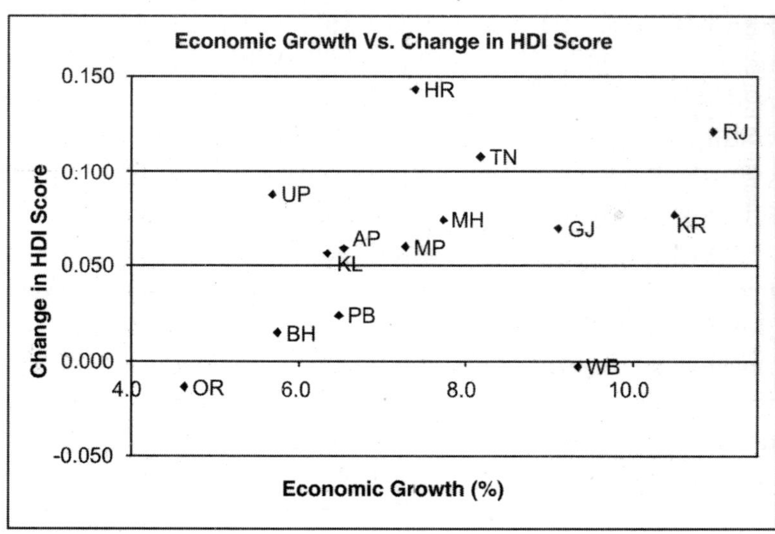

Mukherjee and Chakraborty (2007) report an inverted N-shaped relationship between change in HDI and change in EQI over the two periods (Figure 4.9). It is observed that Karnataka, Rajasthan, Tamil Nadu and UP witnessed a comparable level of change in EQI score despite registering various level of changes in HDI. Gujarat and Orissa, on the other hand, achieved a comparable level of change in EQI, but with a positive and negative change in HDI respectively. Although change in HDI is negative for Orissa and West Bengal, changes in EQI are positive for them. The reverse is true for Bihar, Punjab and Haryana. This particular relationship comes from the similarity in incremental HD or EQ benefits experienced by dissimilar States in terms of EQ or HD achievements. In other words, the result indicates that State-specific factors dominate in determining the EQ.

Figure 4.9: Change in HDI Score Vs. Change in EQI Score

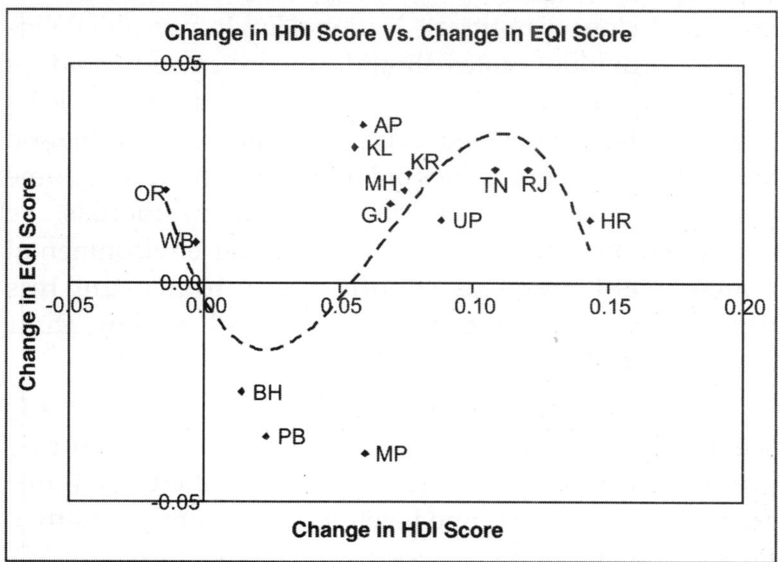

5 Policy Conclusions

The focus on growth in the short run may overshadow the environmental and natural resource concerns of a country or the regions within it. India, which entered the phase of economic reform since 1991, has grown quite commendably over the years, but there is a need to check whether this indeed may lead to 'sustainable development' in the long run. The current study attempts to analyse the cross-State environmental quality (EQ) in India by constructing a composite index by using 63 environmental indicators over two periods: Period A (1990–1996) and Period B (1997–2004).

The results indicate that different States possess different strengths and weaknesses in managing various aspects of EQ. For instance, despite a good overall show, and commendable performance on INDOOR and NPSP, Maharashtra fared poorly on ENERGY and FOREST. On the other hand, the bottom State Punjab is a topper for INDOOR. It seems the States can learn from each other about different aspects of environmental management. In other words, instead of a 'one-size-fits-all' National Environmental Policy (NEP), individual States should be encouraged to adopt environmental management practices based on their local (at the most disaggregated level) environmental information within the broad guidelines of NEP. Furthermore, over time performance of an individual State varies across the environmental criteria, which shows that environmental management practices should take into account this dynamic nature of environment, and review their environmental status or achievement regularly.

The relationship between economic growth and EQ seems non-linear in nature. The economic restructuring process has influenced the EQ of the States differently. While Maharashtra has performed well on both counts, growth in Punjab and Haryana has occurred mostly at the cost of EQ. On the other hand Orissa, despite being a low-income State, has performed reasonably well during both time periods on EQ front. It is observed that even the laggards, like Bihar and MP, have achieved their economic growth at the cost of their EQ. On the contrary, Karnataka and Rajasthan witnessed economic growth and simultaneously managed their environment. The findings re-emphasise the fact that individual States should adopt the environmental measures based on their local environmental impacts assessment of major economic activities, in order to ensure sustainable economic growth.

The relationship between the EQI score and HDI score is observed to be slanting N-shaped. The concentration of several States in low HDI-Low EQI category (Bihar, UP), high EQI-mid HDI category (Gujarat, Karnataka), presence of outliers like Orissa (high EQ-Low HDI) on the one hand and Punjab and Haryana (high HDI-Low EQ) on the other cause this overall relationship. The result indicates the need to review the NHDR methods for calculating the HD achievements of the States. Perhaps, the HD ranking of States like Punjab and Haryana has been influenced too much by their high per capita consumption expenditure. Broadbasing the HD index by incorporating other social achievements might reveal interesting results.

A few policy issues are worth mention. First, contrary to popular belief, industrial pollution is not the only source of all the problems in India—over emphasis on agriculture in Punjab and Haryana seems equally disastrous. Second, it is difficult to comment on the choice of optimal level of income and its composition for a State, which would be in line with the objective of sustainable development. For instance, relatively poorer State Orissa achieves a high level of EQ, which clearly is a result of unutilised natural resources. Third, there is an urgent need for enforcing improved governance in environmentally vulnerable States (e.g. Supreme Court intervention) and ensuring people's participation for guaranteeing sustainable development.

Finally, one limitation of the study is that it is confined only to 14 major Indian States, the constraint being the availability of various secondary environmental information for both the time periods (1990–96 and 1997–2004) under consideration. The analysis may be extended to the excluded States for arriving at a complete picture, when the requisite data series for them are made available. Second, we have focused only on the economic growth of the Indian States during the two periods and look into its relationship with EQ. However, income inequality varies widely across Indian States and has increased in the post-reform period (Deaton and Dreze, 2002). An area of future research could be to analyse the relationship between income inequality of the States, their EQ and HD achievements.

References

Antony, M. J. (2001), *Landmark Judgements on Environmental Protection*, Indian Social Institute, New Delhi.

Appasamy, P., P. Nelliyat, N. Jayakumar and R. Manivasagan (undated), *Economic Assessment of Environmental Damage: A Case Study of Industrial Water Pollution in Tiruppur*, Environmental Economics Research Committee Working Paper Series: IPP-1, available at http://coe.mse.ac.in/eercrep/fullrep/ipp/IPP_FR_Paul_Appasamy.pdf.

Baldwin, R. (1995), 'Does Sustainability Require Growth' in Goldin, I. and L.A. Winters (Eds): *The Economics of Sustainable Development*, Cambridge University Press, Cambridge UK, pp. 19–47.

Bhandari, Laveesh and Aarti Khare (2002), 'The Geography of Post-1991 India Economy', *Global Business Review*, Vol. 3, No. 2, pp. 321-340.

Bhullar, A.S. and R.S. Sidhu (2006), 'Integrated Land and Water Use: A Case Study of Punjab', *Economic and Political Weekly*, December 30, pp. 5353-5357.

Boyce, J.K. (2003), *Inequality and Environmental Protection*, Working Paper Series No. 52, Political Economy Research Institute, University of Massachusetts, Amherst.

Bruvoll, A. and Medin, H. (2003) 'Factors Behind the Environmental Kuznets Curve: A Decomposition of the Changes in Air Pollution', *Environmental and Resource Economics*, Vol. 24, No. 1, pp.27–48.

Carson, R.T., Y. Jeon, and D.R. McCubbin (1997) 'The relationship between air pollution emissions and income: US data', *Environment and Development Economics,* Vol. 2, No. 4, pp.433–450.

Centre for Science and Environment (CSE), "State of India's Environment", various issues of Citizen's Reports, New Delhi.

Chakraborty, P. and J. Singh (2005), *Leather Bound: A Comprehensive Guide for SMEs,* The Energy and Resources Institute, New Delhi.

Costantini, V. and M. Salvatore (2006), *Environment, Human Development and Economic Growth,* The Fondazione Eni Enrico Mattei, Note di Lavoro Series, February 2006, available at http://www.feem.it/NR/rdonlyres/CA11D877-29E3-462F-A591-7738FBD508F0/1891/3506.pdf

Dasgupta, P. (2001), *Human Well-Being and the Natural Environment,* Oxford University Press, New Delhi.

Deaton, A. and J. Dreze (2002), 'Poverty and Inequality in India: A Re-Examination', *Economic and Political Weekly,* September 7, pp. 3729-3748.

Dinda, S. (2004), 'Environmental Kuznets Curve Hypothesis: A Survey', *Ecological Economics,* Vol. 49, No. 4, pp. 431–455.

EPWRF (2003), 'Domestic Product of State of India: 1960–1961 to 2000–2001', Database Software, Mumbai.

Esty, D.C. and M.E. Porter (2001-02), *Ranking National Environmental Regulation and Performance: A Leading Indicator of Future Competitiveness?,* Chapter 2.1, p. 78-100, Harvard Business School, Global Competitiveness Report, http://www.isc.hbs.edu/GCR_20012002_Environment.pdf

Ginkel, Hans van, Brendan Barrett, Julius Court and Jerry Velasquez (edited) (2001), *Human Development and the Environment: Challenges for the United Nations in the New Millennium,* United Nations University Press, Tokyo, Japan.

Gamper-Rabindran, Shanti and Shreyasi Jha (2004), *Environmental Impact of India's Trade Liberalization,* available at http://unpan1.un.org/intradoc/groups/public/documents/APCITY/ UNPAN024230.pdf

Garg, A. and P.R. Shukla (2002), *Emission Inventory of India,* Tata McGraw-Hill Publishing Company Limited, New Delhi.

Government of India (2002), Planning Commission, *National Human Development Report, 2001,* March 2002.

Grossman, G.M. and Krueger, A.B. (1995), 'Economic Growth and the Environment', *Quarterly Journal of Economics,* Vol. 110, No. 2, pp.353–378.

Guha, A. and D. Chakraborty (2003), 'Relative Positions of Human Development Index Across Indian States: Some Exploratory Results', *Artha Beekshan*, Vol. 11, No. 4, pp. 166-181.

Jadavpur University (2006), *Groundwater Arsenic Contamination in West Bengal*, School of Environmental Studies, available at http://www.soesju.org/arsenic/wb.htm

Jha, R. and Bhanu Murthy, K.V. (2001). 'An Inverse Global Environmental Kuznets Curve', Departmental Working Papers: 2001–2002, Division of Economics, RSPAS, Australian National University, Canberra, Australia.

Kadekodi, Gopal K. and L. Venkatachalam (2005), *Human Development, Environment and Poverty Nexus in India*, Institute for Social and Economic Change, Report prepared for the United Nations Development Programme, New Delhi.

Kathuria, V. and T. Sterner (2005), *Monitoring and Enforcement: Is Two-Tier Regulation Robust?—A Case Study of Ankleshwar, India*, April 2005, available at http://www.hgu.gu.se/files/nationalekonomi/personal/thomas%20sterner kathuria.sterner%20ecological%20 economics%20revised.pdf

Kathuria, Vinish (2004), 'Informal Regulation of Pollution in a Developing Country: Empirical Evidence from India', Working Paper No. 6-04, South Asian Network for Development and Environmental Economics (SANDEE), Kathmandu, Nepal.

Kathuria, V. (2002) 'Vehicular Pollution Control in Delhi, India', *Transportation Research, Part D: Transport and Environment*, Vol. 7D, No. 5, pp. 373–387.

Kathuria, V. (2004) 'Impact of CNG on Vehicular Pollution in Delhi: A Note', *Transportation Research, Part D: Transport and Environment*, Vol. 9D, No. 5, pp. 409–417.

Kaushik, A and M. Saqib (2001), *Environmental Requirements and India's Exports: An Impact Analysis*, RGICS Working Paper No. 25, New Delhi.

Melnick, D., J. McNeely, Y. K. Navarro, G. Schmidt-Traub and R.R. Sears (2005), *Environment and human well-being: a practical strategy*, UN Millennium Project, Task Force on Environmental Sustainability, available at: http://www.unmillenniumproject.org/documents/Environment-complete-lowres.pdf

Mukherjee, S. and D. Chakraborty (2007), 'Environment, Human Development and Economic Growth of Indian States after Liberalisation', presented at the National Conference on '*Making*

Growth Inclusive with Special Reference to Imbalance in Regional Development' jointly organised by Department of Economics, Jammu University and Indian Institute of Advanced Studies, March 12, 2007.

Mukherjee, S. and P. Nelliyat (2006), *Groundwater Pollution and Emerging Environmental Challenges of Industrial Effluent Irrigation in Mettupalayam Taluk, Tamil Nadu,* Paper presented at the Ninth Biennial Conference of International Society for Ecological Economics (ISEE) on 'Ecological Sustainability and Human Well-being', December 15-18, 2006, New Delhi.

Mukherjee, S. and V. Kathuria (2006), 'Is Economic Growth Sustainable? Environmental Quality of Indian States After 1991', *International Journal of Sustainable Development,* Vol. 9, No. 1, pp. 38-60.

Murty, M. N., S. C. Gulati and A. Banerjee (2003), *Health Benefits from Urban Air Pollution Abatement in the Indian Subcontinent,* Institute of Economic Growth Working Paper, New Delhi.

NCERT (2002), *7th All India Educational Survey (AIES): All India School Education Survey (AISES),* National Council of Educational Research and Training (NCERT), Ministry of Human Resource Development (MHRD), Government of India, New Delhi, available at: http://www.7thsurvey.ncert.nic.in/

Nagar, A.L., and S.R. Basu (2001), *Weighing Socio-Economic Indicators of Human Development: A Latent Variable Approach,* Working Paper, National Institute of Public Finance and Policy, New Delhi.

Nair, G.K. (2006), *Water scarcity hits parts of south Kerala, The Hindu,* March 31, available at http://www.thehindubusinessline.com/2006/03/31/stories/2006033101981900.htm

Nelliyat, P. (2005), *Industrial Growth and Environmental Degradation: A Case Study of Industrial Pollution in Tiruppur,* Unpublished Ph D Thesis, submitted to the University of Madras, Chennai.

Parikh, J. (2004), *Environmentally Sustainable Development in India,* available at http://scid.stanford.edu/ events/India2004/JParikh.pdf

Raju, S. Durai (Undated), 'Measurement of Human Development – District Level', available at http://www.undp.org.in/hdrc/events/Aggregates/Goa/2

Rithe, Kishor and Ashish Fernandes (2002), *Maharashtra's tiger troubles,* April, available at www.satpuda.org/maharashtratiger.doc

Sankar, U. (2006), 'Trade Liberalisation and Environmental Protection Responses of Leather Industry in Brazil, China and India', *Economic and Political Weekly,* June 17, pp. 2470-2477.

___ (1998), *Laws and Institutions Relating to Environmental Protection in India*, presented at the Conference on 'The Role of Law and Legal Institutions in Asian Economic Development', Rotterdam, November, available at http://www.mse.ac.in/pub/op_sankar.pdf

Shafik, N. and Bandyopadhyay, S. (1992), *Economic Growth and Environmental Quality: Time-Series and Cross-Country Evidence*, World Bank Policy Research Working Paper No. 904, The World Bank, Washington DC.

Selden T.M and Song D.S. (1994), 'Environmental quality and development: Is there a Kuznets Curve for air pollution emissions? *Journal of Environment Economics and Management*, Vol. 27 pp. 47-162.

Tewari, M. and P. Pillai (2005), 'Global standards and environmental compliance in India's leather industry', *Oxford Development Studies*, Vol 33, No. 2, pp. 245-267.

The Energy Resources Institute (TERI), *TERI Energy Data Directory and Yearbook (TEDDY)*, various issues, New Delhi.

The Hindu (2006), 'Air pollution the bane of Davangere', Wednesday, October 4, available at http://www.hindu.com/2006/10/04/stories/2006100414400300.htm

___ (2005), 'Air quality around hospitals, schools deteriorating: *SPM levels are at least 25 per cent more than standard values'*, November 18, available at: http://www.hindu.com/2005/11/18/stories / 2005111817640300.htm

UNDP, 'Human Development Report', various issues.

Venkatachalam, L. (2005), 'Damage Assessment and Compensation to Farmers: Lessons from Verdict of Loss of Ecology Authority in Tamil Nadu', *Economic and Political Weekly*, April 9, 2005.

World Bank (2006), *India: Strengthening Institutions for Sustainable Growth*, Country Environmental Analysis (India CEA), September 2006, available at http://siteresources.worldbank.org/INDIAEXTN/Resources/295583-1176163782791/complete.pdf

World Bank, *World Development Indicator (2004)*, available at http://www.worldbank.org/data/ countrydata/countrydata.html

Appendix 2: Data sources

Environmental group	Data sources
AIRPOL	*MoEF*: National Ambient Air Quality Monitoring Programme Database
INDOOR	*TERI*: TERI Energy Data Directory and Yearbook (TEDDY) – Various Years *CSE*: State of India's Environment: The Citizens' Fifth Report (Part II: Statistical Database)*RGI*: Census of India 2001—Tables on Houses, Amenities and Assets (Database Software)
GHGS	Garg and Shukla (2002) *RGI*: Census of India 2001 – CensusInfo India 2001 (Version 1.0) – Database Software
ENERGY	*CMIE*: India's Energy Sector – Various Years *TERI*: TEDDY – Various Years *RGI*: CensusInfo India 2001EPWRF (2003)*CSO*: Compendium of Environmental Statistics – 2000 and 2002
FOREST	*FSI*: State of Forest Reports – 1997, 1999 and 2001 *MoEF*: The State of Environment – India: 1999, 2001 *CSE*: Citizens' Fifth Report *RGI*: CensusInfo India 2001EPWRF (2003)
WATER	*MoWR*: Annual Report – Various Years *CMIE*: India's Agriculture Sector – Various Years *MoEF*: National Rivers Water Quality Monitoring (NRWQM) Programme Database *MoA*: Annual Report – Various Years *CSE*: Citizens' Fifth Report
NPSP	*CMIE*: India's Agriculture Sector – Various Years *MoA*: Annual Report – Various Years *DoAHD&F*: Livestock Census Data – 1992, 1997 and 2003*RGI*: Census of India 2001 – Tables on Houses, Amenities and Assets (Database Software)
LAND	*CMIE*: India's Agriculture Sector – Various Years

CMIE: Centre for Monitoring Indian Economy, Mumbai.

CSE: Centre for Science and Environment, New Delhi.

CSO: Central Statistical Organisation, Ministry of Statistics and Programme Implementation, Government of India (GoI), New Delhi.

DoAHD&F: Department of Animal Husbandry, Dairying and Fisheries, MoA, GoI, New Delhi.

EPWRF: Economic and Political Weekly Research Foundation, Mumbai.

FSI: Forest Survey of India, Ministry of Environment and Forest, GoI, Dehradun.

MoA: Ministry of Agriculture, GoI, New Delhi.

MoEF: Ministry of Environment and Forests, GoI, New Delhi.

MoWR: Ministry of Water Resources, GoI, New Delhi.

RGI: Office of the Registrar General, Director of Census Operation, Ministry of Home Affairs, GoI, New Delhi.

TERI: The Energy Resources Institute, New Delhi.

Appendix 3: Descriptions of Environmental Groups (Variables) and Indicators

AIR POLLUTION (12 indicators)

Maximum Concentration of NO_2, SO_2 and SPM in Residential and Industrial Area ($\mu g/m^3$): 1990-1995 and 1996–2000 *

INDOOR AIR POLLUTION POTENTIAL (18 indicators)

Monthly Per-Capita Expenditure (MPCE) on Fuel & Lighting (Rs./month/head) Rural and Urban Areas: 1993–94 and 1999–2000 $

Percentage of Rural Households using Bio-fuels (firewoods and chips, dung cake) as primary source of energy (Traditional & Commercial) for cooking (%): 1993–1994 and 1999–2000 *

Percentage of Urban Households using Bio-fuels (firewoods and chips) as primary source of energy (Traditional) for cooking (%): 1993–1994 and 1999–2000 *

Percentage of Rural and Urban Households do not have Access to Electricity: 1991 and 2001 *

Achievement in Installation of Biogas Plants: Up to 1994–95 and up to 2001–2002 $

Kerosene as a Primary Source of Energy for Lighting for Rural and Urban Households (%): 1993–94 and 1999–2000 *

GREENHOUSE GASES EMISSIONS (12 indicators)

CO_2 Equivalent GHGs (CO_2, CH_4, N_2O) Emissions (kg /Person): 1990 and 1995 *

CO_2 Equivalent GHGs (CO_2, CH_4, N_2O) Emissions (tones/Rs. Lakh of GSDP at Constant 1980-81 Prices): 1990 and 1995 *

CO_2 Equivalent GHGs (CO_2, CH_4, N_2O) Emissions (tones/hectare of Reporting Area of Land Utilisation): 1990 and 1995 *

Other GHGs (NOx, SO_2) Emissions (kg /Person): 1990 and 1995 *

Other GHGs (NOx, SO_2) Emissions (tones/Rs. Lakh of GSDP at Constant 1980-81 Prices): 1990 and 1995 *

Other GHGs (NOx, SO_2) Emissions (tones/hectare of Reporting Area of Land Utilisation): 1990 and 1995 *

POLLUTION FROM ENERGY GENERATION AND CONSUMPTION (12 indicators)

Annual Percentage Increase in Motor Vehicles Number (given geographical area) during 1991–92 to 1995–96 and during 1995–96 to 2000–2001 *

Average Per Capita Consumption of LPG, MG, Kerosene, HSD & LDO (in kg per person): 1993–94 to 1996–97 and 1997–98 to 2000–2001 *

Average Petroleum Consumption (in tonnes) per Rs. Lakh of GSDP (at constant 1993–94 Prices): 1993–94 to 1996–97 and 1997–98 to 2000–2001 *

Average Thermal Electricity Generation as a Percentage of Total Electricity Generation (%): 1990–91 to 1995–96 and 1996–97 to 1999–2000 *

Average Electricity Consumption (in KwH) per Rs. Lakh of GSDP at Constant (1993-94) Prices: 1993–94 to 1995–96 and 1996–97 to 1999–2001 *

Average per Capita Consumption of Electricity (in KwH/Person): 1990–91 to 1995–96 and 1996–97 to 1999–2000 *

DEPLETION AND DEGRADATION OF FOREST RESOURCES (16 indicators)

Change in Forest Cover (Dense and Open Forest) as Percentage of Geographical Area (in percentage points): 1995 to 1997 and 1999–2001 $

Change in Per Capita Forest Cover (Dense Forest, Open Forest, Mangrove, Scrub) (in Hectare): 1995 to 1997 and 1999 to 2001 $

Change in Recorded Forest Area as a Percentage of Total Geographical Area: 1997 to 1999 and 1999 to 2001 $

Change in Common Property Forest Area@ as Percentage of Total Recorded Forest Area: 1997 to 1999 and 1999 to 2001 $

Change in Common Property Forest Area[@] as a Percentage of Geographical Area: 1997 to 1999 and 1999 to 2001 $

Change in Per Capita Availability of Recorded Forest Area (Person/ha): 1997 to 1999 and 1999 to 2001$

Change in Per Capita Availability of Common Property Forest Area (in Person/ha): 1997 to 1999 and 1999 to 2001$

Change in Protected Area (National Park & Sanctuary) as a Percentage of Total Geographical Area: 1997 to 1999 and 1999 to 2001 $

Note: [@] - Common Property Forest Area = Protected + Unclassed Forest Area

DEPLETION AND DEGRADATION OF WATER RESOURCES (24 indicators)

Level of groundwater development (%): 1996 and 2004 *

Percentage of Irrigated Area Irrigated by Surface Water Sources (Canals & Tanks): 1992-93 and 1998–99 $

Inland Surface Water Resources (% of geographical area): 1995 and 2001 $

Major & Medium Irrigation Potential Created (Developed) upto the end of the 8th Plan (1992–1997) as a Percentage of Ultimate Irrigation Potential of the State *

Major & Medium Irrigation Potential Utilised as a Percentage of Irrigation Potential Created up to March 1997 *

Minor Irrigation Potential Created (Developed) upto the end of the 8th Plan (1992–1997) as a Percentage of Ultimate Irrigation Potential of the State *

Minor Irrigation Potential Utilised as a Percentage of Irrigation Potential Created up to March 1997 *

Major & Medium Irrigation Potential Created (Developed) upto the end of the 9th Plan (1997-2002) as a Percentage of Ultimate Irrigation Potential of the State *

Major & Medium Irrigation Potential Utilised as a Percentage of Irrigation Potential Created Upto March 2002 *

Minor Irrigation Potential Created (Developed) upto the end of the 9th Plan (1997–2002) as a Percentage of Ultimate Irrigation Potential of the State *

Minor Irrigation Potential Utilised as a Percentage of Irrigation Potential Created Upto March 2002 *

Average Gross Irrigated as a Percentage of Total Cropped Area (%): 1992–93 to 1995–96 and 1996–97 to 1999–2000 *

Average Area Irrigated more than once as a Percentage of Gross Irrigated Area (%): 1992–93 to 1995–96 and 1996–97 to 1999–2000 *

Average Agricultural Consumption of Electricity (in KwH) per Rs. Lakh of Agricultural GSDP at Constant (1993–94) Prices: 1993–94 to 1995–96 and 1996–97 to 1999–2001 *

Number of Energised Pumpsets Per Hectare of Gross Irrigated Area (No./ha): 1995–96 and 1999–2000 *

Change in Number of Energised Pumpsets per Hectare of Gross Irrigated Area (No./ha)/: 1992–93 to 1995–96 and 1995–96 to 1999–2000 *

NON-POINT SOURCE WATER POLLUTION POTENTIAL (20 indicators)

Population Density (Person per km^2 of Geographical Area): 1991 and 2001*

Percentage of Rural and Urban Households without Latrine: 1993 and 1998 *

Average Fertilisers Consumption (kg/hectare): 1992–93 to 1995–96 and 1996–97 to 2000–01 *

Average Annual Rainfall (in mm): 1990–95 and 1996–2000 $

Pesticides Consumption: (kg/hectare) 1995–96 and 1999–2000 *

Area under Pulses as a Percentage of Gross Cropped Area: 1990–91 and 2000–2001 $

Livestock per Head of Person (No. in Cattle unit per Person): 1992 and 1997 *

Poultry Birds per Head of Person (No. per Person): 1992 and 1997 *

Average Total Cropped Area as a Percentage of Reporting Area of Land Utilisation (%): 1992–93 to 1995–96 and 1996–97 to 1999–2000 *

PRESSURE AND DEGRADATION OF LAND RESOURCES (12 indicators)

Average Forest Area as a Percentage of Reporting Area of Land Utilisation (%): 1992–93 to 1995–96 and 1996–97 to 1999–2000 $

Average Non-Forest Common Property Land as a Percentage of Reporting Area of Land Utilisation (%): 1992–93 to 1995–96 and 1996–97 to 1999–2000 $

Average Non-Forest Common Property Land Per Capita (in ha/person): 1992–93 to 1995–96 and 1996–97 to 1999–2000 $

Average Area Sown more than Once as a Percentage of Total Cropped Area (%): 1992–93 to 1995–96 and 1996–97 to 1999–2000 *

Average Gross Irrigated as a Percentage of Total Cropped Area (%): 1992–93 to 1995–96 and 1996–97 to 1999–2000 *

Land Degradation as a Percentage of Geographical Area: 1994 and 2001*

Note:

* - implies that for the environmental indicator we have used (Maximum – Actual) / (Maximum–Minimum) for standardisation.

$ - implies that for the environmental indicator we have used (Actual–Minimum) / (Maximum–Minimum) for standardisation.

Endnotes

1. For instance, setting up of the Local Area Environmental Committees (LAECs) with the active participation of the local people for inspection, monitoring of day-to-day development in hazardous waste-affected sites; the Supreme Court Monitoring Committee (SCMC) on Hazardous Waste has ensured strict compliance of the Hazardous Wastes (Management and Handling) Rules, 1989 on the part of the industries or any other agency involved in Hazardous Waste generation, collection, treatment and disposal.
2. See Chakraborty and Singh (2005) and Kaushik and Saqib (2001) for details.
3. It is argued that environmental degradation or pollution level cannot be measured merely by actual emissions of certain hazardous materials; other factors influencing its spread and intensity also need to be considered (Kathuria, 2002, 2004).
4. The variables for which these two alternative formulas are used are specified at the end of **Appendix 3**.
5. For instance, a higher index for Orissa as compared to Punjab by merely ranking the forest resources of the two States (by taking the percentage of geographical area under forests land) comes from the fact that Punjab possesses very little of the selected variable to begin with. Therefore the analysis does not imply that forest conservation practices of the former are in any way better than those of the latter. Ranking the change in their forest area (as a percentage of geographical area) during any two periods would be the ideal exercise for comparing their forest conservation practices.
6. In several major Karnataka cities, suspended particulate matter (SPM) and respirable suspended particulate matter (RSPM) are far above the permissible limits (*The Hindu*, 2005, 2006).

7. Rithe and Fernandes (2002) argued that Maharashtra has achieved the current level of industrialisation at the cost of the loss of much of its forests. However, the findings of Kadekodi and Venkatachalam (2005) do not support this.
8. Apart from Government regulations, exporter firms increasingly adopted environment-friendly processes to comply with strict norms in export markets (e.g. marine industries in Kochi), which had a significant positive influence on the environment of the State.
9. Nair (2006) noted that depletion of the groundwater table due to indiscriminate sand mining, shrinkage in natural forest cover and reclamation of wetland and paddy fields are major environmental challenges that Kerala is facing today.
10. While Tamil Nadu needs to control industrial pollution (Mukherjee and Nelliyat, 2006; Nelliyat, 2005; Appasamy et al., undated); tackling groundwater arsenic contamination in the rural belt is a major policy challenge for West Bengal.
11. Bhullar and Sidhu (2006) have also reported disregard to environmental sustainability for short-run income and productivity gain in Punjab, which has resulted in over-exploitation of land and water resources.

Chapter 5

Nutrient Mismanagement and Degradation of Soil: Some Implications on Deceleration of Foodgrain Production in India

Pradip Kumar Biswas

1. Introduction

The adoption of high-yielding varieties (HYV) technology and subsequent rapid growth of foodgrains production has no doubt brought about self-sufficiency in foodgrains, but it has in the long run caused serious damage to soil and ultimately threatens the food security itself. The growth rate of foodgrain production as well as their yield peaked in the 1980s and then started decelerating in the early 1990s in almost all the major states. HYV technology requires substantial use of fertilisers, pesticides, and assured irrigation. Cultivation of the same few crops using the same inputs in the same lands over decades degrades soil quality, depletes the water table and pollutes the groundwater with various chemicals. In the face of deceleration, farmers try to raise productivity by intensifying the use of these inputs and inflict further damage to the soil. The cost of cultivation is going up rapidly and farmers are likely to face a severe threat from the global competition.

The problem, however, is not with the technology but its unscientific application. First of all, repetition of the same crop on the same land causes deficiency of particular types of nutrients in the soil. The common belief was that nitrogen, phosphorus and potassium (N-P-K) are the only nutrients that need be supplemented. Farmers continue to raise the intensity of these inputs without any consideration of their exact requirements or of their ratios. In general, N is over-supplied, while the other two are under-supplied. The inappropriate pricing policy of foodgrains distorts the choice of cropping pattern, and the flawed policy of fertiliser prices led to the sub-optimal or inefficient use of fertilisers mix. Further, groundwater is preferred for irrigation as its availability is more certain. The flawed power tariff policy of providing electricity for irrigation at an almost negligible rate led to over-use of the input which caused, on the one hand, depletion of the water table and on the other, salinity, soil leaching and contamination of groundwater. Deficiency of micronutrients in the soil has been overlooked for long. Monoculture over the years brings about deficiency of different types of micronutrients in different parts of India. This would substantially affect the agricultural productivity. The use of organic fertilisers, which would solve some of the above problems, has not yet been adequately emphasised.

The issue of rapid growth in agriculture and the consequent decline in environment, threatening the sustainability of cultivation and livelihood of a large number of people has not received due importance in the National Environment Policy 2006. The Policy describes the problem as one of 'institutional failure' and recommends the adoption of science-based, and traditional sustainable land use practices, through research and development, extension of knowledge, pilot-scale demonstrations, and large-scale dissemination, including farmers' training, and where necessary, access to institutional finance. However, the required institutions to inculcate a scientific temperament among farmers are yet to be evolved, for which the policy should have made specific recommendations.

In this paper, Section 1 is introductory, Section 2 describes the changes in cropping pattern in the last two decades and the rapid growth in the 1980 and deceleration in the 1990s. Section 3 reveals certain facts about degradation of soil quality as a consequence of unscientific agricultural practices. Finally in Section 4, we shall see how far these problems have been addressed in the recent National Environment Policy 2006.

2. Cropping pattern changes, growth and deceleration in foodgrain production

The Relative Importance of Crops/Crop Groups

Crops under foodgrains are clubbed into three broad groups, namely: fine cereals, (rice and wheat), coarse cereals, and pulses, of which fine cereals is the most important group in terms of share in total foodgrains area and output. Next in the order are the coarse cereals, and then pulses. During the 1980s, the average percentage shares of the three groups in total foodgrains area were 50.61, 31.09 and 18.30 respectively and in terms of output 71.55, 20.18 and 8.27 respectively. During the 1990s, the importance of fine cereals further increased and that of coarse cereals decreased. But the rise in area share of the fine cereals was much larger than the output shares. Table 5.1 displays the state-wise average percentage shares of the crop groups in area and production of foodgrains during the 1980s and 1990s.

Growth Performance in the 1980s and 1990s

During the 1980s, production of foodgrains at the all-India level grew at a compound annual rate of 2.94 per cent per annum, and the corresponding yield growth rate was even

Table 5.1: All India estimates of percentage share of rice, wheat, coarse cereals and pulses in total foodgrains

		Rice		Wheat		Coarse cereals		Pulses	
		A	P	A	P	A	P	A	P
North West Region									
Haryana	1980s	13.97	19.04	43.42	65.84	25.00	9.30	17.61	5.82
	1990s	20.49	19.98	49.65	68.62	19.00	6.86	10.81	3.39
Punjab	1980s	30.62	31.67	58.52	63.55	6.87	3.88	3.99	0.90
	1990s	38.25	35.18	56.25	62.18	3.79	2.28	1.69	0.36
Uttar Pradesh	1980s	26.28	24.95	41.10	54.13	18.03	12.39	14.58	8.53
	1990s	27.51	27.91	44.07	56.04	14.50	9.76	13.96	6.37
Eastern Region									
Bihar	1980s	55.54	52.39	20.12	28.57	11.32	11.07	13.02	7.97
	1990s	55.68	49.23	22.97	33.01	10.45	11.89	10.87	5.95
Orissa	1980s	62.90	73.90			9.19	8.45	27.12	16.12
	1990s	76.74	88.78			4.69	3.29	18.19	7.61
West Bengal	1980s	87.07	89.43	5.04	6.77	1.69	1.43	6.21	2.38
	1990s	90.00	92.38	4.93	5.20	1.19	1.15	3.82	1.25
West/Central Region									
Gujarat	1980s	11.40	14.92	12.90	27.31	59.11	48.00	16.59	9.77
	1990s	15.16	18.89	14.65	28.48	48.20	41.08	21.70	11.65
Madhya Pradesh	1980s	27.75	30.61	19.98	29.15	25.22	22.50	27.04	17.75
	1990s	30.00	30.84	23.56	37.37	18.17	13.75	28.26	17.98

		Rice		Wheat		Coarse cereals		Pulses	
		A	P	A	P	A	P	A	P
Maharashtra	1980s	10.68	21.17	6.64	8.27	61.78	58.55	20.89	12.02
	1990s	11.22	20.11	5.80	8.07	58.34	57.47	24.75	14.44
Rajasthan	1980s	1.11	1.77	14.58	41.35	57.59	40.12	26.72	16.65
	1990s	1.22	1.56	18.23	50.40	51.29	32.72	28.61	15.13
Southern Region									
Andhra Pradesh	1980s	45.41	73.88			36.67	20.68	17.76	5.35
	1990s	53.74	78.01			23.59	16.04	22.64	5.91
Karnataka	1980s	15.61	33.73	3.82	2.36	58.68	55.59	21.90	8.32
	1990s	18.45	36.70	3.28	1.93	54.72	53.74	23.64	7.70
Tamil Nadu	1980s	49.54	75.37			33.55	20.53	16.89	4.10
	1990s	55.37	81.47			25.38	14.36	17.98	3.66
All India	1980s	32.19	40.93	18.42	30.62	31.09	20.18	18.30	8.27
	1990s	35.01	42.62	20.59	33.69	26.03	16.54	18.37	7.15

Note: Percentage shares for 1980s and 1990s refer to the average shares for the period 1980-1990 and 1990-1999 respectively. A-area, P-production.

Source: Ministry of Agriculture, Government of India.

higher at 3.16 per cent per annum (Table 5.2). This sustained growth rate of output over a decade, based only on the growth of yield, is commendable for a country like India and is comparable to any international standard. Even for the low-yield crops like coarse cereals, the yield growth was quite high, at 2.64 per cent per annum. Pulses on the contrary maintained a yield growth at a much lower level of 1.34 per cent per annum. Deceleration, however, started in all-India output and yield of foodgrains by the turn of the decade— the growth figures for output and yield were 1.87 and 1.89 per cent per annum during the 1990s. Decline in the foodgrains area made a negative contribution to the total output, as a consequence output growth rate was lower than that of yield. The foodgrains area has been gradually diverted to oilseeds, vegetables and other non-food crops. It may be recalled that a major source of the growth of foodgrains in the pre-Green Revolution and early Green

Table 5.2 All-India estimates of compound annual growth rates of foodgrains

(per cent per annum)

Items	Foodgrains			Coarse cereals			Total Pulses		
Period	A	P	Y	A	P	Y	A	P	Y
1990–92 over 1980–82	–0.21	2.94	3.16	–1.78	0.81	2.64	0.06	1.41	1.34
1997–99 over 1990–92	–0.02	1.87	1.89	–2.08	–0.62	1.51	–0.42	0.76	1.21
	Rice			Wheat			Maize		
	A	P	Y	A	P	Y	A	P	Y
1990–92 over 1980–82	0.65	3.72	3.05	0.58	3.73	3.15	0.04	2.85	2.81
1997–99 over 1990–92	0.61	2.13	1.51	1.66	3.04	1.35	0.96	3.11	2.14

Note: (i) A, P and Y stand for area, production and yield respectively.
(ii) Compound annual growth rates are measured for the triennium averages of terminal years.

Source: Ministry of Agriculture, Government of India.

Revolution periods was the area expansion (Narain, 1988). Within foodgrains there has been a marked change in the composition of crops; coarse cereals have been losing their share relative to finer cereals, like rice and white. The extension of irrigation made the cultivation of fine and high-yielding cereals possible on the soils that were traditionally rain fed and were growing coarse grains. Continuation of minimum support prices (MSP) much above the costs of production (or the market clearing prices) led to such shifts, particularly in the surplus-generating states. Further, much higher levels of yields per acre of rice and wheat as compared to other foodgrains, to an extent, provided the economic basis of such shifts across country. The area shift was sharper in the 1990s than in the 1980s. Area growth rate of coarse cereals was negative for both the decades, at –1.78 and –2.08 per cent per annum respectively for the 1980s and 1990s, respectively.

In fact, in the 1990s when the yield growth rate for finer cereals was in no way higher than that of coarse cereals, the acreage substitution intensified. Wheat gained maximum in terms of area in spite of its yield growth rate falling

Figure 5.1.a: All-India trends in Coarse cereals production

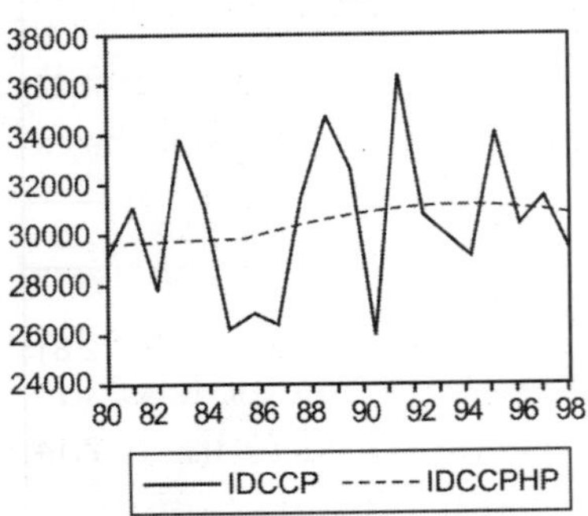

IDCCP - coarse cereals production ('000 mt)
IDCCPHP – smooth series of IDCCP

Figure 5.1.b: All-India trends in coarse cereals yield

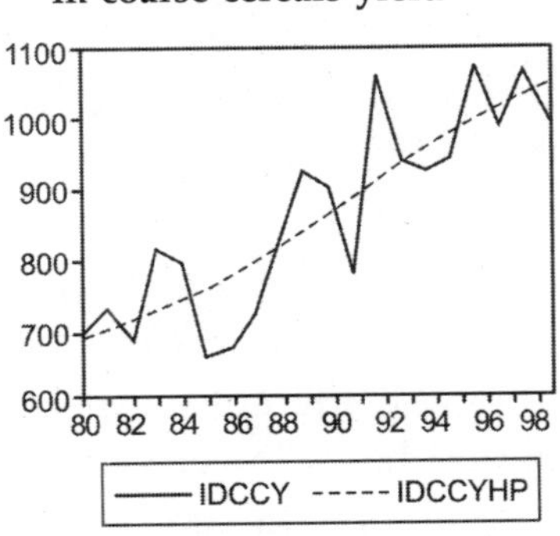

IDCCY – coarse cereals yield (kg/acre)
IDCCYHP – smooth series of IDCCY

Figure 5.1.c: All-India trends in foodgrains production

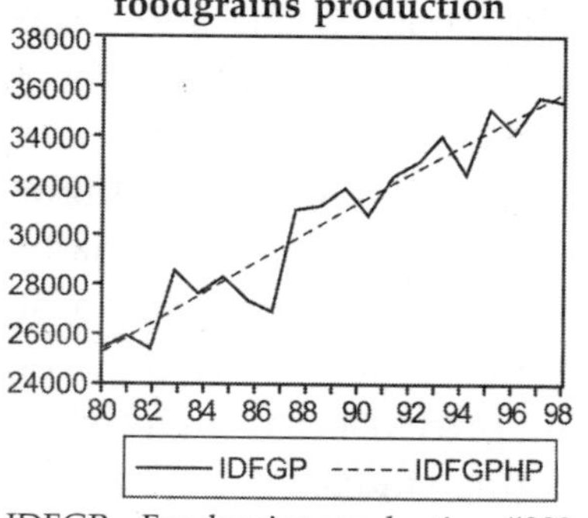

IDFGP - Foodgrains production ('000mt)
IDFGPHP – smooth series of IDFGP

Figure 5.1.d: All-India trends in foodgrains yield

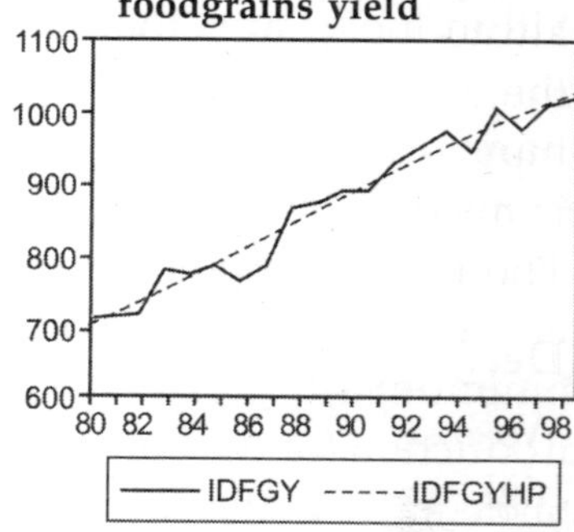

IDFGY – foodgrains yield (kg/acre)
IDFGYHP – smooth series of IDFGY

drastically to 1.35 per cent per annum which was lower than that of coarse cereals (1.51 per cent per annum). The compound annual growth rate of area under wheat in the 1990s was 1.66 per cent, which increased from 0.58 per cent in the 1980s. Rice, however, maintained an area growth rate at a modest level in both the decades (a little higher than 0.6 per cent). Yield growth rate of rice in the 1990s was the same as that of coarse cereals. Among the coarse cereals, maize gained significantly in the 1990s in terms of area. Its yield growth rate in the 1990s was higher than that of rice or wheat. In fact, output growth witnessed acceleration. The area under pulses (in terms of hectares) remained more or less unchanged during the 1980s but declined substantially in the 1990s, having adverse implications in terms of loss of soil fertility.

The deceleration hypothesis may be tested and analysed in three different ways. The first is to divide the time series into two different sub-periods and see whether the later sub-period has a lower growth rate as compared to the earlier sub-period. The second is to test whether the quadratic trend of the data series is significant or not. The third is to see whether there is any diminishing return in yield with respect to the inputs. The diminishing return may not always be reflected in terms of deceleration, especially when the use of inputs grows at a fast rate. But when the use of inputs grows at a constant or declining rate, this deceleration is most likely to set in the presence of diminishing return. These three approaches would be attempted here.

Decline in Growth Rates between the 1980s and 1990s

We have already said that the decade of the 1980s was a high-growth period and the 1990s a low-growth period. The trend break, however, did not take place exactly in 1990, but in the early 1990s, particularly in the all-India context, between 1991 and 1992. Further, across states, the break point varied—clustered around the all-India figure, and, as will be seen later, that there was acceleration or no such break in some states. In order to make the growth rates comparable across states, we have treated the triennium average figure for 1990—1992 as the value at the break point, and the growth rate is estimated on point-to-point basis for two periods, namely, the growth of the 1990-92 figure over the 1980-82 figure and the same for 1997-99 over 1990-92. Further, in view of the heterogeneity of the regions or even states, movements of area, production or yield vary across states. For instance, at the all-India level, the foodgrains area declined, but in the northern states it increased in the 1990s. A disaggregated analysis seems to be useful.

Among the regions, the decline in growth rates was sharper in the north-western region. The three important states, namely Haryana, Punjab and Uttar Pradesh witnessed a substantial decline in the yield growth in the 1990s (Table 5.3). In spite of a rise in the foodgrains area, deceleration of output could not be resisted. Only in Uttar Pradesh the decline in the output and yield growth during the 1990s was relatively low as compared to that of Punjab and Haryana. the rice-wheat cropping pattern dominates

Table 5.3: State-wise estimates of compound annual growth rates of foodgrains

(per cent per annum)

	Period	Rice			Wheat			Foodgrains			Coarse cereals			Total pulses		
		A	P	Y	A	P	Y	A	P	Y	A	P	Y	A	P	Y
North West Region																
Haryana	1980s	3.17	3.92	0.74	1.66	5.68	3.98	-0.42	4.84	5.28	-3.43	-0.32	3.11	-4.31	-0.2	3.6
	1990s	6.12	4.42	-1.53	1.89	2.88	0.96	1.26	2.23	0.91	-1	0.59	1.72	-6.63	-4.87	0.93
Punjab	1980s	5.05	6.19	1.12	1.10	3.79	2.68	1.38	4.11	2.7	-6.7	-4.42	2.42	-8.64	-5.86	3.01
	1990s	2.66	2.76	0.13	0.27	1.68	1.40	0.97	1.98	0.98	-2.61	-0.77	1.93	-5.81	-7.23	-1.5
Uttar Pradesh	1980s	0.47	5.55	5.05	0.77	3.55	2.77	0.01	3.56	3.54	-2.39	1.58	4.07	-0.05	0.65	0.68
	1990s	0.86	3.23	2.35	0.89	2.66	1.75	0.22	2.21	1.98	-1.76	-0.6	1.15	-0.96	-1.16	-0.2
Eastern Region																
Bihar	1980s	-0.27	1.45	1.73	1.45	4.34	2.85	-0.47	2.30	2.79	-3.4	1.93	5.47	-1.63	0.65	2.3
	1990s	0.18	4.61	4.59	0.87	3.40	2.51	0.08	3.40	3.37	0.58	1.66	1.16	-2.61	-1.76	0.92
Orissa	1980s	0.77	4.51	3.72				-0.13	2.83	2.98	-5.28	-4.31	0.45	-0.37	-0.64	-0.6
	1990s	0.06	-0.64	-0.68				-2.64	-2.4	0.29	-11.2	-12.6	-0.8	-11	-15.7	-4.8
West Bengal	1980s	1.23	6.37	5.11	0.34	1.33	1.01	0.80	5.85	5.05	-2.59	3.62	6.65	-4.21	-1.52	2.75
	1990s	0.73	3.01	2.25	4.92	5.18	0.22	0.79	3.06	2.22	-1.06	2.34	3.19	-2.9	-1.92	0.8
West/Central Region																
Gujarat	1980s	1.62	3.24	1.65	-1.43	-0.89	0.62	-0.81	-0.2	0.53	-2.44	-1.64	0.83	3.52	4.36	0.85
	1990s	2.24	2.48	0.28	1.11	2.41	1.06	-1.51	1.56	3.11	-3.71	0.51	4.27	-1.57	-0.13	1.46

Madhya Pradesh	1980s	0.56	3.68	3.10	0.74	4.69	3.92	-0.34	2.90	3.24	-2.42	-0.07	2.41	-0.19	2.18	2.4
	1990s	0.57	-0.06	-0.61	3.11	5.74	2.57	0.26	1.85	1.61	-4.66	-5.11	-0.6	0.84	2.84	1.98
Maharashtra	1980s	0.42	0.05	-0.36	-3.85	-1.50	2.47	-0.12	1.59	1.68	-0.45	2.12	2.53	1.69	4.03	2.17
	1990s	-0.71	1.27	1.99	3.02	4.34	1.02	-0.56	0.03	0.66	-1.48	-1.91	-0.3	1.08	4.26	3.16
Rajasthan	1980s	-0.63	1.49	2.27	0.66	4.34	3.78	-0.36	3.29	3.61	-0.55	3.8	4.27	-0.5	-0.43	-0.1
	1990s	4.07	5.81	1.60	4.42	4.84	0.34	0.76	2.83	2.07	-1.87	-1.52	0.35	2.1	5.57	2.98
Southern Region																
Andhra Pradesh	1980s	0.45	2.08	1.62				-1.85	0.92	2.83	-6.58	-3.87	2.94	1.16	4.26	3.07
	1990s	-0.36	1.10	1.41				-1.26	0.96	2.2	-3.76	1.01	4.89	-0.5	-0.8	-0.3
Karnataka	1980s	1.07	2.22	1.12	-4.47	-2.98	1.55	0.04	1.74	1.69	-0.27	1.74	2.03	0.82	1.05	0.22
	1990s	1.33	2.97	1.64	3.23	2.76	-0.5	0.01	2.58	2.58	-0.96	2.41	3.4	0.95	2.26	1.21
Tamil Nadu	1980s	-0.67	3.75	4.54				-0.66	3.10	3.86	-2.76	-0.19	2.7	3.4	6.59	3.17
	1990s	1.42	2.37	0.94				0.66	2.08	1.4	-1.16	-0.06	1.11	-1.99	-1.54	0.43
All India	1980s	0.65	3.72	3.05	0.58	3.73	3.15	-0.21	2.94	3.16	-1.78	0.81	2.64	0.06	1.41	1.34
	1990s	0.61	2.13	1.51	1.66	3.04	1.35	-0.02	1.87	1.89	-2.08	-0.62	1.51	-0.42	0.76	1.21

Note:

1. Growth rate for 1980s refers to the compound annual growth rate of 1990-92 figure over 1980 -82 figure and 1990s refer to the same 1997-99 over 1980-82.
2. Growth rate for area and yield may not add up to that of output due to triennium averaging of the terminal years.

Source: Ministry of Agriculture,GOI.

this region. Over the last two decades their share in foodgrains area increased at a rapid rate, particularly in Haryana and Punjab. As mentioned earlier this was to a large extent conditioned by the MSP. In spite of a very low growth rate of rice yield, the area under rice in these two states surpassed the growth rates for other crops in the 1980s. The situation became worse in the 1990s when yield growth rate of rice in Punjab became close to zero and in Haryana highly negative, yet the area growth rate in the former remained at a high level with some decline and in the latter doubled.

It is worth noting that a higher rate of yield growth was not associated with a higher rate of acreage expansion or the area shrinkage was not greater when the yield growth rate was lower. In fact, in the north-western region, the deceleration in output of foodgrains and coarse cereals would have been much greater in the 1990s had there been a similar pattern of area response to yield growth, as seen in the 1980s.

Pulses made a rapid decline in the region in both the decades, except in Uttar Pradesh during the 1980s. The rate of yield growth declined and even became negative in Punjab and Uttar Pradesh in the 1990s. Similar to coarse cereals, pulses also witnessed a decline in the share in the foodgrains area.

In the eastern region also, output and yield of foodgrains decelerated, with the exception of Bihar where the growth rate for output and yield increased considerably in the 1990s over the 1980s. West Bengal, which achieved a record growth rate of 5.85 per cent per annum in foodgrains production during the 1980s, witnessed a considerably lower growth rate of 3.06 per cent per annum in the 1990s. Orissa's case is more alarming, where a substantial decline in yield growth was accompanied by a rapid shrinkage in area under foodgrains—the rate of yield growth per annum was 2.98 in the 1980s and 0.29 in the 1990s, and the corresponding area growth rates were –0.13 and –2.64 per cent per annum. The share of wheat declined in West Bengal and increased in Bihar. The growth rate for rice yield accelerated in Bihar during the 1990s whereas in other states it decelerated, and became negative in Orissa.

Deceleration in foodgrains production was also evident in the Central/Western region where all the states, excepting Gujarat, witnessed a decline in the growth rate of area and yield. Gujarat witnessed a negative growth in production and a meagre 0.53 per cent yield growth rate in the 1980s but made a faster comeback in the 1990s—yield increased at a rate of 3.11 per cent per annum. Yet, the area under foodgrains declined at a greater rate in the 1990s, than in the 1980s which led to a moderate growth rate in production.

The southern region as a whole has no strong tendency of deceleration of foodgrains production and yield. There was a substantial rise in the growth rate of output in Karnataka and a marginal rise in Andhra Pradesh in the 1990s, as compared to that in the 1980s. Only in Tamil Nadu was the decline in the rate of yield growth of foodgrains marked. However, yield growth rate accelerated in the 1990s in Karnataka.

Pulses are also quite important in these states sharing a sizeable amount of foodgrains area, although the output share in total foodgrains was quite low, indicating a very low yield. In Andhra Pradesh and Tamil Nadu area and output declined during the 1990s. So was the yield growth. During the 1980s yield growth rate was above 3 per cent per annum and the production growth was much faster in the two states. However, in the 1990s, only Karnataka maintained a moderate growth in area, production and yield.

In brief, the growth rates of production and yield of foodgrains have been significantly reduced in the 1990s over the 1980s in most of the states. The deceleration was much more prominent for rice and wheat than for coarse cereals and pulses. Yet, area expansion continues for the former crops and the area decline continues for the latter crops. This tendency seems to be induced by the pricing policies relating to foodgrains and inputs, and higher yields of rice and wheat.

The Quadratic Trend

In order to establish the hypothesis of deceleration in foodgrains production and yield, we need to conduct statistical tests on the relevant data series. As a first step, we need to find out the trend component in the data series. This is done using an appropriate filter (Hodrick and Prescot, 1997). The smoothed data series is then regressed on time to find a log-quadratic trend by using the following equation:

$log\ (y)=a+bt+ct^2$, where t represents time and y exponentially smoothed series for the period, 1980 to 1999. The sign of c (that is the coefficient of t^2) indicates deceleration or acceleration. When it is negative, deceleration is said to be operative and the positive sign implies acceleration; of course it has to be statistically significant. This test has been conducted for thirteen major states and throughout India.

All-India foodgrains production and yield witnessed deceleration in the period under consideration, as the coefficients of t^2 are significantly negative. The same is true for coarse cereals although the intensity of deceleration was much below that of foodgrains. It may be seen in Table 5.4 that in the northern and eastern regions all the states under consideration witnessed deceleration for both foodgrain production and yield. In the Central region, deceleration in both production and yield was noted only in Madhya Pradesh and only in production in Maharashtra. On the contrary, acceleration in production and yield was noted in Andhra Pradesh. As regards coarse cereals, the growth rate for both output and yield slowed down in the states of Madhya Pradesh, Maharashtra and Rajasthan and accelerated in Gujarat.

In the southern region, only in the state of Tamil Nadu the growth rate of foodgrains output and yield tapered off during the period. In Andhra Pradesh, output and yield followed diametrically opposite growth paths—while the former decelerated, the latter maintained an increasing growth rate. This was due to the rate of area decline being much greater than the rate of yield growth. In Karnataka, both foodgrains output and yield accelerated. As regards coarse cereals, acceleration in production and yield was noted in Andhra Pradesh and Karnataka and deceleration in Tamil Nadu. Overall, majority of the states witnessed a significantly negative second order coefficient for output and yield, as may be seen in Table 5.4. The deceleration however, is more widespread in the case of foodgrains than in the case of coarse cereals and also more prominent in the north-western and eastern regions than in the central and southern regions. A measure of the contribution of coarse cereals, fine cereals, and pulses to growth rate of overall foodgrain yields for the 1980s and 1990s may be useful to locate the sources of deceleration.

Diminishing Return to Fertilisers

We have already noted deceleration in the production and yield of foodgrains. Some of this may be explained in terms of diminishing return to inputs or slowdown in the growth of vital inputs, like fertilisers, irrigation, certified or quality HYV seeds, and pesticides. Here we shall discuss the condition of fertiliser use—whether the growth of its use

Table 5.4: Estimates of exponential trends for production and yield of coarse cereals (CC) and foodgrains (FG)

		Regression coefficients for production				Regression coefficients for yield			
		A	t	t^2	Adj R^2	a	t	t^2	Adj R^2
North West region									
Haryana	(a) CC	6.616	-0.020	0.00119	0.83	6.368	0.029	0.0001@	0.99
	(b) FG	8.616	0.057	-0.00090	1	7.197	0.070	-0.00159	1
Punjab	(a) FG	9.405	0.053	-0.00106	1	7.823	0.037	-0.00081	1
Uttar Pradesh	(a) CC	8.120	0.022	-0.00086	1	6.682	0.047	-0.00101	1
	(b) FG	10.08	0.042	-0.00062	1	7.063	0.043	-0.00069	1
Eastern region									
Bihar	(a) CC	6.907	0.025	0.00001*	0.99	6.605	0.080	-0.00195*	1
	(b) FG	9.044	0.031	-0.00034	0.99	6.780	0.038	-0.00047	1
Orissa	(a)CC	6.361	0.020	-0.00558	0.99	6.693	0.009	-0.00083*	0.86
	(b) FG	8.548	0.050	-0.00222	0.98	6.630	0.044	-0.00126	0.99
West Bengal	(a) FG	8.783	0.066	-0.00122	0.99	7.013	0.059	-0.00121	0.99
West/Central region									
Gujarat	(a) CC	7.837	-0.050	0.00201	0.76	6.727	-0.032	0.00267	0.97
	(b) FG	8.474	-0.031	0.00180	0.87	6.907	-0.023	0.00197	0.96
Madhya Pradesh	(a) CC	8.029	0.027	-0.00247	0.99	6.413	0.038	-0.00134	0.99
	(b) FG	9.403	0.031	-0.00029	0.99	6.497	0.036	-0.00047	1
Maharashtra	(a) CC	8.542	0.035	-0.00102	0.95	6.413	0.038	-0.00134	0.99
	(b) FG	9.117	0.023	-0.00033*	0.97	6.471	0.021	-0.0001@	0.98

Rajasthan	(a) CC	7.930	0.030	-0.00095	0.98	5.962	0.027	-0.00032	0.99
	(b) FG	8.867	0.017	0.00059	1	6.313	0.028	-0.00001@	1
Southern region									
Andhra Pradesh	(a) CC	8.023	-0.067	0.00244	0.99	6.486	0.017	0.00127	0.99
	(b) FG	9.246	0.008	0.00019	0.99	7.005	0.034	-0.00027	1
Karnataka	(a) CC	8.139	0.014	0.00031	1	6.678	0.010	0.00088	0.99
	(b) FG	8.738	0.010	0.00057	1	6.750	0.008	0.00075	0.99
Tamil Nadu	(a) CC	7.247	0.001@	-0.00061*	0.86	6.716	0.034	-0.00068	1
	(b) FG	8.652	0.033	-0.00057	0.99	7.115	0.052	-0.00113	1
All-India	(a) CC	10.279	0.006	-0.00016	0.87	6.501	0.023	-0.000004	0.99
	(b) FG	11.732	0.032	-0.00038	1	6.870	0.037	-0.00048	1

Note:

1. * Confidence level between 95 and 99 per cent, @ insignificant, and the rest are significant at less than one per cent.
2. Regression equation: **log (y)=a+bt+ct**^{2}, where **t** represents time and **y** exponentially smoothed series for the period, 1980 to 1999.

Source: GOI, Ministry of Agriculture.

declined or diminishing return started taking place. The rationale for testing diminishing return with respect to fertilisers is that they have been seen as a key input towards explaining productivity performance of the HYV crops. Consumption of chemical fertilisers (N+P+K) per hectare more than doubled in the 1980s, from 31.83 kg in 1980-81 to 67.49 kg in 1990-91; thereafter it increased at a slower rate and reached 90.04 in 1998-99.[1] Along with this deceleration in the use of this vital input if diminishing return also starts operating to this input, the consequence is any body's guess. In order to test the diminishing return of yield with respect to fertilisers, we assume the following production function:

$$Log\ Y_t = a + b(logF_t) - c(logF_t)^2 + d(logM_t) + et$$

Where Y - yield per hectare, F - fertiliser per hectare, M - manure per hectare, t – time, a, b, c, d and e are parameters.

If the coefficient c is found to be significantly negative, we may accept the hypothesis of diminishing return.

Relevant data are collected from the published reports of the Commission for Agricultural Costs and Prices, Ministry of Agriculture, Government of India. There exist several problems with the data, such as: (i) it covers selected crops in selected states, (ii) the data series is often incomplete covering the years from early 1980s to the mid or late 1990s.

This test has been conducted for rice, wheat, jowar, bajra, urad and gram for selected states where data gaps are relatively low. In some cases, manure is dropped when its application was at a very low level. Regression results are presented in Table 5.5. In the case of rice, the test is conducted in five states, namely, Punjab, Karnataka, Andhra Pradesh, Uttar Pradesh, and West Bengal. The relevant coefficient is found to be negative in three cases, viz. Punjab, Andhra Pradesh and Uttar Pradesh, but nowhere is it highly significant. In the other two cases the coefficient is positive, but is significant only in Karnataka, where in fact there was acceleration in the yield of foodgrains, including rice, as we have noted earlier.

As regards wheat, the test is carried on for four states, namely Punjab, Haryana, Uttar Pradesh and Rajasthan. For the last two cases, the relevant second order coefficient is negative and in the first two cases it is positive. In Punjab, the positive coefficient is highly insignificant whereas in Haryana it is highly significant. The negative second order coefficients for Uttar Pradesh and Rajasthan are not highly significant, nevertheless they indicates at least the beginning of diminishing return.

Jowar in Karnataka and urad in Andhra Pradesh showed a negative value of *c*, Andhra Pradesh's case is highly significant. Bajra in Haryana and gram in Rajasthan showed a positive value of *c* but neither is significant.

It is worth noting that in several cases, the contribution of manure to yield is found to be negative—quite contrary to the desired result. This happened due to the fact that in those cases the use of manure steadily declined when yield continued to rise and this declining trend in *M* could not be eliminated by the coefficient of *t*.

The regression analysis therefore suggests that the diminishing return of yield to fertilisers is not pervasive but it has made a beginning in a number of states and for a number of crops. This, however, cannot explain deceleration in foodgrains production, so widespread in Indian agriculture.

3. Unscientific agricultural practices and degradation of soil

A massive yield growth of the major crops was witnessed throughout the 1980s. It is well known that the HYV plants, given adequate water or moisture of the soil, extract soil nutrients at a much larger rate than their traditional counter-

Table 5.5: Regression results of yield on fertilisers and manure, 1981-1998

Items		Intercept	logF	$(\log F)^2$	LogM	t	Adj R^2
	Rice						
Punjab	Coefficients	44.5358	4.1742	-0.3889	-0.1630	-0.0257	0.31
	t-values	0.50	0.12	-0.12	—1.92	-3.14	
Karnataka	Coefficients	111.1357	-15.6267	1.7653	0.4047	-0.0376	0.27
	t-values	1.68	-1.96	2.00	2.10	-1.36	
Andhra Pradesh	Coefficients	-45.4629	8.5582	-0.8154	-0.0175	0.035	0.84
	t-values	-1.33	0.88	-0.83	-0.14	1.48	
Uttar Pradesh	Coefficients	13.9993	0.6394	-0.0325	-0.1334	-0.0063	0.88
	t-values	0.40	0.24	-0.10	-2.25	-0.41	
West Bengal	Coefficients	-21.2052	-1.2553	0.1957	-0.1033	0.0135	0.40
	t-values	-0.52	-0.45	0.55	-0.27	0.65	
	Wheat						
Punjab	Coefficients	-9.8607	-7.9049	0.7711		0.0169	0.74
	t-values	-0.21	-0.54	0.54		1.79	
Haryana	Coefficients	-1.3328	-10.1984	1.0922		0.0149	0.93
	t-values	-0.07	-2.58	2.62		2.32	
Uttar Pradesh	Coefficients	-61.7548	10.3341	-1.0900	0.0460	0.0203	0.87
	t-values	-1.57	1.24	-1.21	1.29	1.45	
Rajasthan	Coefficients	-59.4700	2.9235	-0.3528		0.0285	0.68
	t-values	-3.65	1.16	-1.12		4.09	

	Bajra						
Haryana	Coefficients	52.4333	-0.4503	0.1908		-0.0254	0.47
	t-values	1.02	-0.70	1.42		-0.98	
	Jowar						
Karnataka	Coefficients	-12.2411	2.1623	-0.2855	0.2467	0.0500	0.20
	t-values	-0.21	0.91	-0.78	1.10	0.17	
	Urad						
Andhra Pradesh	Coefficients	-210.991	-0.6788	-0.2000	0.4456	0.1073	0.84
	t-values	-4.71	-5.57	-5.00	4.82	4.75	
	Gram						
Rajasthan	Coefficients	-21.2401	-0.0158	0.0649	0.1308	0.0117	0.54
	t-values	-0.95	-0.16	0.68	2.86	1.03	
Uttar Pradesh	Coefficients	11.8523	-0.0192		0.0378	0.0048	-0.34
	t-values	0.33	-0.20		0.66	-0.27	

Note: Yield and Manure in quintals and Fertilisers in kilograms.

Source: Data based on Commission for Agricultural Costs and Prices.

parts and thus the soil cannot replenish nutrients on its own within a short span of time. Multiple cropping and repeated cultivation of the same combination of crops most often aggravates the problem of deficiency of particular nutrients in particular soils. Nutrient supplementation is required to be made, which causes another kind of problem, namely that of soil nutrient management.

In India, three kinds of fertilisers that have been commonly used are: nitrogenous (N), potash (K) and phosphate (P), called primary nutrients. N is required for a quick plant growth and therefore needs to be supplemented on the basis of its deficiency in the soil. The other two nutrients, required for metabolism, growth of root and plant resistance, are also required to be supplemented for their deficiency in the soil.[2] On the basis of soil testing agro-scientists observed that, for the country as a whole, given the cropping pattern, N, K and P are to be supplemented in the ratios of 4: 2: 1. It is common practice in Indian agriculture that the application of N is much higher than K and P, so that the ratio for N in total fertiliser far exceeds the recommended one. During the 1970s and 1980s, N:P:K ratios hovered around 6:1.9:1 and 5.9:2.4:1 respectively, thereafter with the decontrol of phosphatic and potassic fertilisers, their prices soared in comparison to nitrogen. The N:P:K ratios shot up to 9.5: 3.2:1 in 1992-93 and thereafter the ratios fluctuated around this figure. Thus the distortion in fertiliser use was further aggravated in favour of N and against K and P in the 1990s.

The requirement of nutrient supplementation, however, varies across regions, depending on, among others, soil conditions, cropping intensity and cropping pattern. Different crops in different regions are therefore recommended varied amounts of these primary nutrients as supplements. Several studies have highlighted that these

Table 5.6: Farmers practice and soil test based trials on farmers field: state wise results, 1992-94

State	Crop	Farmers practice				Recommended dose			
		Nutrient (kg/ha)			Yield	Nutrient (kg/ha)			Yield
		N	P	K	q/ha	N	P	K	Q/ha
Maharashtra	Blackgram	0	50	0	11.9	20	50	50	15.7
Uttar Pradesh	Blackgram	5	14	0	8.3	15	40	0	9.5
Bihar	Maize	138	0	0	32.1	120	75	50	61.1
Karnataka	Maize	72	32	32	60.2	150	75	40	67.6
Rajasthan	Maize	45	20	10	15.3	90	30	11	19.8
Gujarat	Pearl millet	60	36	0	14.8	80	40	0	16.4
Rajasthan	Pearl millet	40	15	0	12.3	90	30	0	14.3
Andhra Pradesh	Rice	73	28	14	46.8	80	49	30	49.7
Haryana	Rice	160	52	4	55.4	139	62	62	58.8
Orissa	Rice	67	32	34	42.7	78	39	39	46.5
Punjab	Rice	152	31	0	58.8	125	30	29	62.5
Tamil Nadu	Rice	105	40	50	50.6	125	53	53	56.9
Uttar Pradesh	Rice	87	18	3	37.5	120	60	29	47.7
West Bengal	Rice	108	51	45	50.1	88	44	44	50.3
Haryana	Wheat	149	58	0	45.3	134	60	27	48.1
Madhya Pradesh	Wheat	57	31	8	30.7	95	58	30	39.6
Punjab	Wheat	146	64	8	44	125	61	30	47.2
Rajasthan	Wheat	95	37	2	39.5	113	38	14	44
Uttar Pradesh	Wheat	92	30	7	34.3	120	60	31	42.9

Source: Govil and Kaore (1999).

recommendations are rarely adhered to and consequently actual yield remained far below the potential. Govil and Kaore (1999) made a summary report of several studies. In Table 5.6 we reproduce a part of their observation:

It may be seen in Table 5.6 that in most of the states and for most of the crops, potash is often under-used. In Punjab and Haryana, for both wheat and rice, nitrogen is applied in excess of the recommended dosage and the application of potash is negligible, although the amount recommended for its use is sizeable. Imbalances in fertiliser uses also vary across size classes. Velrasu and Singh's (1999) study of Erode district in Tamil Nadu revealed that marginal farmers (less than 1 ha) apply N below the recommended dosage and other categories exceed it. However, the maximum amount is applied by small farmers (1–4 ha). As regards P and K, small farmers exceed the recommended doses and other categories apply below the recommended doses. Under-application is maximum for marginal farmers and then come medium and large, in that order. In the study area as a whole, N is used in excess and P and K are below what is recommended. The average percentage deviation of N is smaller than that of P and K. For rice, N exceeds the recommended dosage, P is at the recommended dosage and K is below the recommended dosage. Overdose of one nutrient and under-dose of another or disproportionate use of nutrients have many implications, such as reducing yield much below the potential of the crop variety, or raising cost of production, harming long-term soil fertility through nutrient-fixation, toxicity or salinity.

The state policy of subsidising fertilisers favoured N, as it was made relatively cheaper compared to P and K, and further, overemphasis on the use of NPK led to relative neglect of secondary or micronutrients, like iron, zinc, sulfur, molybdenum, manganese, silicon chlorine, etc. which the plant requires and extracts from the soil. These elements are required in small quantities, and in the pre-Green Revolution period, their deficiency in the soils was not felt. Introduction of HYV crops, rising cropping intensity, and use of high analysis fertilisers raised production, but in the course of time, deficiencies of micro and secondary nutrients became critical for sustaining optimum productivity of crops in several soils. Deficiency of zinc (Zn) was first noticed in rice in tarai soils and in wheat on the sandy soil of Punjab in 1970 and then in most of the intensively cultivated soils where these crops are grown. Later, deficiency of iron (Fe) in rice, sugarcane, chickpea, groundnut, etc. in sandy soils, of manganese (Mn) in wheat in rice-wheat system on the sandy soils of Punjab, Boron (B) in chickpea in highly calcareous soils of Bihar, and later S became critical for sustaining high soil productivity in many areas.[3] These micro and secondary nutrients therefore became critical to the growth of the plants. Several studies indicate that the soil of almost every region of India is deficit in some nutrient or other. Supplementing these kinds of nutrients in the soil has not yet been started to any considerable extent. Singh (2001a), while analysing 1.48 lakh soil samples data of recent years for different agro-ecological zones and soil types, found that 45, 8.3, 4.5 and 33 per cent mean deficiency of Zn, Fe, Cu, and B respectively. Soils of Maharashtra, Karnataka, Haryana, Tamil Nadu, Andhra Pradesh, Orissa and Bihar tested Zn deficiency more than 50 per cent. Among micro nutrients, Zn deficiency was found most widespread in various states. Its deficiency leads to reduction in crop productivity. Boron deficiency is widespread in Bihar, Karnataka, West Bengal, Orissa, Uttar Pradesh, Madhya Pradesh, Tamil Nadu and Punjab.[4]

Deficiency of sulphur is also widespread in the soils of India. It reduces plant growth, as does N deficiency. In another study, Singh (2001b) revealed that S deficiency of soils of various states varied from 5 to 83 per cent with an overall mean of 41 per cent. Most of the Indo-Gangetic alluvial plains, red and lateritic hill soils are prone to S deficiency. Severe S deficiency was reported in Punjab, Haryana and Uttar Pradesh soils with low organic matter. This deficiency is on the increase due to intensive cropping. S deficiency is widely reported in many districts of the central, eastern and southern states. Several manmade factors are responsible for S deficiency, besides soil properties. The management practices such as increased cropping intensity, introduction of HYVs, cultivation of high sulphur demanding crops, use of sulphur-free fertilisers like urea, DAP, MOP, lack of organic manure addition, lesser crop residue recycling, and irrigation with canal water having low sulphur content are causing many adverse effects on the availability of sulphur and have led to greater removal and depletion of sulphur from the soil. This has adversely affected productivity of cereals, pulses and oilseeds. In fact, a recent study by Rao et al. (2001) established the fact that sulphur deficiency in the soils was one of the key factors for very slow yield growth of pulses in the predominantly pulse-growing states like Uttar Pradesh, Rajasthan, Madhya Pradesh, Maharashtra, Andhra Pradesh, Karnataka and Orissa. Pulses in general receive very little mineral fertilisers and their yield, though very low, are sustained mainly by soil nitrogen and nitrogen fixed by rhizobium. All the nutrients present in the soil are not available to plants as these elements mostly occur in forms which cannot be absorbed by the plant. Proper dosage of fertiliser containing deficit nutrients can be applied. Problems still arise in other ways: (i) infiltration of the nutrients in the deep soil with rain water, (ii) conversion of nutrient in the form not absorbable by the plant (fixation of nutrients), and (iii) difficulty in maintaining a balance between sufficiency and toxicity in the soils throughout the farming period. Agricultural extension service seems to be highly necessary for regularly testing the soil nutrient contents and recommending the required doses of different fertilisers required for specific crops.

In the rice-wheat cropping system, which is a predominant cropping system in India, S deficiency is chronic—increased use of S-free mineral fertilisers, intensive cropping with high yielding crop cultivars without recycling their residues and restricted supply of organic manure have aggravated S-deficiency (Sakal et al., 1999). Increasing levels of S progressively raise the uptake of N, P, K and S in rice and also to a lesser extent in wheat. S is also very important for its active role in protein synthesis. The choice of cropping pattern is unscientific in many cases. For instance, upland rice cultivated in many parts of Bihar, Uttar Pradesh, West Bengal, Madhya Pradesh and Maharashtra and in other states, sharing 17 per cent of rice area and contributing 10 per cent of the rice production of the country. Poor fertility, land degradation, and lack of availability of plant nutrients to crop are important bottlenecks in increasing and stabilising the productivity of upland rice. Upland rice soils are deficit or marginally sufficient in N content. However, only 20–30 per cent of the applied N fertiliser is utilised by crops due to various kinds of losses such as through high permeability, infiltration, and percolation loss of water. These soils are also deficit in P and K (Mishra, 1999). In Punjab, due to intensive agriculture, the soil became deficient in N, P, Zn and S, the watertable lowered at the rate of 23 cm per annum, underground water became polluted due to more and more use of chemical fertilisers, pesticides, weedicides, monoculture of rice-wheat rotation led to higher incidence of pests and diseases attack, etc. (Sidhu and Dhilon 1997). Out of 118 blocks in the state, 67 per cent were found to have acute inadequate N in the soil, 44 per cent had inadequate P, and although the level of K is generally sufficient in the soil, due to intensive cropping it declined over the years. The availability of Zn declined by 50 per cent between 1981–1991. Over exploitation of underground water sources led to a decline in the watertable in most areas of the state, especially the central plains where paddy is the main kharif crop. The expansion of the area under paddy crop has been associated with the lowering of the watertable. It is further noted that, apart from being a water-intensive crop, paddy is often over-irrigated in Punjab, compared to its requirement (ibid).[5]

Toxicity of micronutrients has now become an important issue. In soils having a pH below 5.5, this problem arises. In strong acidic soils, toxicity of Mo and Al associated with Ca and Mn may occur (Mishra 1999). When Al concentration becomes very high in the soil, it adversely affects the uptake of most of the plant nutrients. In highly weathered lateric and sandy soils with low pH, Si deficiency occurs. Zn deficiency occurs in alkaline soils and its intensity increases with rising pH value of soils, and Fe deficiency occurs primarily in the alkaline soils (ibid). All this indicates, that adoption of more scientific methods of cultivation is required, which in turn requires greater involvement of researches and agricultural extension activities.

The environmental problems associated with rapid agricultural growth based on HYV technology are found to be widespread and they seem to have ultimately brought down the growth rate itself, making the technology unsustainable. The problem, however, arose due to the unscientific use of the technology. The various types of chemical fertilisers including micronutrients are not being applied according to their requirements in the soil for specific crops. Over-irrigation of crops often causes soil leaching when fertlisers infiltrate to the watertable, and excess draft of groundwater over what is recharged causes depletion of watertable. Both make cultivation unsustainable in the long run. Deceleration seems to be a reflection of the mismanagement of the various inputs and resultant degradation of soils at a much wider level. Although the New Environmental Policy 2006 recognised these problems, it did not come out with specific remedial policies.

4. Some implications of the National Environmental Policy 2006

In the Policy statement the problem has been recognised as: The degradation of land, through soil erosion, alkali-salinization, water logging, pollution, and reduction in organic matter content has several proximate and underlying causes. The proximate causes include ...excessive use of irrigation (in many cases without proper drainage, leading to leaching of sodium and potassium salts), improper use of agricultural chemicals (leading to accumulation of toxic chemicals in the soil), diversion of animal wastes for domestic fuel (leading to reduction in soil nitrogen and organic matter). These proximate causes of land degradation in turn, are driven by implicit and explicit subsidies for water, power, fertilizer and pesticides...The absence of conducive policies and persistence of certain regulatory practices reduces people's incentives for afforestation, and leads to reduced levels of green cover" (p.22).

It further says that tariff policies for irrigation systems, which, through inadequate cost recovery, provide incentives for overuse near the headworks' of irrigation systems, and drying up of irrigation systems at the tail-ends. This results in excessive cultivation of water intensive crops near the headworks, which may lead to inefficient water use, water logging and soil salinity and alkalinity. The irrigation tariffs also do not yield resources for proper maintenance of irrigation systems, leading to loss in their potential. In particular, resources are generally not available for lining irrigation canals to prevent seepage loss. These factors result in reduced flows in the rivers. Pollution loads are similarly linked to pricing policies leading to inefficient use of agricultural chemicals" (p.29). Pollution of groundwater from agricultural chemicals is also linked to their improper use, once again due to pricing policies, especially for chemical pesticides, as well as agronomic practices, which do not take the potential environmental impacts into account. While transiting through soil layers may considerably eliminate organic pollution loads in groundwater, this is not true of several chemical pesticides. The pesticides themselves may become a source of pollution when it leaches into the ground water.

The policy rightly diagnosed that the direct causes of groundwater depletion have their origin in the pricing policies for electricity and diesel. In the case of electricity, wherever individual metering is not practised, a flat charge for electricity connections makes the marginal cost of electricity effectively zero. Subsidies for diesel also reduce the marginal cost of extraction to well below the efficient level. Given the fact that groundwater is an open access resource, the user then 'rationally' (i.e. in terms of his individual perspective), extracts groundwater until the marginal value to him equals his now very low marginal cost of extraction. The result is inefficient withdrawals of groundwater by all users, leading to the situation of falling water tables. Support prices for several water-intensive crops with implicit price subsidies aggravate this outcome by strengthening incentives to take up these crops rather than less water-intensive ones.

The policy thereby recommends that the efficient use of groundwater would, accordingly, require that the practice of non-metering of electric supply to farmers be discontinued in their own enlightened self-interest. It would also be essential to progressively ensure that the environmental impacts are taken into account in setting electricity 'tariffs, and diesel pricing. The policy suggests that the relevant fiscal, tariffs, and sectoral policies should take explicit account of their unintentional impacts on land degradation' (p.22). It then recommends to encourage 'adoption of science-based, and traditional sustainable land use practices, through research and development, extension of knowledge, pilot scale demonstrations, and large scale dissemination, including farmer's training, and where necessary, access to institutional finance... Encourage agro-forestry, organic farming, environmentally sustainable cropping patterns, and adoption of efficient irrigation techniques'. (p.23)

It has been recommended to take an explicit account of impacts on groundwater tables of electricity tariffs and pricing of diesel, and to promote efficient water use techniques, such as sprinkler or drip irrigation, among farmers. There is a need to provide necessary pricing, inputs, and extension support to feasible and remunerative alternative crops which may be raised by efficient water use. Excessive use of fertilisers, pesticides and insecticides is the main non-point source of the pollution. These pollutants contribute to the pollution of the ground water as well as surface water. The optimal utilisation of fertilisers, pesticides and insecticides should be encouraged for improving the water quality. (p.32)

The policy report introduced the problem as 'institutional failures, referring to unclear or insufficiently enforced rights of access to, and use of, environmental resources, result in environmental degradation' (p. 4). In fact, several institutions that are essential for dealing with the management of natural resources or environmental components under consideration do not exist at all. For instance, the culture of scientific farming practices of choosing a cropping pattern and inputs based on their sustainability, frequent interaction between farmers and the local scientific community regarding soil conditions, input uses, choice of crops, plant diseases, etc. interaction between policy makers and the scientific community in determining prices for inputs and agricultural products so as to influence cropping pattern and input consumptions in a socially optimal way. Inculcation of this culture spanning the central to the decentralised level, requires both macro- and micro-level planning and concerted efforts for its implementation. The sooner these institutions evolve, the better prospect of the security of livelihood of the millions of farmers who are facing the danger of unsustainable cultivation.

Bibliography

- Government of India, Ministry of Agriculture, (2001), 'Reports of the Commission for Agricultural Costs and Prices', New Delhi.
- Govil, B.P and S.V. Kaore, (1999) 'Fertiliser Industry in Promotion of Balance Fertilisation', *Fertiliser News*, Vol.44, No. 2, February.
- Haque, T. (1996) (Ed), '*Small Farm Diversification: Problems and Prospects*', NCAER, New Delhi.
- Hodrick, R.J. and E.C. Prescot (1997), 'Postwar U.S. Business Cycles: An Empirical Investigation', *Journal of Money, Credit and Banking*, Vol. 29, No. 1.
- Jha, B.K. and D. Jha (1996), 'Constraints in Small Farm Diversification – A Study in Kurukshetra District of Haryana

(India)', in Haque (ed). *Small Farm Diversification: Problems and Prospects'*, NCAER, New Delhi.

- Mishra, G.N. (1999), 'Fertiliser Use in Upland Rice', *Fertiliser News*, Vol.44, No 7.
- Narain, Dharm (1988), *Studies on Indian Agriculture*, Oxford University Press, Delhi, pp. 140-42
- Pal, S. and A. Singh (1997), *Agricultural Research and Extension in India: Institutional Structure and Investments*, Agricultural Economics and Policy Research, New Delhi.
- Patnaik, Utsa (1975), 'The Process of Commercialisation in Agriculture in Colonial Condition: A Hypothesis', *Occasional Paper No. 14*, Centre for Economic Studies and Planning, Jawaharlal Nehru University.
- Rao, C.S., K.K., Singh and M. Ali (2001), 'Sulfur: A Key Nutrient for Higher Pulse Production', *Fertiliser News*, Vol.46, No.10, December.
- Sakal, R, A.P. Singh, N.S. Bhogal and Md. Ismail (1999), 'Impact of Sulfur Fertilisation in Sustaining the Productivity of Rice-Wheat Cropping System', *Fertiliser News*, Vol.44, No.7.
- Sakal, R, R.B. Saha, A.P. Singh and N.S. Bhogal (1999), 'Effect of Boron and FYM alone and in Combination on Boron Nutrition of Crops in Maize-Lentil Cropping System', *Fertiliser News*, Vol.44, No.11.
- Saleth, R.M. (1996), Diversification is a Strategy for Small Farm Development: Some Evidence from Tamil Nadu', in Haque (ed) '*Small Farm Diversification: Problems and Prospects*', NCAER, New Delhi.
- Sidhu, R.S. and M.S. Dhilon (1997), 'Land and Water Resources in Punjab Agriculture: Their Degration and Technologies for Sustainable Use', *Indian Journal of Agriculture Economics*, Vol.53, No.3.
- Singh, M.V. (1999) 'Current Studies of Micro and Secondary Nutrients Deposits and Crop Response in Different Agro-Ecological Regions', *Fertiliser News*, Vol. 44, No. 4.
- Singh, M.V. (2001a), 'Evaluation of Current Micronutrient Stocks in Different Agro-Ecological Zones of India for Sustainable Agriculture', *Fertilise News*, Vol.46, No. 2 February.
- Singh, M.V. (2001b), 'Importance of Sulfur in Balanced Fertiliser Use in India', *Fertiliser News*, Vol.46, No.10, October.

- Velrasu, P. and P. Singh (1999), 'Fertiliser Use Pattern and its Impact on Crop Productivity: A Case Study of Erode District in Tamil Nadu', *Fertiliser News,* Vol.44, No.10, October.
- Vig, A.C., G.S. Bahl and M. Chand (1999), 'Phosphorus – its Transformation and Management under Rice-Wheat system, *Fertiliser News,* Vol.44, No.11.

Endnotes

1. Government of India, Ministry of Agriculture, (2001) Agricultural Statistics in Brief, p.109.
2. It is observed by Vig, Bahl and Chand (1999) that inherent P fertility status of most of the soils under rice-wheat cropping system (Indo-Gangetic alluvial plains of India) falls under the low to medium category. Transformation of naturic P is more important for rice than wheat. On field trial, it was found that application of P fertiliser to wheat gave a better residual effect to the following rice crop than vice versa et al. noted that many parts of Indo-Gangetic alluvial soils are sodic i.e. alkaline deficit in Zn and N. Iron deficiency is coming up. Further, loss of N of chemical fertiliser through seepage is also severe. It was found that organic manure, neem cake or prilled urea are useful in preserving N in the soil through slow release of N of chemical fertiliser. All this indicates that agricultural extension service has a greater role to play in cultivation in a scientific manner.
3. Singh, M.V. (1999) 'Current Studies of Micro and Secondary Nutrients Deposits and Crop Response in Different Agro-Ecological Regions', *Fertiliser News,* Vol. 44, No. 4.
4. A study by Sakal, Saha, Singh and Bhogal (1999) on Bihar found that boron has emerged as second limiting micronutrient after Zn in soils of Bihar. In north Bihar, calcarious soil have exhibited boron deficiency of around 48 per cent. It also has S deficiency around 20 per cent.
5. The fall in the watertable during 1984–94 has been estimated to be about 5.1 metres in Sangrur, 4.8 metres in Patiala, 4.5 metres in the paddy-growing areas of Ferozepur and Faridkot, and 2.5 metres in the Jalandhar district (ibid).

Chapter 6

Climate Change Response: Equity or Nothing

C.E. Karunakaran

The Intergovernmental Panel on Climate Change (IPCC) released on February 2, 2007 its Summary for Policymakers[1] of the physical science basis of climate change as part of its Fourth Assessment Report, an assessment exercise it takes up once in five years. The operative part of the summary that captured attention was the use of the phrase 'very likely' (indicating greater than 90 per cent probability) to attribute global temperature increase to increase in anthropogenic greenhouse gas (GHG) concentrations,[2] a change from the use of the word 'likely' (greater than 66 per cent probability) in its Third Assessment Report.[3] The very structuring of the IPCC, established in 1988 as a joint subsidiary of the United Nations Environment Programme and the World Meteorological Organization—where government representatives of all countries too play an important role in finalising the reports—implies that its findings tend to tone down the scientific consensus on the gravity of global warming and the imperative for action and thus represent the 'lowest common denominator science', as pointed out by Tim Flannery[4] in his book, *The Weather Makers*.

It is because mainstream science has been pointing out this level of certainty for very many years now and had been forecasting the ruinous impacts of the GHG increase, that many environmentalists, and even some policy makers, have gone to the extent of describing climate change as the ultimate weapon of mass destruction[5] and a threat to the world, worse than terrorism or nuclear war. To understand why it is so, one should look at some basic facts. Global warming is caused primarily by the very foundation on which modern civilisation is built—burning of coal, oil and gas. So much so, the real solution to the problem would include lifestyle changes,[6] something that goes against the grain of the consumer culture—and the socio-economic system built on it. The earth has not seen anything like this build-up of carbon dioxide for at least 650,000 years, as seen from ice core analyses.[7] If this continues, by the end of the century, the earth will be hotter than at any other time in the past two million years.[8]

It is already too late to avoid major consequences because of the inertia of the ecosystem—even if no more CO_2 or other greenhouse gases are emitted by mankind from tomorrow, the earth will continue to warm up for some decades, the sea will continue to rise for some centuries and the ice sheets will continue to adjust for thousands of years.[9] The world is already facing increasing sea intrusions, floods, storms, droughts, heat waves, disease transmissions and environmental refugees. The proportion of the world's population affected by weather disasters has doubled between 1975 and 2001[10] and the World Health Organisation (WHO) estimates that climate change already accounts for 150,000 deaths and 5.5 million life years lost due to premature mortality and disability, in a year.[11]

The coming years—resulting from what has already been done, not to speak of further emissions—will be worse. An insurance specialist estimates that insurance losses due to extreme weather events are increasing by 10 per cent a year, against the world economic growth of 3 per cent, and that, even by 2010, insurance companies could be charging annual rates of 12 per cent, forcing many to drop out.[12] There is the additional threat of runaway warming, because of crossing some threshold and triggering positive feedback—such as frozen peat bogs thawing and releasing huge quantities of methane, which is twenty times more powerful than carbon dioxide in causing global warming,[13] even if its atmospheric life is shorter. The magnitude of the impending crisis is best expressed in the words of Dr. James Hansen, the highly regarded director of the NASA Institute of Space Studies that we are 'near a tipping point, a point of no return, beyond which the built in momentum and feedbacks will carry us to levels of climate change with staggering consequences for humanity and all of the residents of this planet.'

> He also points out: The Earth's history suggests that with warming of 2-3°C the new equilibrium sea level will include not only most of the ice from Greenland and West Antarctica, but a portion of East Antarctica, raising sea level of the order of 25 meters (80 feet).[14]

Given this background, how does one assess the future of this planet and where does one draw the line and say, this far we can pollute our atmosphere and somehow manage its consequences—keeping fingers crossed about positive feedbacks—but beyond this would be unacceptable chaos? Does the developing situation provide a window and a plausible timeframe for mankind to mend its ways and step back before this line? Scientists, activists and policy makers have been grappling with this issue and have now come to a broad understanding of where this line is to be drawn, taking into account all relevant factors. This consensual limit is a 2°C warming, over and above the pre-industrial global average temperature.[15] Of this, the earth has already reached the 0.8°C mark and is currently warming up by 0.2°C per decade.[16] The latest IPCC report forecasts that by the end of this century, the probable temperature rise will be in the range 1.8 to 4.0°C, if we go along business as usual.[17]

There cannot, obviously, be business as usual. Even the Kyoto Protocol—which was agreed to in 1997, but came into force only in 2005, with the USA and Australia opting out, that asked the industrial North to bring down its emissions from its 1990 levels by 5.2 per cent average before 2012, with the developing South exempted—and extensions of it on the same lines, cannot be the solution. One scenario, which holds a risk of 9 to 26 per cent of crossing the 2°C mark, demands that the total manmade emissions should start declining from 2010 and should reach one quarter of the starting level within 30 years and keep declining further—a challenge of a very high order.[18] Other, less demanding, scenarios are also being propagated, as in a report submitted to the UK government by Sir Nicholas Stern, former Chief Economist of the World Bank—but hold a higher risk of breaching the target limit.[19]

It is in this overall context of the urgent need for phase-shift action that one should assess the direction in which multi-national and multi-dimensional action to contain climate change can move forward and be effective. But first, the question whether it is technically, socially and economically feasible to achieve such a drastic reduction in energy use in such a short time needs to be addressed. The consensus on this is, it is feasible—efficient use of energy, renewable energy, hybrid, hydrogen and electric cars, revised taxes and incentives and lifestyle changes, like more use of public transport and many others are possible, if there is a will.[20] The much-quoted Stern Review mentioned earlier, makes the point that it would cost much less to address climate change now than to live with it.[21]

The real question, however is: who is to bear this cost of lessening as well as living with climate change, as the world must? It is obvious that the atmosphere can take only so much more pollution by greenhouse gases if the warming is not to cross the two degree mark, and that scarce space is currently being filled up 60 per cent of the time by the carbon dioxide of industrialised countries that hold less than 20 per cent of the world's population.[22] It is in the interest of those countries to see a grandfather approach prevail, by which, all countries would be required to reduce emissions from their current levels by similar proportions. The developing countries with large populations do not accept this approach. The industrial countries have reached their level of development riding on low-cost fossil fuels, while the developing countries need to do the same to reduce their poverty levels and cannot afford the higher cost of climate-friendly technologies in the short to medium term. Their economic growth rates, and therefore their emissions growth, are also of a higher order compared to industrial countries.[23] Grandfathering does not suit them. It is this dichotomy between luxury emissions and survival emissions that led to the principle of 'common but differentiated responsibilities and respective capabilities' that was agreed to by the nations of the world—the USA included—in the historic 1992 UN Framework Convention on Climate Change (UNFCCC), the first collective step the nations of the world took to deal with this issue.

There is then the responsibility approach, based on the 'polluter pays' principle. Emission estimates show that the average American consumed and emitted 46 times as much carbon as the average Indian during the 40 years after 1950.[24] There is also the capability approach, by which those who are more capable of handling mitigation and adaptation, take on the larger share of the burden. Then there is the global commons approach—the atmosphere, the dumping ground of carbon dioxide—is a common resource that belongs to all citizens of the world. Free riding on it is possible only if it has unlimited capacity. This certainly is not the case, as far as carbon dioxide parking is concerned. Most countries of the world have laws that recognise equal rights to common pool resources.

To understand the dimensions of this principle of equal rights to global commons one can do a simple exercise of allocating today what is available of the remaining atmospheric dumping space equally among the 6.5 billion people on this earth and allowing each person to park her or his emissions there for the rest of the century. This is because the world will run out of its available carbon emission budget for this century—under the precautionary scenario of not exceeding 2°C warming—in a quarter of that time, under normal circumstances.[25] Such an exercise immediately reveals the huge disparity in consumption levels and the immense problem that would be faced if one were to implement this approach. If such boxes are allocated—and by some means, all on this earth stick to their current levels of emissions and do not increase them any further—the average American will run out of her box in far less than a decade while the average Chinese will take less than half a century and the average Indian more than a century to fill up their boxes; the Bangladeshi can go on for a good five centuries with just that one box.[26]

This gives rise to the possibility that certain countries can rent their entitlements to other large emitters since the world as a whole may turn away from carbon-based energy inside of half a century. An NGO based in the UK—the Global Commons Institute—has been pushing forward the concept of Contraction and Convergence[27], by which, the international community first agrees to a single per-capita allowance to converge on, and the time for convergence. The northern countries will have to contract their emissions and the South can increase their emissions and converge to that agreed level within the agreed time; thereafter, all together will reduce the per capita emissions uniformly to levels needed to avoid climate disaster. The formula also envisages trading of entitlements that developing countries may not need for the given year.

Even this approach, which appears to be reasonable from the perspective of developing countries, may prove to be inadequate to populous and rapidly developing countries like China, India, Brazil and South Africa. India and China are growing at three times the rate of the developed nations and need to carefully weigh the likely GHG target they could get allotted against their ballooning future emission requirements. At the recent 12th Conference of Parties at Nairobi, Christian Aid pointed out in a paper that the South's rapid growth trajectory is such that its emissions alone will cross the line of total global emissions allowable under the precautionary two-degree scenario by year 2020, even if the North's emissions suddenly dropped to zero at that time.[28] In other words, the situation has reached such a stage, mainly because of the historic emissions—and also because of lack of vigour in addressing this issue by the nations of the world since 1992—that high-growth developing nations might find themselves with no room to manoeuvre and be forced to choose between their development and saving the planet. Their resources are far too limited to take on the challenge of switching to a low carbon-energy path in such a short time.

It is this scenario that has given rise to a vigorous examination of the responsibility approach leading to the concept of carbon debt—meaning just recompense for occupying scarce global commons in disproportionate manner. An assessment made by Christian Aid in 1999 pointed out that the debt owed on this count by the rich nations to the highly indebted poor countries—evaluated in terms of wealth generated through excess occupation of atmospheric CO_2 space—is of the order three times the conventional financial debt the latter owe the former and the rich countries continue to incur a debt of $13 trillion a year to poorer nations.[29]

Eco-Equity, the campaigner for a realistic approach to devising a climate framework, considers the basket of equity principles relevant to mitigation sharing to be—equal rights to global commons, polluter pays, capacity-based burden sharing and need-based resource sharing.[30] Based on these principles, they have developed the concept of Greenhouse Development Rights, by which the first priority of a southern nation will be its development and all mitigation action taken by it will be paid for by the North, until the former reaches a certain level of development. The North will, of course, drastically cut down on its own emission as also pay for the emission reductions all over the globe. A point to note here is that the present connection between emissions and economic development needs to be broken as quickly as possible, for developing countries like India and China cannot hope to reach anywhere near the current levels of per-capita North emissions before peaking in a couple of decades and getting on to the down slope. Investment in such drastic shift to newer technologies is beyond the capacity of developing countries and can only be borne by the developed world.

This is but one possible framework. But the real question is: how is any framework to be agreed upon and implemented in a world where the largest polluting nation is still outside the Kyoto framework, other developed countries do not look like reaching up to the Kyoto targets—

which themselves are just a beginning of much steeper reductions to follow thereafter—while demanding that developing countries start shouldering responsibility and the developing countries steadfastly refuse to do so. There is a big divide between the G-77 countries and China on the one side and the lead group among the developed nations, the European Union on the other; the former feel that the latter has not done enough to bring the USA on board and to aggressively address the mitigation issue in their own countries[31] and are wary of entering into any discussions that might ultimately land them with grandfathered commitments.[32]

From their perspective, considerations of equity in sharing global commons is nowhere in the picture in serious negotiations between countries under the UNFCCC, even as several environmentalists have been actively campaigning for equity in broad terms. (India's Centre for Science and Environment was among the earliest to flag the issue by pointing out the inequity in per-capita emissions.[33]) The National Environment Policy 2006 of India stresses that 'equal per-capita entitlements of global environmental resources to all countries' will be an essential element of India's response to climate change.[34]

There are signs that the northern countries could be gradually coming round to accept that anything less than equal rights to the atmosphere will not take the negotiations anywhere. There is a realisation that time is running out—according to one projection, if global emissions are peaked in 2010 and taken downward thereafter, a 2.6 per cent per annum reduction will help to keep within 2°C warming; if, however, this peaking is delayed by 10 years, the reduction will have to be by a drastic 6.7 per cent per annum to achieve the same result.[35]

Some activists hope that this exigency would now drive all countries to move towards an acceptable solution and practical wisdom suggests that it can only be equity based to be endurable. It is this realisation that made the Royal Commission on Environmental Pollution of the UK recommend, as early as in 2000, the contraction and convergence strategy in its report to the government[36] and the government's Energy White Paper of 2003 to implicitly accept it in projecting future UK emissions.[37] The insurance industry is the earliest among the business community to recognise the seriousness of global warming and the most concerned to find a quick solution, as it impacts its bottom line directly. Looking for a real-world solution that will truly work, the Chartered Insurance Institute of the UK had no hesitation in accepting per-capita emission convergence.[38]

There is also a more pressing concern—the threat of social disruptions and warfare. Large sections of population will get displaced by the impacts of climate change and will have to compete for resources. A 2003 report commissioned by the Pentagon warned that nuclear arms will proliferate as people fight for resources as a result of global warming. It pointed out that 'disruption and conflict will be endemic features of life,' and cautioned the US administration that climate change could 'challenge United States national security in ways that should be considered immediately'.[39]

It is in this context of global urgency to address the issue and the fast track on which policy makers of many countries have been put by their societies, that there cannot be any more laidback attitude by India in international negotiations. India will now be forced to go beyond what it has been saying so far: our emissions are small, our per capita emissions are very small and we cannot divert attention from other priorities.[40] China and India cannot now afford to factor in any consideration of grabbing as much atmospheric space for as long as possible, so long as the USA does not come on board; they should reckon with the fact that the advanced countries are grabbing much larger spaces at the same time, making future drastic reductions a fait accompli for one and all. Besides, their economies get more and more entangled in path dependence—the more the investment one makes in fossil technologies, the less viable it becomes to switch to a non-carbon path.

More important, their people stand to lose heavily if climate change gets out of hand. The fate of a third of the world's population cannot be left to meandering and unworkable climate negotiations where the lead is taken primarily by the European nations. The time has come for China, India and a few other countries to take on a proactive role in shaping the future order of greenhouse emissions and climate adaptations. They should work to build a consensus among G-77 to put on the table a radically new approach—an approach based on equity—to emissions mitigation and adaptation, as compared to Kyoto. Some experts have already divided the southern countries into four different groups, ranging from newly industrialised to least developed—with China in group 2 (rapidly industrialising developing) and India in Group 3 (other developing)—to devise differential commitments and resources for them.[41] It is in the interests of India and China to work out a formula that is seen as equitable by the poorest nations of the world. That would be the only workable solution in the medium term and would also preserve the unity of the South in future parleys.

There is no escaping that any future parley should focus not merely on emission reductions and sharing of reductions, but very largely on managing the domino effects of the damage already done to the atmosphere which is presently hurting the poor nations of the world through increased droughts and storms and vector diseases and will, unavoidably, hurt them more in the future.[42] This can only be addressed on the basis of responsibility and capacity principles and cannot be regarded as charity or aid. It follows from this that any fund set up for climate change adaptation should contribute directly to poverty reduction and development of the affected nations and not overly try to establish complex connections to climate change-related damage or adaptation need. Development is the best instrument to build adaptation capability.

Countries like India also face another sensitive issue because of the very large income disparities among their people. The huge social divide that exists in India and several other countries is also a huge carbon divide. It is not very difficult to imagine the emission disparities between the urban upper middle-class Indian who runs two cars and four air conditioners and the rural working woman who treks for hours to fetch head loads of shrub every day for the kitchen fire and whose daughter uses a flickering kerosene lamp to pore over her school work. A study by the Indira Gandhi Institute for Development Research, Mumbai, showed that in 1989-90, the per-capita carbon emission of the bottom half of the rural population was one-thirteenth of the top 10 per cent of the urban population in India.[43] Equity is an indivisible concept. If one seeks justice based on inter-national per-capita emission rights, it cannot but be applied in the intra-national context too. The NEP 2006 foregrounds equity as one of its guiding principles in these terms: 'Equity, in the context of this policy refers to both equity in entitlements to, and participation of, the relevant publics, in processes of decision-making over use of environmental resources'.[44] State and society in India need very carefully to assess how this is to be applied to climate change response within the country.

The 2007 G8 Summit of the rich and powerful nations of the world[45] conformed to the existing pattern of the US's refusing to accept emission reduction commitments and the European Union's compromising with that stand. The final consensus report did not mention the crucial 2°C warming limit and had no specific target for emission reduction, both sought by the EU.[46] The G5 developing nations[47] that were also invited to the meet re-emphasised the concept of 'common but differentiated responsibilities and respective capabilities' and sought transfer of technologies at 'affordable costs'.[48] In fact, India, which made an aggressive sounding statement at the Summit with the Prime Minister's pointing out 'we are not here as petitioners',[49] watered down its stand from the 2005 G8 Gleneagles Summit where it sought removal of patent protection for energy technologies for use in developing countries with no licensing fees,[50] to asking in 2007 for clean technologies to be 'made affordable for developing countries'.[51] The silver lining in the cloud is that the G5 countries—called BICSAM countries, using a newly coined acronym—have decided to consult each other on a regular basis and to coordinate their positions.[52]

India has its job cut out if it decides to change course from being a reactive participant in multilateral discussions to a proactive campaigner for the equal rights to development of its more than one billion people.

Endnotes

1. IPCC, 2007: Summary for Policymakers. In: *Climate Change 2007: The Physical Science Basis. Contribution of Working Group I to the Fourth Assessment Report of the Intergovernmental Panel on Climate Change* Solomon, S., D. Qin, M. Manning, Z. Chen, M. Marquis, K.B. Averyt, M.Tignor and H.L. Miller (eds.). Cambridge University Press, Cambridge, United Kingdom and New York, USA.
2. Carbon dioxide, Methane and Nitrous oxide are the major greenhouse gases that cause global warming by trapping part of the heat radiated back from the earth. Carbon dioxide which gets generated when fossil

fuels coal, oil and gas get burnt, accounts for the major part of global warming.

3. IPCC, 2001: *Climate Change 2001: The Scientific Basis. Contribution of Working Group I to the Third Assessment Report of the Intergovernmental Panel on Climate Change* Houghton, J.T.,Y. Ding, D.J. Griggs, M. Noguer, P.J. van der Linden, X. Dai, K. Maskell, and C.A. Johnson (eds.). Cambridge University Press, Cambridge, United Kingdom and New York, USA, p. 881.
4. Flannery, T. 2006. *The Weather Makers,* Allen Lane, Great Britain.
5. Michael Fish, Britain's longest serving TV weather forecaster says this: *http://www.bbc.co.uk/wear/content/articles/2007/03/28/climate_countdown_michael_fish_feature.shtml*
6. Working Group III of IPCC Fourth Assessment Report states that there is 'high agreement' that 'changes in lifestyles and consumption patterns that emphasize resource conservation can contribute to developing a low-carbon economy that is both equitable and sustainable.' IPCC, 2007: Summary for Policymakers. In: *Climate Change 2007: Mitigation of Climate Change. Contribution of Working Group III to the Fourth Assessment Report of the Intergovernmental Panel on Climate Change*
7. IPCC 2007. *op cit. supra* note 1.
8. Ibid.
9. The Chairman of the Intergovernmental Panel on Climate Change is quoted on this in AFP, 2005. *Climate Change is Already Here,* Agence France Presse, reproduced in *http://www.countercurrents.org/cc-afp020205.htm*
10. Muller, B. 2002. *Equity in Climate Change: The Great Divide,* Oxford Institute for Energy Studies, Great Britain.
11. McMichael, A.J. et al., 2003. *Climate Change and Human Health—Risks and Responses,* World Health Organisation, Geneva.
12. Lohmann, L. 2006. *Carbon Trading—A Critical Conversation on Climate Change, Privatisation and Power,* Development Dialogue No.48, September 2006, The Dag Hammarskjöld Centre, Sweden.
13. Connor, S. 2006. 'Our Worst Fears are exceeded by Reality', *The Independent,* UK, 29 December 2006.
14. Hansen, J.E. 2006, *Can we Still avoid Dangerous Human-Made Climate Change?* Presentation on February 10, 2006 at New School University, New York City, *http://www.columbia.edu/~jeh1/newschool_text_and_slides.pdf* (accessed May 2007).

15. Athanasiou, T. 2007. *An Inconvenient Truth, Part II, an EcoEquity discussion paper, www.ecoequity.org/docs/InconvenientTruth2.pdf.*
16. IPCC 2007, *op cit. supra* note 1.
17. Ibid.
18. Athanasiou, T., Kartha, S., Baer, P. 2006. *'Greenhouse Development Rights: An approach to the global climate regime that takes climate protection seriously while also preserving the right to human development,* EcoEquity and Christian Aid.' *http://www.ecoequity.org/GDRs/*
19. Stern, N. 2006. *Stern Review: The Economics of Climate Change.* HM Treasury, London.
20. IPCC, 2007: Summary for Policymakers. In: *Climate Change 2007: Mitigation of Climate Change. Contribution of Working Group III to the Fourth Assessment Report of the Intergovernmental Panel on Climate Change*
21. Stern, N. 2006. *op cit. supra* note 19.
22. Roberts, J.T. and Parks, B.C., 2007. *A Climate of Injustice: Global Inequality, North South Politics, and Climate Policy,* MIT Press.
23. IPCC, 2007 *op.cit. supra* note 20 states that two thirds to three quarters of increase in energy CO_2 emissions during the years 2000–2030 is projected to come from non-Annex I (i.e. developing) regions, even if their per-capita emissions will still be only a third of that of the developed regions by 2030.
24. As established from the historic emission statistics in *Marland, G., T.A. Boden, and R. J. Andres. 2006. Global, Regional, and National Fossil Fuel CO_2 Emissions.* In *Trends: A Compendium of Data on Global Change.* Carbon Dioxide Information Analysis Center, Oak Ridge National Laboratory, U.S. Department of Energy, Oak Ridge, Tenn., USA.
25. Based on current emission levels and growth rates, vide IPCC 2007, *op.cit. supra* note 20 and low-risk emission pathway total emissions for current century, vide Baer, P. and Mastrandea, M. 2006, *High Stakes: Designing Emissions Pathways to reduce the Risk of Dangerous Climate Change,* Institute for Public Policy Research, London, www.ippr.org
26. By extrapolation, using estimated 2006 per-capita emission levels of countries and the low-risk emission pathway for the world, vide Baer, P. and Mastrandea, M. 2006, *op cit. supra* note 25.
27. Global Commons Institute, 2007, GCI briefing: Contraction and Convergence, *http://www.gci.org.uk/*
28. Kartha, S. et al. 2005, *Cutting the Knot: Climate Protection, Political Realism, and Equity as Requirements of a Post-Kyoto Regime.* Paper

presented at a side event at COP 10, Argentina. *http://www.ecoequity.org/docs/CuttingTheKnot.pdf*

29. Simms, A. 1999, *Who owes Who? Climate Change, Debt, Equity and Survival*, A collaborative report by Christian Aid, Global Commons Institute and the International Institute for Environment and Development. *http://www.jubileeresearch.org/ecological_debt/Reports/Who_owes_who.htm*
30. Kartha, S. et al. 2005, *op.cit. supra* note 28.
31. The European Environment Agency estimates that EU's existing policies will bring down greenhouse-gas emissions in the EU-15 by only 0.6 per cent in 2010 against the Kyoto target of 8 per cent, vide European Environment Agency 2006. *Greenhouse gas emission trends and projections in Europe 2006, EEA Report No.9, 2006*, European Environment Agency, Copenhagen.
32. For more on how the South is cautious about a possible 'abate and switch' strategy of the North, see Kartha S., et al., 2004, *Cutting the Gordian Knot, Adequacy, Realism and Equity, http://www.gtinitiative.org/documents/gordianknot%20-%20color.pdf* Also see Ott, H E 2003, *Global Climate*, Yearbook of International Law, Vol.13 (2002), Oxford University Press.
33. Agarwal, A. and Narain, S., 1991. *Global Warming in an Unequal World: a Case of Environmental Colonialism*, Centre for Science and Environment, New Delhi, India.
34. Government of India, 2006. *National Environment Policy—2006*, Ministry of Environment and Forests, Government of India.
35. Kartha, S. et al., 2005, *op.cit. supra* note 28.
36. Royal Commission on Environmental Pollution, 2000. Energy—*The Changing Climate, The Royal Commission on Environmental Pollution's 22nd Report*, Royal Commission on Environmental Pollution, London.
37. Upham, P. 2003, *Climate Change and the UK Aviation White Paper, Tyndall Briefing Note No.10*, Tyndall Centre for Climate Change Research, Norwich, UK.

38 Chartered Insurance Institute, 2001. *Climate Change and Insurance*, Chartered Insurance Institute, London.

39. Schwartz, P. and Randall, D., 2003. *An Abrupt Climate Change Scenario and Its Implications for United States National Security, http://www.environmentaldefense.org/documents 3566_AbruptClimate Change.pdf*

40. This position is implied as well as stated in several negotiating fora. See Vajpayee, A.B., 2002. *Speech of Prime Minister Shri Atal Bihari Vajpayee at the High Level Segment of the Eighth Session of Conference of the Parties to the UN Framework Convention on Climate Change New Delhi—30 October 2002. http://unfccc.int/cop8/latest/ind_pm3010.pdf.* Also Ram,N. 2005. Greenhouse Gas Emissions, G8 and India, The Hindu, July 9, 2005. and the foreign ministry note referred in Madhavan, N. 2007. 'China and India set to make common cause on global warming', The Hindusthan Times, June 6, 2007.
41. Brouns, B and Ott, H. E. 2005. *Solidarity understood Correctly,* D+C—Magazine for Development and Co-operation, Vol. 46, No.5, pp. 207–209.
42. More on this in IPCC, 2007: Summary for Policymakers. In: *Climate Change 2007: Impacts, Vulnerability and Adaptability,. Working Group II Contribution to the Fourth Assessment Report of the Intergovernmental Panel on Climate Change,* IPCC Secretariat, Geneva. The report warns that a quarter of the world's plant and animal species could face extinction with temperature rises exceeding 1.5–2.5°C and that hundreds of millions of people will be exposed to increased water stress.
43. Murthy, N. S., Panda, Manoj K. and Parikh, Jyoti K. 1996. *Consumption Pattern, Energy Demand and Carbon Dioxide Emissions in India: 1990–2020,* in Structural Transformation Processes towards Sustainable Development in India and Switzerland, INFRAS, Zurich, May 1996.
44. Government of India, 2006. *op. cit. supra* note 34.
45. Canada, France, Germany, Italy, Japan, Russia, the UK and the USA.
46. G8 Summit, 2007. *Growth and Responsibility in the World Economy, Summit Declaration (7 June 2007), G8 Summit 2007, Heiligendamm. http://www.g-8.de/Content/DE/Artikel/G8Gipfel/Anlage/2007-06-07-gipfeldokument-wirtschaft-eng,property=publicationFile.pdf.* The report states: 'In setting a global goal for emissions reductions in the process we have agreed today involving all major emitters, we will consider seriously the decisions made by the European Union, Canada and Japan which include at least a halving of global emissions by 2050.'
47. Brazil, China, India, Mexico and South Africa.
48. Government of India, 2007. Joint Position Paper of Brazil, China, India, Mexico and South Africa participating in the G8 Heiligendamm Summit, June 2007, 8/6/07, Ministry of External Affairs, *http://www.mea.gov.in/*

49. *Reuters*, 2007. 'India seeks Direct Role in tackling Climate Change', Reuters Video News, June .07, 2007. *http://rtv.rtrlondon.co.uk/2007-06-08/3f34293a.html* Soundbite of Shiv Shankar Menon, India's Foreign Secretary: 'The idea was put I suppose also very clearly by PM when he said that we are partners and participants, we are not here as petitioners and I think, I think that idea was very, also very, very strong.'
50. Government of India, 2005. *Dealing with the Threat of Climate Change, India Country Paper. The Gleneagles Summit*, Ministry of External Affairs, *http://www.mea.gov.in/*
51. Government of India, 2007. *PM's intervention on Climate Change at the Heiligendamm meeting, 08/06/2007*, Ministry of External Affairs, http://www.mea.gov.in/
52. Varadarajan, Siddharth 2007. Brazil proposes G5 Summit—G8 Forum has limited utility for Third World, *The Hindu*, June 11, 2007 and Government of India, 2007. Joint Press Release issued after the meeting of O-5 leaders in Berlin, 07/06/2007, Ministry of External Affairs, *http://www.mea.gov.in/*

Chapter 7

Linkage Between Biodiversity and Cultural Diversity: An Exploratory Analysis

M.S. Bhatt and A.K.M. Nazrul Islam

Introduction

In the last few decades, the man–environment relationship has become fragile. Trade-off between the existing patterns of development and protection of environment is the main cause behind this deterioration. Dangerous consequences of this situation have been widely demonstrated. These include global warming, thinning of the ozone shield, dumping of nuclear waste, contamination of water bodies, destruction of biodiversity,[1] deforestation, desertification, soil erosion, air pollution, water pollution, melting of icebergs, drying of rivers, etc. For example, one-third of the world's cropland is losing topsoil at a rate that is undermining its long-term productivity. Almost 50 per cent of the world's rangeland is overgrazed and deteriorating into deserts. Two-third of oceanic fisheries is now being fished out or beyond their capacity. Watertables are falling alarmingly in three large food-producing countries—China, India and the USA. In North China, which produces 25 per cent of its grain harvest, the watertable is falling roughly by 1.5 metres (5 feet) per year. The same thing is happening in the Punjab region, which is the bread basket of India.

The world's forests have shrunk by about half since the dawn of agriculture and continue to do so. Worldwide, forests are shrinking by over nine million hectares per year, an area equal to Portugal. Deforestation is causing unprecedented flooding and diminishing the recycling of water inland, thus reducing rainfall. The IUCN (The World Conservation Union) shows up that one out of eight of the world's 9,946 bird species is in danger of extinction, as is one in four of the 4,763 mammals species, and nearly one-third of all 25,000 fish species.

Due to increased temperature, the world is experiencing devastating storms which cause losses worth millions of dollars every year. Rising temperature is the worst consequence due to current patterns of growth. Over the last three decades, the ice covering of the Arctic Sea has thinned by 42 per cent. In the near future, a large lower part of the world land will go under seawater, like the whole southern parts of Bangladesh (Brown, 2001).

Ecological history also tells us that many ancient civilisations collapsed due to environmental degradation. The builders of Avebury and Stonehenge seem likely to have caused massive deforestation, leading to soil erosion, climate change and probable famine (Wall, 1994). The early Sumerian civilisation of the fourth millennium BC, which was blessed with its irrigation system, created a high productive agriculture, but due to an environmental flaw in the design of the irrigation system, the food production declined and this civilisation collapsed due to lack of food. The Mayan Civilisation of present-day Guatemala developed a sophisticated highly productive agriculture. Due to heavy deforestation and soil erosion, their system, was undermined, leading to scarcity of foods and collapse of the civilisation. Another ancient civilisation which collapsed due to environmental degradation was Easter

Island. Due to large-scale deforestation for making canoes to catch dolphins from the sea, there was shortage of large tress and subsequently lack of dolphins for food. Whether it was salting of the Sumerian land, the soil erosion of the Mayans or the loss of distant-water fishing capacity of the Easter Islanders, collapse of all these early civilisations appears to have been associated with a decline in food supply. Today the addition of 80 million people a year to the world's population at a time when watertables are falling suggests that food supplies again may be the vulnerable link between the environment and the economy (Brown, 2001). The scene in India is equally frightening. The pressure of an overgrowing population, demands for foods, clothing, housing and excessive consumerism are degrading the key stone functions of our environment beyond repair. The IUCN report for 2002 contains 366 species of animals found in India categorised as 'threatened', which is approximately 6.73 per cent of the world's total number of threatened faunal species (5,453 species). This includes 148 species of mammals, 138 birds, 32 reptiles, 3 amphibians, 17 fish and 28 invertebrate species. The analysis of the data on threatened species by threat categories reveals that out of 366 species listed as threatened, 18 are critically endangered (Cr), 52 species are endangered (En), 150 species are vulnerable (Vu), while 10 species are at lower risk or conservation dependent (LR/cd), 101 species are at lower risk but near threatened (LR/nt), and the data for 35 species is not available (DD). On analysis of the threat categories by groups (at the global level), it is found that out of 366 species of threatened Indian fauna, 220 species are at significant risk, of which 88 species are of mammals, 72 of birds, 25 of reptiles, three of amphibia, nine of fishes and 23 species of invertebrates.

Survival of mankind cannot be separated from the survival of other species. Numerous plants, animals, micro-organisms provide indispensable human food, fibre, firewood, medicine, industrial raw materials, etc. They interact with their physical environment to form an ecosystem[2] which regulates energy flows on the earth and sustains natural cycling of materials and, as a consequence, influence concentrations of gases in the atmosphere, determine the properties of soil and regulate various natural cycles like the hydrological cycle, carbon cycle, nutrition cycle, etc. Complex and diverse life forms and their combinations that are the earth's biodiversity and its physical environment together constitute all-encompassing life-support systems on which we humans depend.

The rate of degradation of our natural resources has left us no option but to take necessary steps to check it without delay. Despite the number of national and international treaties/ conferences/ seminars, very little has been achieved in terms of effective policies and their implementations. This does not mean that nothing positive has happened till now. Development vis-à-vis environment has become a serious research agenda in the recent years. This in turn has raised the levels of understanding of the issues involved. The world has become more educated about the environmental concerns. Experts from all the major disciplines of knowledge are engaged in analysing these problems from their perspectives. Ways and means of integrating ecology and development have been discussed in a number of international seminars and conferences such as United Nations Conference on Human Environment (Stockholm, 1972), World Conservation Strategy (1980), Our Common Future (1987), Caring for the Earth (1991), UN Conference on Environment and Development (1992), and the Rio Summit (1992). These international conventions, agreements, and treaties inspire member countries to formulate their own policies for fighting environmental degradation. The Government of India too has made many policy documents to maintain and protect our environment. For example, the Indian Forest Act, 1865; Elephant Preservation Act, 1879; Wild Birds Animal Protection Act, 1912; Motor Vehicles Act, 1938; Factories Act, 1948, River Boars Act, 1956. As a prelude of the 1972 Stockholm Conference, the Government of India had formed the National Commission on Environmental Planning (NCEP) in the year 1972 (Murty, 2001). Then followed the Wildlife Protection Act, 1972; Water (Prevention and Control of Pollution) Act, 1974; Water (Prevention and Control of Pollution) Cess Act, 1977; Forest (Conservation) Act, 1980; Air (Prevention and Control of Pollution) Act, 1981; Environment Protection Act, 1986; National Forest Policy, 1988; National Conservation Strategy and Policy Statement on Environment and Development, 1992; Policy Statement on Abatement of Pollution, 1992; National Water Policy, 2002; National Environment Policy, 2006 and Forests Right Act, 2007 which are some of the policies and acts formulated by the Government of India.

The concerns of the present-day environmental problem in most of the developing countries, like India are of relatively recent origin. Even the recent awareness and concern for environmental protection at the policy level is donor induced. At the grass roots level, it is due to the efforts by individuals and NGOs. In the absence of ground-level awareness and concern, environmental aspects have remained more or less peripheral to the contemporary social movements in India. Some of the contemporary movements, in retrospect, however, acquired the status of ecological or environmental movements as they have widened their focus from basic survival needs to ecological concerns (see Sethi, 1993 and Gadgil and Guha, 1994). In general these movements are often grouped under tribal or peasant movements (Shah, 1990) and also under new social movements (Omvedt, 1993 and Sethi, 1993). Some even title them as middle class or elitist movements (Shah, 1990; Sethi, 1993). The reason is that the ecological aspects are linked with the problems associated with peasants and tribals whose survival is attached to the status of natural resources. The urban middle class and elite often articulate the problems or demands of tribals as well as of the non-tribal poor. In the context of a coalition between affected people and the middle-class spokespersons, the real issues tend to get clouded as the debate is drawn into different forums in order to attract national and international attention (Sethi, 1993). Many of our environmental movements, like the Chipko Movements, Appiko Movements, Narmada Bachao Andolan, etc. have attracted worldwide attention and paved the way for people's awareness about environment. Another area where significant development has taken place is exploration of new research paradigms such as cultural ecology, causality between cultural diversity and biodiversity and rediscovering conventional ways of environment, reducing the trade-off between development and ecology, among others.

To address a problem effectively requires a proper understanding of the nature, extent; its links with others associated areas. Then, one can find an effective and implementable way in the light of past experiences. As mentioned above, the loss of biodiversity in India is indeed a serious and complex problem. Among other things, it requires multi-dimensional scientific studies for an informed understanding of the issues involved. We have to explore every possible preservation technique that may suit our economic development, both for equity and efficiency considerations. Experts have identified many such techniques. Of these, preservation of cultural diversity[3] has been identified as one of the critical inputs to preserve biodiversity. It is argued by ecological historians that biodiversity and cultural diversity (which means diversity of different languages, traditions, religions, faiths, beliefs, rituals, folklores, etc. in a particular society) are closely linked and if sublime aspects of cultural diversity generate a number of positive externalities, which can automatically preserve biodiversity without any additional cost or effort.

Many societies or cultures have traditionally developed strategies of conserving and managing nature and natural resources. These strategies are highly congruent to the traditional life-styles of the perspective societies. The sacred conservation practices followed by the local people have come into focus of late due to their importance in protecting several delicate ecosystems and threatened species, the explicit connections between cultural diversity and biodiversity and their potential of people-oriented conservation efforts. This has recently been drawing attention from academicians, policy planners, politicians, researchers. It is argued that the biodiversity crisis should be nuanced on the basis of the interaction among a wide range of social, cultural, economic, political and ecological variables (WWF International-Terralingua, 2000). Many researchers have tried to identify such linkages (Posey, 1999a, 1999; Maffi et al., 1999, Maffi, 2001, 2005a, 2005; Laird, 1999; Harmon, 1992, 1995, 1996, 2002; Harmon & Loh, 2004; UNESCO, 1996, 2001; Zent and Zent, 2004; Moore et al. 2002; etc.).

The received literature further suggests that there is indeed a definite causality between cultural diversity and biodiversity, although its direction seems interactive and in some cases fuzzy. Given the state of knowledge of this causality and complex nature of the issues involved, there is a need for systematic studies to identify this causality and generate replicable research designs. This would indeed go a long way to generate empirical evidence, which in turn can be used as a building block for policy formulation and its effective implementations. 'The Global Source Book on Biocultural Diversity[4]' by Terralingua[5] is a source of vast knowledge on links between linguistic, cultural and biological diversity and shows how working for integrated conservation can foster sustainable livelihoods in local communities. The most important study that has tried to develop quantitative measurement of biocultural diversity based on certain indicators of both biodiversity and cultural diversity at global level is by Harmon and Loh (2004, 2005). A quantitative understanding of the linkage between biodiversity and cultural diversity could certainly give us an edge over a theoretical justification of such linkage but the nature of the study requires comparable quantitative data at national or regional level on different indicators of biodiversity and cultural diversity.

However, these aspects have not been researched on an extensive scale. Economists, in particular, surprisingly have glossed over these issues. A number of studies has been conducted to evaluate and assess the various aspects of biodiversity, but not much literature has come out related to cultural diversity–biodiversity linkages, particularly in India. Many studies on biodiversity have been attempted but these lack replicable and objectionable methodologies. No separate full-fledged study has been conducted in India. This is disquieting, in view of the vast potential for research in the biodiversity–cultural diversity perspective. India is culturally so rich that there are 4,635 ethnic communities, 325 languages from 12 major language families, six major religions, and 3 racially distinct groups. There are hundreds of tribes, having different cultures and lifestyles. 'Complex juxtaposition of biological and cultural diversity in India is no coincidence. The widespread traditional practices of protecting patches of natural habitat or "sacred groves" and certain species, the miracle, myths and folklore representing animals and plants as ancestors and totems, lifestyles which were incredibly fine-tuned to be rhythms and limits of their natural surrounds, hunting practices which protected nature's limits: in all these and other ways local communities have expressed a profound understanding of their relation ship with nature' (Kothari, 1997). Through religions, folklores and traditions, the village communities had drawn a protective ring around the forests, like the Chipko Movement (Guha, 1989). Culture itself is the customary manner in which the human groups learn to organise their behaviour and thought in relation to their environment (Howard, 1989). India being a mega-diversity country[6] in terms of both cultural diversity and biodiversity requires a much in-depth study to find out such possibilities (Islam, 2007).

This paper has been arranged into four sections. Section-I briefly analyses a profile of both biodiversity and cultural diversity in India. Linkage between cultural diversity and biodiversity has been discussed both generally and in the Indian context in Section-II. Section-III deals with current trends of the loss of both biodiversity and cultural diversity at global as well as Indian level. The last section discusses the need for the preservation of cultural diversity and biodiversity for greater benefits.

A Profile of Biodiversity and Cultural Diversity in India

India is a global hub of both biodiversity and cultural diversity. According to the *National Biodiversity Strategy and Action Plan* (NBSAP, 2005) report, India, with 2.4 per cent of the world's area, has over 8 per cent of the world's total biodiversity, making it one of the 12 mega-diversity countries in the world. The country's forest cover has been assessed to be 20.55 per cent of the country's geographical area; of this, dense forest areas cover 4,16,809 sq km (12.68 per cent) and open forests cover 2,58,729 sq km (7.87 per cent). About 45,000–47,000 plant species are reported to occur in India, representing 11 per cent of the known world flora wherein about 33 per cent flowering plant species and 29 per cent of the total Indian flora are endemic. The reported number of species for the angiosperms or flowering plants varies between 16,500 and 19,395 taxa (including intra-specific categories) under 247–315 families and represents roughly 7 per cent of the described species in the world. India is also one of the world's 12 Vavilovian Centres of origin and diversification of cultivated plants. The Indian region has approximately 107 species of aquatic angiosperms, which represent nearly 50 per cent of the total aquatic plant species of the world. The reported number of species for the angiosperms or flowering plants varies between 16,500 and 19,395 taxa (including intra-specific categories) under 247–315 families and represents roughly 7 per cent of the described species in the world, India alone represents 17,672 species. Nearly 17 per cent (193 species) of the pteridophyte flora of the world are endemic to India. Out of the total 750 species (53 genera) of gymnosperms found in the world, 48 species are found in India itself. Nearly 90,000 species of fauna have been reported from India, a little over 7 per cent of the world's reported animal diversity. In the case of mammals, worldwide there are 4,629 recorded species. India harbours 390 species or about 8.42 per cent. The number of microbial cells is estimated to be about 4–6 × 10^{30}, containing nearly half the total carbon and 90 per cent of the nitrogen and phosphate on this planet. (For more details, refer to the National Biodiversity Strategy and Action Plan, 2005).

In terms of cultural diversity, India is a 'cultural laboratory' for the globe. It is a land of enormous genetic, cultural and linguistic diversity (Majumder, 2001). History suggests the coexistence of many faiths, beliefs, ideologies, life-styles, languages, customs, and traditions. People from ancient times have been living here with these cultural variations. Internationally, it has been recognised that India is a culturally mega-diversity country. It has as many as 4,693 distinct ethnic communities, 751 SC communities, 461 ST communities, who speaks about 750 dialects, 8,650 *gotras,* 13,156 clans, 12,057 surnames, 1963 lineages, 2,994 titles, 775 sub-groups, three distinct racial communities, six major religions, besides many other smaller and indigenous faiths and 325 languages. There are 584 Muslim, 130 Sikh, 100 Jain, 93 Buddhist, 339 Christian, seven Jew, nine Parsi, 411 Tribal religious communities live in India, besides numerous Hindu communities. These communities derive their identity mostly from their environment, the places of their settlement, the hills and mountains, the forests and plains, and the occupations based on local resources. There are 96 eco-cultural zones, defined by ecology, language, culture, history and administration. Ecology determines crop production and food habits, which are also influenced by culture. Sixty per cent of the Indian communities is purely vegetarian. India is a land of ethnic diversity too. It has six major world religions and many more smaller and indigenous faiths. It is the home of three distinct racially ethnic communities (Singh, 1994). These diversities of languages, communities, cultural habits represent India's cultural richness and its diversity.

Linkage between Biodiversity and Cultural Diversity: As mentioned earlier, the debate on linkages between cultural diversity and biodiversity and their possible directions has recently emerged as a new research paradigm in the domain of environmental studies. Many researchers have attempted to identify such linkages (Posey, 1999a, 1999; Maffi, 1999; Laird, 1999; Harmon, 1992, 1995, 1996, 2002; Harmon & Loh, 2004; Moore et al., 2002; UNESCO, 1996, 2001; Zent and Zent, 2004; etc.). Harmon (1996) points out that there is significant correlation between the richness of countries with respect to biodiversity and cultural diversity. The latter has arisen, at least partly, by human adoption to the former, and has in turn nurtured it. The diversity of natural ecosystems and species is one basis of cultural diversity, as different human communities have devised a remarkable range of practices, rituals, technologies and social relations in response to the challenges and opportunities presented by their natural surroundings (Kothari, 1997). This relation can either be positive or negative. Harmon and Loh (2004, 2005) in their work on global biocultural diversity index conclude '...these relationships affect each other, in certain cases are closely linked, and sometimes may even be constitutive of each other in important ways'. In many cases, cultural diversity plays a vital role in preserving natural diversity, especially biological diversity. There are many communities and groups which, because of their food habits, rituals, lifestyles or other cultural traditions damage biological diversity. For instance, if we over-use our commons for performing rituals or observe religious practices, it is bound to affect the quality of our environment. We cannot presume that traditional societies are by definition eco-friendly and sustainable.

Attempts to romanticise societies and cultures are fraught with serious consequences. For example, in the domain of biodiversity it can result in irreversible intra- or inter-generational damages (like the loss of endemic biodiversity). (For an excellent exposition refer to: 'Forests and Settlements' by Romila Thapar in "Environmental

Issues in India—A Reader" edited by Mahesh Rangarajan, Pearson-Longman, New Delhi, India, 2007). On the sample analogy, the present division of economy-environment debates into stereotype (like 'North-South', 'Ecological Sustainability versus Economic Growth') is by presumption flamed and needs to be dispelled outright. But the overall changes in the socio-economic and cultural life of the people and threats from global mono-cultures start to eradicate the smaller and indigenous cultures globally. In India too the situation is no different. These declining trends of diversity in languages and cultures have been documented in literature (Krauss, 1992; Harmon, 1995; Mühlhäusler, 1996; UNESCO, 1996, etc.). This is indeed a matter of worry for us.

Many Indian cultural habits are rooted in its nature. At the same time, culture influences nature through its various manifestations. Our customs, beliefs, traditions, languages, religions, faiths, folklore, lifestyles are influenced by our surrounded nature and natural elements. Nature shapes and moulds our behaviour and our behaviour too influences our environment. Popular cultural expression cuts across religions; 775 traits in India relating to ecology, settlement, identity, food habits, marriage patterns, social customs, social organisation, economy, occupation and impact of change and development have been identified by experts, which reveal a sharing of cultural traits across religious categories. Clans bearing names of animals, plants or inanimate objects cut across religions, language and region, indicating their intimate relationship with nature and the living diversity of plant and animal life. Such clans include those with the names of crocodile, corn, panther, buffalo, salt, tree, shrub, tortoise, leopard, water chestnut, chrysanthemum, jiggery, goat, vermilion, rice, flower, tiger, tiger's claw, kite, herb, horse, radish, cobra, pearl, gold, silver, bird, monkey, ape, peacock, rabbit, pepper, fish, jasmine, porcupine, food, brush, red lotus, dead tree, musk, cumin seed, broomstick, tamarind seed, horse gram, milk, cow, bell, gigantic lizard, sandal paste, areca nut, bear, leaf plate, bamboo, moon, frog, banyan tree, cart, plantain, butter onion, pipal tree, elephant, sacred rice, castor seed, etc. (Singh, 1994). The indigenous and local communities of India represent various cultural values. Their languages, food habits, customs, traditions, rituals, beliefs, etc. are very closely related to nature as they too live mostly near nature. Their lifestyle, socio-cultural norms and traditions are formed by the surrounding nature. Being closer to nature gives them an edge over others in understanding and its elements. They are more aware of nature's organising principles, sometimes describes as entities, spirits or natural laws. On the one hand, nature shapes their socio-cultural identity and on the other, their cultural diversity has much influence over its state. Although conservation and management practices are pragmatic, indigenous people generally view this knowledge as emanating from a spiritual base. All creations are sacred and sacred and secular are inseparable. Spirituality and spiritual consciousness is the highest form of awareness. Cultural beliefs, values, knowledge, and behaviour have taken shape and have been transmitted through human–environment interactions. The widespread traditional practices of protecting patches of natural habitat 'sacred groves' and certain species, like myriad myths and folklore representing animals and plants as ancestors and totems, lifestyles which were incredibly fine-tuned to the rhythms and limits of their natural surroundings, hunting practices which respected nature's limits, in all these and other ways local communities have expressed a profound understanding of their relation with nature. The sacred conservation practices followed by local people have come into focus of late due to their importance in protecting several delicate ecosystems and threatened species, the explicit connections they show between cultural and biological diversity, and their potential of people-oriented conservation efforts (Gokhale, 2001). The relationship between human language[7] and biodiversity has been an important manifestation of ethnology. Human language is also greatly influenced by the physical and biological world around. There are also evidences from Indian communities that, due to their cultural and food habits, they depend much on nature. They depend on hunting and gathering, trapping of birds and animals, practising shifting cultivation, etc. which cause much environmental degradation. This suggests that people are closely related to their natural surroundings for their sustenance and growth. This intrinsic relationship among culture, languages and biodiversity gets affected by the changing environment or change in cultural elements as they are linked to each other and get affected due to each other's deterioration.

The National Biodiversity Strategy and Action Plan Report (NBSAP, 2005) mentions that for millennia, biodiversity has supported the livelihoods and life of the people of India, shaping a diversity of cultures in which respect for nature and its myriad life forms has enjoyed a central place. Animals and plants have been revered and worshipped. Forests, rivers, mountains and lakes have been seen as abodes of the gods. The significant tradition of protecting patches of forests dedicated to deities and/or ancestral spirits as sacred groves by many Indian communities is another reflection of the reverence for nature in their religious and socio-cultural life. Many of the sacred groves still provide a safe refuge to several endangered and threatened species of flora and fauna. The *'People of India'* series by the Anthropological Survey of India (ASI) identifies that the lives and livelihoods, occupations, dress, songs and settlement patterns of 85 per cent of the Indian communities are rooted in local ecosystems and local natural resources. It also finds that the names of 1018 communities suggests a close correlation of the nomenclature of a community with its occupation based on locally-available resources, place of origin or village or territory, and deity or religious association. A majority of the communities of India (about 55 per cent) derive their names from the traditional occupations they pursue. These occupations relate mainly to land (agriculture, forestry, fishing, etc.) or to labour (craftsmanship, transport, cottage industry, etc.). There are also occupations relating to other economic pursuits, and finally, there are miscellaneous occupations related to the needs of peasants, folk communities, and urban dwellers. Next to these occupations is a series of community names (about 14 per cent) associated with geographical locations, hills and mountains, plains and valleys, rivers and streams, deserts and forests. Lastly, there are names based on religions and sects.

Loss of Biodiversity and Cultural Diversity

Recent trends of the loss of global, as well as Indian biodiversity and cultural diversity are a cause for worry. Maffi (2005) does not believe that the loss of biodiversity and ethno-linguistic loss (a proxy to cultural diversity) are merely parallel trends: rather, she thinks of it as an interlinking process. There is surely some correlation between these trends. A closer look at the current facts and figures will help us appreciate the gravity of the situation. Comprehensive studies show that presently 24 per cent of mammals and 12 per cent of bird species are threatened by extinction. The situation in India is not also very encouraging. India figures fairly high up in global assessments of threatened species using the revised IUCN threat criteria and categories across taxa. The IUCN report for 2002 contains 366 species of animals categorised as 'threatened', this is approximately 6.73 per cent of the world's total number of threatened faunal species (5,453 species). This includes 148 species of mammals, 138 birds, 32 reptiles, three amphibians, 17 fish and 28 invertebrate species. The analysis of the data on threatened species by threat categories reveals that, of the 366 species listed as threatened, 18 are critically endangered (Cr), 52 species are endangered (En), 150 species are vulnerable (Vu), while 10 species are at lower risk or conservation dependent (LR/cd), 101 species are at a lower risk but near threatened (LR/nt), while the data for 35 species is not available (DD). On analysis of the threat categories by groups (at global level), it is found that out of 366 species of threatened Indian fauna, 220 species are at significant risk, of which 88 species are of mammals, 72 birds, 25 reptiles, three amphibia, nine fishes and 23 species of invertebrates. Further analysis of the critically endangered list reveals that, out of 18 species, five are mammals, seven are birds, four reptiles, and one insect. Of these 366 species, the populations of only three species are stable, 101 species are showing distinct deteriorating trends and the population status of 75 species is uncertain, while the population trends of 187 species have not been assessed. None of the species is, however, showing an upward population trend. At least 166 species of crops (6.7 per cent of total crop species in the world) and 320 species of wild relatives of cultivated crops are believed to have originated in India.

The reasons behind this extinction and threats can be attributed to many factors, ranging from natural as well as human-caused-like current model of development and economic progress, increasing social, political and economic inequities, changes in cultural, ethical and moral values, lack of recognition of the full values of biodiversity, inappropriate and contradictory laws and policies, demographic changes, etc (NBSAP, 2005). If we look at the pace of biodiversity loss in India, even a layman can understand the situation is really alarming. If this trend of losing biodiversity continuous, we will end up losing our rich biodiversity. This will obviously affect our economic, medicinal, cultural, social, as well as life-support systems. The pace of biodiversity loss is alarming in India which is indeed a matter of concern as human survival is largely dependent on biological resources. Over 70 per cent of Indians are still dependent on agriculture and allied sectors and others too are directly or indirectly dependent on nature for their sustenance. Once the biodiversity is eroded it will have an immense effect on human interests—food security, medicinal needs, life-saving supports, recreational needs, nature-based research and innovation, etc.

Human cultural diversity is also threatened on an unprecedented scale. Linguists estimate that between 5000 and 7000 languages are spoken today on the five continents[8]. On an average, one language disappears every 14 days and with this language a unique culture, unique tales, myths and unique ways of thinking disappears as well (Schauer, 2002). A UNESCO (1996) report states that within a century, half of the smaller and indigenous languages will disappear. Powerful and dominant cultures are first marginalising and then destroying traditional cultural diversity. Several factors can be responsible for the depletion of cultural diversity: the changing structure of lifestyles, consumption patterns, adoption of global powerful cultures, influence of media and communication technologies, disintegration of the joint family system, lesser esteem for knowledge in primary school curricula, transition from oral to written culture, etc. Schauer (2002) identifies that the spread of the dominant

'Western culture' which is facilitated by the new media and by industrial products is one of the main reasons for this decrease in diversity of cultures. Losing this cultural diversity will have a strong negative effect on biological diversity, which in turn will affect human society. India has an advantage over many other countries by virtue of being a mega-diversity country in terms of both biological and cultural diversity. A better understanding and systematic maintenance of our cultural diversity will not only preserve them from depletion, but help us in the preservation of our endangered biological diversity and socio-economic balance. For the preservation of cultural diversity, understanding and respecting the local and traditional cultures, a more holistic lifestyle and awareness must be taken into consideration in policy as well as implementation.

Preservation of Biocultural Diversity—An Urgent Need:

Maffi (2005) points out that, like other species, humans are an intrinsic part of the natural environment. In the course of their history, humans have modified the environment through their cultural beliefs, values, knowledge and behaviour; at the same time, human cultures have been moulded by the process of adaptation to the natural environment. Many of the indigenous communities in India, because of their cultural habits, practise a lifestyle which is more eco-friendly. Although many scientists are skeptical about the traditional and indigenous communities' role towards conservation of nature, almost everybody agrees that they are more likely to employ environmentally sustainable practices when they enjoy autonomous territorial security and local autonomy.

The conservation of biological diversity (CBD) recognises the role of traditional ecological knowledge by indigenous and local communities in the conservation of biodiversity and calls for its protection. So, any loss of cultural diversity or biodiversity will have a definite effect on the other. We are undoubtedly dependent upon biodiversity for our survival; our nourishment, our development, our research and inventions, etc. Any loss of these precious natural elements will surely affect our own interests. On the other hand, loss of cultural diversity means loss of associated knowledge, historical experiences, aspirations and world views. Experts rightly assert, that deprive a people of their language, culture and spiritual values and they will lose all sense of direction and purpose of life. The natural environment and human societies should be seen as inextricably inter-related within their complex system. The breakdown of these connections underlines many of the environmental and social problems that humanity is facing today. Therefore, any action to protect, maintain, and restore the ecological health of natural environment should be combined with actions to protect, maintain, and restore the social, cultural, spiritual, and biophysical health of human societies and vice versa (Maffi, 2005). A WWF-Terralingua (2000) study shows that success in conserving biodiversity may well be inter-related to the maintenance of cultural diversity, and that conversely, the loss of cultural diversity is part and parcel of the same socio-economic and political processes, leading to biodiversity loss.

Many experts around the world have proved that extinction of cultural diversity has a negative effect on the management of biodiversity. Atran et al. (1997, 2002) find that a better understanding to intricate cultural patterns, favouring environmental maintenance, may enhance the value of biodiversity and reduce its chance of extinction. They also give importance to cultural diversity in the formation of people's behaviour and management of commons. The Universal Declaration on Cultural Diversity[9] (UNESCO, 2001) rightly identifies the importance of maintaining biodiversity and cultural diversity: '...as a source of exchange, innovation and creativity, cultural diversity is as necessary for humankind as biodiversity is for nature'. So, there are linkages between the two and preservation of biodiversity, as well as cultural diversity is very important. International bodies like the UNEP, IUCN, UNDP, UNESCO and many non-governmental organisations, like the Terralingua, are tirelessly working to identify such linkages and make proper policy for their preservation. Growing concentration on wealth, skewed control over access to natural resources (sources energy and arable land) and marginalisation of the poor, both within and across the countries, is bound to exacerbate the conflict between development and environment. The biggest challenge before today's world is fulfilment of increasing demand by its vast and growing population and simultaneously keeping the earth livable and keeping a balance between the present consumption of and entitlements of the future generation to environmental assets. One suggested solution for maintaining natural resources, which are increasingly becoming scarce, is sustainable use.[10] Human society must preserve biodiversity and maintain a sustainable use of these resources to get long-tern benefits. However, preservation of biodiversity while maintaining a certain pace of economic growth along with vast poverty, landlessness, unemployment, illiteracy is not an easy task. We must explore every possible way that will save our biodiversity, keeping the pace of growth, and will not require much of the already scarce resources. In that direction, preservation of cultural diversity may, in turn preserve biodiversity and generate greater benefits out of these resources.

Experts believe that conserving biodiversity without conserving associated knowledge systems is like building and maintaining a library without a catalogue. Maintaining diversity in nature is also important, not only for economic benefits but for our own survival. Therefore, a necessary condition for life on earth as well as cultural diversity is the presence of required biodiversity. It is now not unusual to read prominent (though often rather superficial) declarations of the importance of preserving biological and cultural diversity as a central conservation goal (Harmaon & Loh, 2004, 2005). Our development should be informed by and designed with respect to our cultural, spiritual, ethical values along with experiences of great socio-economic transformation.

Reference

Atran, Scott; Medin, L. Douglas; Ross, Norbert; Lynch, Elizabeth; Vapnarsky, Valentina; Ucan Ek', Edilberto; Coley, John; Timura, Christopher; Baran, Michael (2002). 'Folkecology, Cultural Epidemiology and the Spirit of the Commons', *Current Anthropology*, Vol. 43.

Atran, Scott and Medin, L. Douglas (1997). *Knowledge and Action: Cultural Models of Nature and Resource Management in Mesoamerica*, The New Lexington Press, San Francisco.

Brown, Lester R. (2001). *'Eco-Economy'*, Earthscan Publications Ltd., London, UK.

Coombe, Rosemary J. (2000). *Preserving Cultural Diversity through the Preservation of Biodiversity: Indigenous Peoples, Local Communities, and the Role of Digital Technologies*, University of Toronto, Canada.

Gadgil, M. and Guha, Ramachandra, (1994). Ecological Conflicts and Environmental Movements in India, *Development and Change*, Vol. 25, No. 1.

Gokhale, Yogesh (2001), 'Biodiversity as a sacred Space', *The Hindu*, 20 May, New Delhi.

Guha, Ramachandra (1989). *The Unquiet Woods—Ecological Change and Peasant Resistant in the Himalaya*, Oxford University Press, New Delhi.

Harmon, David (1992). 'Indicators of the World's Cultural Diversity', Paper presented at the Fourth World Congress on National Parks and Protected Areas, 10–21 Februray 1992, Caracas, Venezuela.

Harmon, David (1995). 'The Status of the World's Languages as Reported in Ethnologue', *Southwest Journal of Linguistics*, 14, 1–33.

Harmon, David (1996). 'Losing Species, Losing Languages: Connections between Biological and Linguistic Diversity', *Southwest Journal of Linguistics*, Vol. 15, Nos. 1 & 2.

Harmon, David (2002). *In Light of our Differences: How Diversity in Nature and Culture Makes us Human*, The Smithsonian Institute Press, Washington DC, USA.

Harmon, David & Loh, Jonathan (2004). *IBCD: A Measurement of the World's Biocultural Diversity*, in Borrini-Feyerabend et al. '*History, Culture and Conservation;*, IUCN, Switzerland pp. 271–280.

Howard, M.C.(1989). 'Contemporary Cultural Anthrocopology' (3rd Ed.), Glenview, II: Scott, Foresman and Company.

Islam, A.K.M. Nazrul (2007). Ethno-cultural Diversity and Environment: An Indian Perspective, *Indian Journal of Psychology & Mental Health*, Vol.1, No. 1, pp. 121–130.

Kothari, Asish (1997). *Understanding Biodiversity: Life, Sustainability and Biodiversity*, Orient Longman, New Delhi.

Krauss, M. (1992). 'The World's Languages in Crisis', *Language*, 68(1): 4–10.

Laird, S.A. (1999). *Forests, Culture and Conservation*, in *Cultural and Spiritual Values of Biodiversity*, Intermediate Technology Publications, London.

Loh J. and Harmon, D. (2005). 'A Global Index of Biocultural Diversity', Ecol. Indic. 5(3): 231-241.

Maffi, Luisa (2001). (Ed.) *On Biocultural Diversity: Linking Language, Knowledge and Environment*, Smithsonian Institution Press, Washington DC.

Maffi, Luisa (2005a). 'Linguistic, Cultural and Biological Diversity', *Annual Review of Anthropology*, 34:599–617.

Maffi, Luisa (2005). *Report on the Global Source Book on Biocultural Diversity*, Terralingua, USA.

Maffi, Luisa, T. Skutnabb-Kangas and J. Andriyanarivo (1999). *Linguistic Diversity", in "Cultural and Spiritual Values of Biodiversity—A Complementary Contribution to the Global Biodiversity Assessment"*, Intermediate Technology Publications/UNEP, London/Nairobi.

Majumder, Partha P. (2001). 'Ethnic Population in India as seen from an Evolutionary Perspective', *Journal of Bioscience*, Vol. 26, No. 4, p.534.

Millennium Ecosystem Assessment (2005). *Ecosystems and Human Well-being: Biodiversity Synthesis*, World Resources Institute, Washington DC.

Moore, Joslin L., Lisa Manne, Thomas Brooks, Neil D. Burgess, Robert Davies, Carsten Rahbek, Paul Williams and Andrew Balmford (2002). *The Distribution of Cultural and Biological Diversity in Africa*, The Royal Society, UK.

Mühlhäusler, P. (1996). *Linguistic Ecology: Language Change and Linguistic Imperialism in the Specific Rim*, Routledge, London.

Murty, M.N. (2001). *Environmental Regulations and the Economics of Environmental Policies*, in Rabindra N. Bhattacharya (Ed.), *Environmental Economics—An Indian Perspective*, Oxford University Press, New Delhi, pp. 101–102.

National Biodiversity Strategy and Action Plan (NBSAP, 2005). Ministry of Forests and Environment, Government of India, New Delhi.

Omvedt, G. (1993). *Reinventing Revolution: New Social Movements and the Socialist Tradition in India*, East Gate Book, London.

Posey, Darrel Addison (1999a). "*Cultural and Spiritual Values of Biodiversity—A Complementary Contribution to the Global Biodiversity Assessment*", Intermediate Technology Publications/UNEP, London/ Nairobi.

Posey, Darrel Addison (1999). *Cultural and Spiritual Values of Biodiversity*, Intermediate Technology Publications, London.

Rangarajan, M. (2007). *Environmental Issues in India—A Reader*, (Ed.), Pearson-Longman, New Delhi.

Schauer, Thomas (2002). "*Biological and Cultural Diversity in a Globalised Information Society*", Universitätsverlag Ulm, Germany.

Sethi, Harsh, (1993), 'Survival and Democracy: Ecological Struggles in India' in Ponna Wingnaraja (Ed.), *New Social Movements in South: Empowering the People*, Vistaar Publications, New Delhi, pp. 122–148.

Shah, Ghanshyam, (1990) *Social Movements in India: A Review of Literature*, Sage Publications, New Delhi.

Singh, K. S. (1994), 'People of India', (Ed.), *Anthropological Survey of India*, New Delhi.

UNESCO (1996). 'Sacred Sites—Cultural Integrity, Biological Diversity', Programme Proposal, Nov. 1996, Paris.

UNESCO (2001). 'Universal Declaration on Cultural Diversity', Approved by UNESCO General Conference, 31st Session, Paris.

Wall, Derek (1994). *Green History—A Reader in Environmental Literature, Philosophy and Politics*, Routledge, London.

WWF International-Terralingua (2000). *Indigenous and Traditional Peoples of the World and Ecoregion Conservation-An Integrated Approach to Conserving the World's Biological and Cultural Diversity*, Gland, Switzerland.

Zent, Stanford and Zent, Eglee L. (2004). *On Biocultural Diversity from a Venezuelan Perspective: Tracing the Interrelationships among Biodiversity, Cultural Change and Legal Reforms*, Depto. de Anthropologia, Instituto Venezolano de Investigaciones Cientificas, Apartado 21827, Caracas, Venezuela.

www.iucn.org/redlist

www.terralingua.org

www.unep.org

Endnotes

1. According to the *Convention on Biological Diversity* (CBD), "Biodiversity means the variability among living organisms from all sources, including, inter alia, terrestrial, marine, and other aquatic ecosystems and ecological complexes of which these are a part, this includes diversity within species, between species and of ecosystems".
2. An interdependent system of living and inanimate, but biologically active, components: designates terrestrial and aquatic systems of widely different sizes, from woods to tropical forests, from meadows to prairies, from ponds to oceans.
3. *Cultural diversity* is the variability of human expression and organisation, including that of interactions among cultural groups and between the groups and the environment. It can be thought of as the totality of the cultural richness present within human species (Harmon, 1998). Harmon (1992) identifies the following indicators for *cultural diversity*: local language; ethnic affiliation; forms of social organisation; subsistence practices; land management; dietary habit; medicine; aesthetic manifestation; and religious manifestation.
4. 'Biocultural diversity' is the diversity of life on earth both in nature and culture, or the diversity of life forms that has been jointly shaped by both natural and cultural forces through co-evolutionary processes (Maffi, 2005).

5. 'Terralingua' is a non-governmental organisation working on the area of cultural diversity-linguistic diversity-biodiversity related areas in close collaboration with many international organisations, like the IUCN, International Resource Institute, WWF, etc. It is basically a US-based organisation and has offices in other countries, like Canada.
6. Mega-diversity countries are those which are likely to contain the highest percentage of global species richness.
7. Human language is a very important element of cultural diversity. Many studies have used language diversity as a proxy to cultural diversity as human culture depends much on the language they speak.
8. Posey, Darrel A. (1999a). "Cultural and Spiritual Values of Biodiversity", Intermediate Technology Publication, UNEP.
9. The UNESCO Universal Declaration on Cultural Diversity was adopted unanimously in the 31st Session of the UNESCO General Conference in Paris, 2nd November 2001.
10. 'Sustainable use' means using some resources without harming the balance between present needs and future needs and maintaining the environmental balance too.

Chapter 8

Marginalised People and the National Environment Policy, 2006

Archana Prasad

I

Marginal People and the Political Economy of Natural Resource Dependence

This paper attempts to evaluate the National Environment Policy, 2006 (NEP) in the light of the concerns of the marginalised people of India. Here the term 'marginalised people' refers to the Scheduled Tribes (ST) and Scheduled Castes (SC) of India, who form a substantial portion of the population that is dependent on natural resources for their daily survival. Their marginalisation is historic in character because it reflects that long-term processes have dislocated these people from fertile and potentially prosperous tracts to the margins of the agricultural society. At the same time, the legislative and social regime that regulated the access to commons has also had an extremely adverse impact on the daily lives of these people, most of whom depended on seasonal livelihoods for their subsistence. The restructuring of the economy in the colonial times itself had led to conversion of most of these people from producers to casual labourers.[1]

These developments of the colonial times structured both the natural resource management and social welfare functions that were meant to solve the problems of the SCs and STs in independent India. The Nehruvian consensus that evolved in the early years after Independence was based on two contradictory trends: on the one hand, the state continued to exercise monopoly control over common property resources like forests, minerals, pastures and water; on the other hand, it also carried out a social welfare programme to assist in the development of these people. So, even while the first Five-Year Plan resolved to help the 'backward classes' to pursue a path of self development, especially in health, education and economic life,[2] the impact of the macro-economic policies was completely the opposite. Large developmental projects, without adequate care for the rights of the marginal people, ensured that the inequities within the system increased substantially.

This was reflected in the occupational status of the Scheduled Castes and Tribes from the Census figures from 1961 onwards, which reveal the picture depicted in Table 8.1.

Table 8.1: Occupational Profile of Schedule Caste and Tribe People, 1961 – 91

Year	Cultivator		Agricultural Labourer		Household Industry		Other Workers	
	SC	ST	SC	ST	SC	ST	SC	ST
1961	37.76	68.18	34.48	19.71	6.56	2.47	21.20	9.64
1971	27.87	57.56	51.74	33.04	3.33	1.03	17.06	8.37
1981	28.17	54.43	48.22	32.67	3.31	1.42	20.30	11.84
1991	25.44	54.50	49.06	32.69	2.41	1.04	23.08	11.76

Source: Census of India, *Tables of Scheduled Castes and Tribes* for years 1961 – 1991.

Table 8.1 shows the increasing conversion of the SCs and STs from agricultural on-farm producers to labourers. This is largely due to the fact that there was land alienation in the marginal areas, both for public purposes as well as in distress. Since many of these areas were resource rich, they felt the direct impact of the strategy of harnessing for natural resources for national development. According to one estimate, nearly 213 lakh people were displaced through development projects, application of forest laws and other factors from these areas between 1952 and 1991.[3] Further, land alienation was a common feature and took place through benami transactions due to indebtedness and distress sales. Even though there was legal protection against these actions, the nature of economic stagnation was such that marginalised people were unable to resist these processes. The first factor that reflected this trend was the pattern of land ownership and alienation amongst the SC and ST population. By 1986, though the operational holdings of the SCs and STs had increased marginally to 12.4 and 7.9 per cent respectively, the rate of increase hovered around 1 per cent in comparison to 9.3 per cent for other social groups, thus revealing the inequities within the system.[4]

These inequities have in fact accentuated in the period after the reforms, when the policy of state withdrawal from social and economic sectors gathered momentum. The crisis resulting out of structural changes within agriculture and repeated attempts to open up common property resources like water and forests to industry has also had an adverse impact on the status of SCs and STs people. For example, externally-aided poverty-reduction programmes, like the Velugu project in Andhra Pradesh promote mono-culture commercial crops on the one hand, and even as they purchase food from the open market and lend it to the poor people as credit, on the other. This is seriously undermining the capacity of the state-owned Public Distribution Systems and co-operatives to expand their capacities to cater to tribals' basic needs in a socially viable way. Similarly, the proposal to set up SEZs in forest areas and introduce industry as a third party in joint forest management has resulted in leasing out forested and wasteland areas to big companies at nominal rates in the marginal regions. These trends are reflected in the patterns of employment and work in the 1990s, represented in Table 8.2.

Table 8.2: Patterns of Scheduled Caste & Tribe Employment (percentage of Population)

Category of work	Scheduled Tribes		Scheduled Castes	
	1991	2001	1991	2001
Total workers	49.3	49.0	39.2	40.4
Main workers	42.02	33.8	36.08	29.5
Marginal workers	7.02	15.2	3.1	10.9
Non-workers	50.7	51.0	60.8	59.6

Source: Census of India, 1991 – 2001.

One of the most significant features of Table 8.2 is the increase in marginal workers and a sharp decline in the number of dalits and tribals who get about 100 – 180 days of work. Another thing to keep in mind is the fact that the 2001 figures also include 'unpaid' work as 'work' in the framework of its analysis. If this is so, we do not know whether the rise in the total number of workers includes unpaid women labourers and other form of bonded labour who belong to the SC and ST communities and whose labouring conditions are inhuman. Thus the increase in total workforce in this case may not reflect the true picture about the nature and conditions of work in the post-reforms era.

This is seen in the occupational profile of the SCs and STs in the 1990s represented in Table 8.3.

Table 8.3: Occupational profile of SCs and STs in the 1990s

Category of workers	Scheduled Tribes		Scheduled Castes	
	1991	2001	1991	2001
Cultivators	54.5	44.7	25.44	20.0
Agricultural labourers	32.69	36.9	49.04	45.6
Agriculture workers	87.9	90.6	73.48	75.6
Non-Agriculture workers	12.1	9.4	26.62	34.4

Source: Census of India, 1991 and 2001.

If we compare these trends with that of the dalits, we find that the dependence on agriculture and allied activities by tribals is much higher. Dependence on forests needs to be seen in this context of increasing dependence on the agricultural sector by the dalits and tribals. Though the data records a decline in dependence on forestry and livestock related activities from 2.27 per cent in 1971 to 1.2 per cent in 1991, this does not reveal the true picture, as it does not record the seasonal and supplementary dependence on forests. Literature from all over the world suggests that people with marginal holdings and those with irregular work usually get an important supplementary income from forests.[5]

In studies conducted on different field sites, it has been shown that more than 85 per cent of the land-owning dalits and tribals are in fact marginal land holders who cannot meet their livelihood only through the practice of agriculture. In fact, what is revealed is the fact that marginalised communities are still following a system of livelihood that is borne out of the combination of dependence on agriculture, forests and labour. Micro studies from the field in the post-agrarian crisis period show that dependence on casual labour is increasing in proportion to the dependence on forests in the post-economic reforms period.[6] At the same time, the increasing pace of industrialisation in the marginal areas has underscored the precariousness of livelihood systems and their (often unrecorded in macro-economic analysis) increasing dependence on forests.

In the context of the foregoing analysis, it is important to reiterate the main issues that plague the economic system of the Scheduled Caste and Scheduled Tribe people today. The first main problem experienced by these marginal people is the lack of access to land and 'common and public resources'. This is best reflected in the land-holding figures of the SC and ST people. The average access to land has been declining since the 1980s, with the average holdings of the Scheduled Tribe family declining from 2.25 hectares in 1985-86 to 1.74 hectares in 2000-2001 and for Scheduled Caste families declining from 1.15 hectares to 0.85 hectares.[7] More important, one must note that most of the tribal people have marginal holdings of less than two hectares and at least a third of them are landless. As far as the SCs are concerned, the situation is even worse as the proportion of landless people is higher than even the STs. This has created a contradiction between the SCs and STs in some of the areas like Jharkhand where the SCs themselves work on tribal lands. In this situation, any national policy has to address the diverse circumstances that structure the access of SCs and STs to both land and natural reources by keeping in mind the following context:

- Lack of power sharing and control in management of resources (lack of implementation of existing laws and policies; lack of decentralisation of power);
- Displacement due to large development and other industrial projects, lack of control and capability to deal with both corporate capital and market forces, especially in a situation where the government is

looking for greater partnerships with and investments from the private sector;

- And Last but not least, the environmental policy of the government has to ensure that the goals of sustainable development are met by ensuring the livelihood security and development of all marginalised people. This means that the goals of social justice and equity should go hand-in-hand with the programme of conservation and development.

If these concerns are to be addressed, the national environmental policy must be evaluated not only in terms of the policy prescriptions it espouses, but also the implications of these measures for the path of development, which profoundly affects the future of marginalised people.

II

Marginalised Areas and People in the National Environment Policy, 2006

As is evident at first glance, the National Environment Policy attempts to provide an overall inter-sectoral framework for introducing environmental concerns into mainstream developmental policy. Many of the objectives it sets out for itself have a direct bearing on the lives of the marginal people of the nation. The most important of these is inter- and intra-generational equity,[8] which should guide the process of socio-economic development. It is well known that the people living in resource-dependent areas have made and continue to make sacrifices for the nation's overall development. Following the tenor of the National Forest Policy, 1988, which states that 'ecological and economic security'[9] are two sides of the same coin, the NEP stresses the need to have human-centred development which ensures a healthy and productive life, in harmony with nature. Further, in order to achieve this objective, the policy lays down the principle of equity, right to development, decentra-lisation and the public trust doctine as basic principles by which its aim is to be met.

It is significant to note that the principles mentioned above are quite significant from the point of view of the resource-dependent people, and need to be considered in the context of the rest of the policy, according to which the principle of equity 'refers to both equity in entitlements to, and participation of, the relevant publics, in processes of decision-making over use of environmental resources'[10]. This is supplemented with the doctrine of decentralisation, which 'involves ceding or transfer of power from a Central Authority to State and Local Authorities, in order to empower public authorities to address these issues'. Both these principles should have necessitated that the policy spell out the ways in which the local people's decision-making powers are strengthened. But, far from this, the policy has actually resulted in creating an opening for making diversion of forest land easier through dilution of environmental impact assessment procedures, as enunciated in the new guidelines of 2006.[11]

At another level, it is equally important to consider the public trust doctrine espoused by the NEP. The policy states that the State is not an absolute owner, but a trustee of the resources meant for public use. This understanding is essential to protect both the national interest and the legitimate interest of a large number of people. For example, the section on forests underlines the need to give legal recognition of the traditional entitlements to forest dwellers under the Panchayat (Extension to Scheduled Areas) Act.[12] The Scheduled Tribes and Other Traditional Forest Dwellers (Recognition of Forest Rights) Act, 2006[13] can be seen as fulfilling this principle in a different perspective. In this context, the policy also claims to provide long-term incentives to these communities to conserve the forests by giving a direction to universalise Joint Forest Management.[14] However, the main difficulty with this proposition is that it is unable to provide a credible analysis of how the state can disinvest in common lands when it is only a 'trustee' and not the 'absolute owner' of these resources. This is precisely what the Environmental Impact Assessment Rules, 2006 aim to do.[15]

It would appear that the principles espoused by NEP would lead to a more equitable path of development. But, the framework developed by the policy for the analysis of the causes for and solutions to the current environmental problems leaves much to be desired. For instance, the NEP states that 'the loss of the environmental resource base can result in certain groups of people being made destitute, even if overall, the economy shows strong growth'.[16] However, the measures it advocates fail to offer any succour to the poor to give better access to resources and capabilities. Many of the proposed regulatory reforms will be dependent on market-based instruments and would not provide a solution to those problems of environmental degradation that affect the livelihood security of the poor.[17]

It is significant that the NEP pursue the current path of development rather than making it more ecologically and socially just. Even though the revised draft makes some effort to discuss the problems of tribals, women and urban poor, their access to resources is only discussed in forests and wildlife areas. It fails to address their problems regarding access to land, water and biomass resources. Further, the discourse of rights is replaced by an emphasis on 'traditional entitlements' which cannot take care of the problems of the rural and urban poor. The NEP's vision of structural, institutional and policy interventions is embedded in the ideology of ecology of affluence. It is pertinent to note that though the NEP talks of the unsustainable consumption patterns in the industrialised countries that are having serious adverse impacts on the environment, both local and global, and are accentuating poverty in the developing countries, it is unconcerned how its own policy prescriptions for environmental management feed into the interests and lifestyle of the rich. Because of this, the prescriptions the NEP promotes are neo-liberal in character and have disastrous implications for dalits and tribals.[18]

III

The Ideological Faultlines of the National Enviromental Policy

The key propositions of the NEP regarding strategies and actions straddle a conventional conservationist agenda (excluding people's rights over the use of their local resources) and a strategy that seeks to open up the natural resources sector to market forces.[19] The policy also seeks to expand the control of wildlife conservators in other areas where endangered species exist. Clearly, these provisions go against the spirit of the Wildlife Protection Amendment Act, 2006, which attempts to provide a basis for recognition of some local rights in protected areas. At the same time, it chooses to transform the role of the Indian State in the direction of facilitating the market forces to self-regulate their activities for environmental concerns and confining largely its own direct interventions to the application of price and taxation instruments.[20] For example, in order to deal with the problem of groundwater pollution, the NEP talks of tackling the pollution of groundwater from agricultural chemicals by changing the current pricing policies for chemical pesticides, which do not take the potential environmental impacts into account. It hopes to curb the use of agrochemicals by opting for a mild measure like an optional utilisation of fertilisers, pesticides and insecticides to cause an improvement in the water quality. By doing this, the NEP severely dilutes the development and welfare role and responsibility of the Indian state by proposing to rely on the practice of regulation through the codes of environmental management in a big way.

It is pertinent to note that the NEP recognises 'when water tables are very deep the capital costs may be relatively high, with no assurance that water would actually be found, and in such a situation, a user who may be a marginal farmer able to borrow the money only at usurious rates of interest, may, in case water is not found, find it impossible to repay his debts'.[21] But it only suggests promotional measures like: 'a) intensive water and moisture conservation through practices based on traditional and science based knowledge, and relying on traditional infrastructure, b) enhancing and expanding green cover based on local species and c) reviewing the agronomic practices in these areas, and promoting agricultural practices and varieties, which are well adapted to the desert ecosystem'. It consciously chooses to remain silent on what type of institutional forms and rights would be created in order that rights of access and ownership accrue to natural resource-dependent people in a sustainable way (area-based land use and water management, area planning for sustainable cropping patterns). Needless to say, the role of the state has to be developmental in nature to provide sustainable access to resources to the marginalised sections of the Indian people.

The NEP attempts to bring issues of the settlement of ownership and control under the rubric of 'institutional failures' or the failure to settle rights. But it is not successful in doing this because it does not get into the establishment of a policy framework for what would constitute the legitimate framework of 'rights of use' on different resources. Like other donor agencies, this policy also stresses the importance of community-based natural resource management as the only alternative in the current scenario. But it is totally silent on the way in which these activities will be integrated and related with the larger mechanisms of conservation. While the policy states that it will follow the Public Trust Doctrine[22] where the state is 'not an absolute owner, but merely a trustee of all natural resources, which are by nature meant for public use and enjoyment, subject to reasonable conditions, necessary to protect the interests of a large number of people', this assertion of the state being not an absolute owner may have been welcome but for the fact that it does not limit the state power when it comes to the privatisation of natural resources. If the state is not an absolute owner, can it privatise something it does not own? But the policy is silent on how the state's power to denotify these areas for commercial or large developmental projects will also be delimited through this review of legislation. Instead, it provides a justification for the Environmental Impact Assessment Rules[23] which dilute the controls over-diversion of resource-rich areas for commercial purposes, and whose implementation will lead to large scale displacement of people.

Further, the policy does not specify how the ownership and right of access shall be established over these resources. The policy also states that deregularisation is just another term for privatisation, by encouraging environmental investments that will be made by corporate houses since the resource-dependent poor people are not in a position to make such investments. For example, the policy makes claims about the rationalisation of Coastal Regularisation Zone (CRZs) notifications. It specially talks of decentralising 'the clearance of specific projects to State Environmental Authorities, exempting activities which do not cause significant environmental impacts....' This formulation is clearly detrimental to the livelihood of fisherfolk and can easily lead to the destruction of the coastal ecology. Yet again, it fails to outline the inalienable rights of the fisherfolk or a mechanism by which these rights can be settled. The impression given is that the market forces will necessitate multi-stakeholder partnerships through which rights will be settled once again, leaving the people to the mercy of corporate shipping and fishing companies. The recent protests by grass roots movements on changes in the CRZ rules and the post-tsunami experience of the coastal areas have shown that the influx of private capital in the development of coastal areas has only destroyed the livelihood of local fishermen and small farmers.

The improvement of people's access to resources cannot be done only through the grant of 'traditional entitlements' and 'community-based management'. It needs to have co-evolved structures for promotion and regulation to be in a position to protect the knowledge of local people and develop that knowledge in a way that the local communities can get the maximum benefit out of it. This would be possible only if the policy makes explicit affirmative proposals for the development of 'marginalised people-friendly' mechanisms of project evaluation, benefit sharing and decentralisation of natural resource management. Such types of proposals are missing from the document. For example, the document talks of universalisation of joint forest management (JFM). However, it is quite well known that even in those areas where JFM is already in practice, the forest-dependent people have been gradually marginalised on account of the lack of co-evolved institutional structures for the promotion and regulation of markets in non-timber forest produce trade. The revised policy, too, does not take this into account and ignores the ground reality on the implementation of PESA laws whose spirit has been consistently violated by both central and state governments. Further, even though it attempts to talk of devolution of powers to Van Panchayats and other customary institutions, it ignores the processes by which their authority is being continuously eroded in the wake of economic reforms.

The draft NEP had reiterated the age-old principle of 'non-regularisation' of encroachments in the post-1980 period. This has been deleted in the revised policy, especially in the wake of the heated debate over the Scheduled Tribes and Other Forest Dwellers (Recognition of Forest Rights) Bill 2006.[24] But like the draft NEP, the revised policy also forcefully argues for the rationalising of EIA principles and solving the problems of delays in clearances. The policy document fails to make a clear commitment towards making EIA statements open and public and subject to evaluation and contestation by the affected parties and supports the changes brought about by the Draft EIA Rules 2005. The democratisation of decision-making structures and expertise demands that all stakeholders have the equal opportunity to evaluate the impact of the project on their livelihood security and environmental futures. Further, all major development and industrial projects should have the prior informed consent of the affected parties and should make public their benefit sharing plans and agreements. At another level, it should also be mandatory to provide information to local committees and bodies, and hold public hearings at periodic intervals.[25]

It is significant that NEP limits itself largely to creation of market-friendly regulatory regimes to facilitate the introduction of private players in natural resource management. It does not pay sufficient attention to the creation of knowledge, expertise, technology and infrastructure for the development and diffusion of environment friendly innovations. For example, while outlining the scope of institutional changes needed at the cross-sectoral level in the section on strategies and actions, it fails to acknowledge the need to fill the gaps in respect of the development of critical Science and Technology infrastructure.[26] This infrastructure is required to undertake the generation of scientific expertise for environmental monitoring, enforcement, development and diffusion of technologies needed for the sustainable development. There is an urgent need to make interventions in the direction of upgrading the skills of marginalised people through appropriate technological interventions, if their livelihood is to be ensured. Measures for this are totally missing from the current NEP.

The policy interventions outlined in the document are informed by an extremely narrow approach to regulatory reforms. Its propositions are mostly focused on the problems of transaction costs arising out of state regulation. The remedies it proposes lie essentially in the shift to the use of soft civil law, fiscal instruments and use of economic principles in economic decision-making. The revised policy talks of capacity building for the implementation of environmental management principles, but the marginalised groups are not an explicit target of the efforts to be made by the government. The policy document is completely oblivious of the obvious fact that environmental transformation is not an activity performed in isolation but one which involves a variety of actions within the system. This is especially true with respect to democratisation and access to natural resources by marginalised communities. By ignoring the systemic issues raised in this essay, and further providing the basis for opening up of the natural resource sector, the NEP has ensured that the future of dalits and tribals will be more jeopardised than it already is in the current scenario.[27] For the environmental, dalit and tribal movements, which have been fighting for more equitable solutions to problem of marginalised, the struggle will become more challenging and difficult if the full implications of this policy are realised.

Endnotes

1. For a detailed analysis of the historical processes leading to marginalisation of the Scheduled Castes and Scheduled Tribes, see Archana Prasad, *Tribal Development and Globalisation: Exploring the potential of non-timber forest produce in Central India*, Nehru Memorial Museum and Library 2005.
2. Government of India, *First Five-Year Plan*, Planning Commission, New Delhi, 1952, Chapter on Backward Classes.
3. Walter Fernandes and Vijay Paranjape (eds.) (1997). *Rehabilitation Policy and the Law in India*, Indian Social Institute, New Delhi, p.15.
4. Ministry of Agriculture, *2001 Agricultural Statistics of India 2000 – 2001*, New Delhi, pp. 8-9.
5. See, for instance, N.C. Saxena, *Saga of Participatory Forest Management*, Centre for International Forestry Research (CIFOR) Publications, Bogor, Indonesia, 1997. Several literature surveys by CIFOR between 1997-2001 show this.
6. Archana Prasad, *Tribal Development and Globalisation, Exploring the Potential of Non-timber Forest Produce in Central India*, Nehru Memorial Museum and Library 2005, pp. 31 – 40.
7. Ministry of Agriculture, *Agricultural Census 2000-2001*, pp. 8-9.
8. Ministry of Environment and Forests, *National Environment Policy 2006*, New Delhi (hereafter NEP), p.8.
9. Ministry of Environment and Forests, *National Forest Policy 1988*, New Delhi, see preamble.
10. NEP, p.12.
11. Ministry of Environment and Forests, *Environmental Impact Assessment Notification 2006*, New Delhi.
12. NEP, p.25.

13. Archana Prasad, 'Survival at Stake', *Frontline*, December 30, 2006–January 12, 2007.
14. NEP, 24.
15. Archana Prasad, 'Re-Engineering Environmental Impact Assessment' *People's Democracy*, September 24, 2006.
16. NEP, pp.5 – 6.
17. NEP, p.8.
18. Dinesh Abrol and Archana Prasad, 'Liberalising Environmental Regulation', *People's Democracy*, 30 July 2006.
19. NEP, p. 26.
20. NEP, p. 20.
21. NEP, p. 30.
22. NEP, p. 9.
23. Archana Prasad, 'Re-Engineering Environmental Impact Assessment', *People's Democracy*, September 24, 2006.
24. Archana Prasad, 'Forest Conservation and Tribal Development', *Social Scientist*, Decemeber 2005.
25. Archana Prasad, 'Re-Engineering Environmental Impact Assessment', *People's Democracy*, 30 July 2006.
26. Dinesh Abrol and Archana Prasad, 'Liberalising Environmental regulation'.
27. See, for example, Memorandum submitted to Ministry of Environment and Forests by Kalpavriksh Environmental Action Group on the Draft National Environment Policy, 2005.

Chapter 9

Legal Parameters of the National Environment Policy, 2006

Furqan Ahmad

I. Introduction

A policy provides broad guidelines for planners for governance in any particular sphere. It is an instrument of transformation which is expected to identify the problem, and provide a choice among alternatives on the basis of goals expressed to ensure effective implementation. In India, attention has been paid right from ancient times to the present age in the field of environmental protection and improvement. However, the approaches may have been different. Historically speaking, the laws relating to environment were simple and quite effective and the people were aware of the necessity of environmental protection. The present-day laws in India are the outcome of growing industrialisation and developmental projects due to population pressure.

In ancient India, protection of environment was the essence of Vedic culture. There was a philosophy of environmental management enshrined in many scriptures and *Smritis*. The environmental ethics of nature conservation was not confined to the common man; the rulers and kings were also bound by them. Similarly, in medieval India, environment conservation was given great importance. The establishment of British colonial rule brought about many changes in the religiously oriented indigenous system. Thus, the environmental policy during British rule was not directed at the conservation of nature but at the appropriation and exploitation of common resources, with the primary objective of earning revenue. Neither were there effective laws for the protection of environment. Further, these laws had a narrow scope and limited territorial reach. The present paper will briefly discuss the environmental policy in post-Independence India and analyse the laws made by the Ministry of Environment and Forest, Government of India in order to implement these policies. The paper will examine the present policy of 2006 and the legal measures suggested therein.[1]

II. Post-Independence Policy

As far as environment policy in independent India is concerned, the Indian Constitution did not deal with the subject of 'environment' till the 1976 amendment in 1972, when a national body was established to bring about greater coherence and coordination in environmental policies and programmes. In February 1972, a National Committee on Environmental Planning and Coordination (NCEPC) was established in the Department of Science and Technology. The NCEPC was an apex advisory body in all matters relating to environmental protection and improvement. The committee was to plan and coordinate, but the responsibility for execution remained with the various ministries and governmental agencies. Over time, the composition of the committee changed significantly and it became unwieldy, and decision making more complex.

The Fifth Five-Year Plan (1974 – 79) stressed that the NCEPC should be involved in all major industrial designs and a link and balance between developmental planning and environmental management be maintained. In this context, the minimum needs programme (covering rural education, health, nutrition, drinking water, etc.) received a fair priority, and was expected to minimise environmental pollution and degradation in rural areas. In the Sixth Five-Year Plan (1980 – 85), an entire chapter on 'Environment and Development' was included, which emphasised sound environmental and ecological principles in land-use, agriculture, forestry, mineral, extraction, energy production, etc. It provided environmental guidelines to be used by administrators and resource managers and laid down an institutional structure for environmental management by the Central and State Governments. The basic approach taken by the Seventh Plan (1985 – 90) was to emphasise sustainable development in harmony with the environment, as the federal government had recognised the negative effects that development programmes were having on the environment. The Plan called for the government and voluntary agencies to work together to create environmental awareness. The Seventh Plan recognised that the nation's planning for economic growth and social well-being in each sector must also work to secure improvement in environmental quality. The leaders of the country had realised that poverty and under-development, as opposed to development activities, had led to many of the country's environmental problems. The Eight- Five-Year Plan (1992 – 1997) gave important place to the environment by moving it to the fourth category of subjects examined in the text. The Plan stated:

> Systematic efforts have been made since the Sixth Plan period to integrate environmental considerations and imperatives in the planning process in all the key socio-economic sectors. As a result of sustained endeavour, planning in all major sectors like industry, science and technology, agriculture, energy and education include environmental considerations.[2]

The Ninth Plan (1997 – 2002), while re-emphasising 'Growth with Social Justice and Equity' focused on joint forest management and community forestry. The Tenth Plan (2002 – 2007) by and large endorsed the same.

In 1992, the Union Government adopted a 'National Conservation Strategy and Policy Statement on Environment and Development' (NCS). It adopted the policy of 'sustainable development' and declared the government's commitment to re-orient policies and action 'in unison with the environmental perspective'. Special sections in the NCS deal with the rehabilitation of persons ousted; the role of NGOs; and the special relationship between women and environment. Again, in 1992, the Union Government came out with 'Policy Statement for Abatement of Pollution'. This statement declares the objective of the government to integrate environmental consideration into decision making at all the levels. To achieve this goal, the statement adopted four fundamental guiding principles, viz. (i) Prevention of pollution at source; (ii) Adoption of the best available technology; (iii) 'Polluter pays' principle; and (iv) Public participation in decision making.

The policy statements, though unenforceable in a court of law, represented a broad political consensus and the duties of the government under the Directive Principles of State Policy contained in Part IV of the Constitution. In the *State of H.P. versus Ganesh Wood Product,*[3] the Supreme Court relied upon the National Forest Policy and the State Forest Policy of Himachal Pradesh to invalidate a decision taken by the State Industrial Project Authority. The court cautioned government departments against ignoring the forest policies and warned that disregard of these policies would imperil government decisions.

III. Legislation enacted from Stock Home till Date

Based on the aforementioned policy statements and commitments, various legislations were passed. The special laws on environmental protection enacted by the Central Government in India were:

Water (Prevention and Control of Pollution) Act, 1974. This provides for the establishment of pollution control boards at the Centre and States to act as watchdogs for prevention and control of pollution. Similarly the Air (Prevention and Control of Pollution) Act, 1981, aimed at checking air pollution via pollution control boards.

Environment (Protection) Act, 1986. This is a landmark legislation which provides for single focus in the country for protection of environment and aims at plugging the loopholes in existing legislation. It provides mainly for pollution control, with stringent penalties for violations. After the Bhopal tragedy the Factories Act, 1948 was also amended in 1987 in order to introduce the legal matters to control the pollution form 'Hazardous Process'.

Public Liability Insurance Act, 1991. This provides for mandatory insurance for the purpose of providing immediate relief to persons affected by accidents occurring while handling any hazardous substance. The National Environment Tribunals Act, 1995, was formulated in view of the fact that civil litigations take a long time (as happened in the Bhopal case). The Act provides for speedy disposal of environment-related cases through environment tribunals.

National Environment Appellate Authority Act, 1997. This provides for the establishment of National Environment Appellate Authority (NEAA) to hear appeals with respect to restriction in areas in which any industries, operations or process shall not be carried out or shall be carried out subject to certain safeguards under the Environment (Protection) Act, 1986.

In the area of forest and wildlife, the *Forest (Conservation) Act, 1980* aimed to check deforestation, diversion of forest land for non-forestry purposes, and to promote social forestry. Similarly, the Wildlife (Protection) Act, 1972, aimed at rational and modern wildlife management. It was amended in 2006. The Biological Diversity Act, 2002, is a major legislative intervention effected in the name of the communities supposed to be involved in the protection of biodiversity around them. The Act intends to facilitate access to genetic materials while protecting the traditional knowledge associated with them.

Critical Evaluation of Major Environmental Legislations

The law is a reflection of a policy statement. However, the examination of these legislations reveal that they were prepared in haste, did not effect objectives enumerated in policies, and had a number of lacunae. For this reason, the National Environment Tribunal Act, 1995 has not yet been enforced though its objectives have again been reflected in the National Environment Policy (NEP) 2006. We will discuss here the inherent flaws in the major legislations.

Environment (Protection) Act (EPA), 1986

The EPA is known as an umbrella legislation to provide a framework for the Central Government to coordinate the activities of various Central and State authorities established under previous laws, such as the Water Act and Air Act. It is also an 'enabling' law, which articulates the essential legislative policy of environment protection and change in emphasis from the narrow concept of pollution control to the wider aspect of environment protection. The exhaustive definition of environment, which includes water, air and land and the inter-relation which exists among and between water, air and land, other human creatures, plants, micro-organisms and property gave a very good impression. However, deeper scrutiny discloses that the legislation has several missing links. Only a wide net is drawn over the pollution laws not covered by other laws. All that the Act has done is to arm the Central Government with comprehensive control of environment pollution by industrial and related activities of man.

Section 24 of the EPA makes the penal provision redundant by limiting the scope and propose of the Act. This Section postulates that where an offence under this Act is also an offence under any other Act, the offender is to be punished only under the other Act. The Factories Act, the Wildlife Act, the Forest Act, the Insecticides Act and a host of other legislations covering almost every aspect of environment are available; thus, where an offender under the EPA is also an offender under any of such Act(s), he shall be punished only under such Act, thereby rendering the EPA to be meaningless. Though the Water Act and the Air Act do not give the common man a locus standi to initiate action but Section 19 of the Environment Protection Act gives a right to move the court though it requires that notice of not less than 60 days be given to the offender of alleged offence and intention to prosecute. The provision defeats the purpose through prior intimation.

The Supreme Court has viewed that the purpose of the EPA was only to create authority u/s 3(3) with adequate powers to control and protect the pollution.[4] In general, the courts have also interpreted other provisions together with subordinate legislation made thereunder to enforce abatement of pollution.[5] The law of environment is thus in a stage of evolution in India.

Factories Amendment Act, 1987

The extensive amendments in the Factories Act, 1948

mentioned above are intended to strengthen safety measures and are particularly related to pollution control from a hazardous process. However, the new legislation has also some flaws which sometimes may nullify its objective.

There are several welcome features in this amendment, such as Enquiry Committees. The Central Government now has power to appoint Enquiry Committees, to enquire into the standards of safety and health in the factory. However, it is disappointing that the recommendations of such committees shall be only advisory in nature. Section 41-G of the Factories Act envisages the setting up of a safety Committee by the occupier of every factory. Where a hazardous process takes place or a hazardous substance is used or handled, safety committees with workers participation will be formed in different factories. However, in Section 41-G it is stated:

> 'Provided that the state government may, by order in writing and for reasons to be recorded, exempt the occupier of any factory or class of factories from setting up of such committees'.

Thus the State Government has been given power to exempt any occupier of a factory or class of factories from setting up such a committee for reasons to be recorded. This exemption was criticised by many members in both the houses of Parliament while debating the legislation.

In addition to their participation in the Safety Committee, workers of a factory have right under Section 41-H of the Act to warn against the eminent dangers in the factory. This is a good provision. However, labourers can give warning to the Inspector of the locality, to the management and the like. It is important to take into account as to how that warning will be complied with and how that will be respected or ignored. Compliance should be time bound. It should be immediately attended to. The labour machinery, which is charged with the implementation of these labour laws in the States, is very inadequate. It does not have a sense of urgency. In many places, even if warning is given to them, they do not take it very seriously. Such warnings are given to them also. It is will help, promote or improve things if there is no time limit prescribed in this provision.

Section 41-E mentions emergency standards. The Government knows that the Factory Inspector is required to know whether some hazardous process is involved in the factory. If some hazardous process is involved, naturally certain special safety measures are to be provided for to meet the requirement of that hazardous process in manufacture.

How can one set of measures be categorised as safety measures for regulating process and another set of measures as emergency standards. Instead, there should be a fool proof arrangement to provide for adequate safety measures in all these units, in all those establishments where hazardous processes are involved, and nothing should be left for emergency standard.

The Factories Amendment Act provided a new Chapter 4-A, by which all efforts have been made take precautions against any future hazards. However, in the new chapter there is no provision for control or design of a unit. One of the causes of the gas leakage at Union Carbide Factory was its faulty design. The Union Carbide factory at West Virginia does not have the same design as in Bhopal. Computerised safety systems were installed at Virginia to prevent the accidents whereas no such control systems existed at Bhopal.

Another major lacuna is that the pollution of rivers is not covered here. Whenever there is an industry or factory, its effluents are spoiling the nearby rivers. The result of its flowing into the rivers is that the water in these rivers becomes unfit for drinking purposes. Apart from that, the creatures living in these rivers, like the fishes, die due to the poisons created by the industrial effluents. Therefore, there should have been a provision against the same also in the Amendment Act.

The most important thing is occupational disease. The government has not covered this problem. This disease a very damaging effect on the health and employability of workers employed. Those who work in the chemical industries are the first victims of diseases like cancer. That is why there occupational diseases should have also been covered through this amendment.

Public Liability Insurance Act, 1991

The legislation excludes public undertakings from its ambit. In an era where distinction between public and private sector industries is disappearing, it is anomalous that while the public sector industries cannot have immunity from liability to give relief in terms of section 3 (1), they are exempted from the insurance requirement. Government companies and statutory corporations have to set up adequate funds to meet payments under Section 3. However, there will be no justification for exemption to make to the Environment Relief Fund. Several members opposed this exemption in both the houses during the passage of the Bill.[6] Another shortcoming of the Act is that it makes it compulsory for owners to take out insurance for public liability but casts no such obligation on the insurance companies to insure such liability. The basic principle and terms of insurance should be spelt out in the Act itself rather than leaving them to the insurance companies, to avoid exploitation of owners and general public.

It may be noted that the concept of hazardous substance and handling needs to be widened for better protection of victims. Clause (c) of Section 2 defines 'handling 'as covering almost every conceivable operation which follows Section 2 (d) of Environment (Protection) Act. 1986. It is debatable whether, it covers personal injury or adverse effects on health or damage to property from the disposal of hazardous substances.

According to Section 2 (e) of the EPA, 'hazardous substance' means any substance or preparation which, by reason of its chemical or physio-chemical properties or handling, is liable to cause harm to human beings, other living creatures, plants, micro-organism, property or environment. The hazardous wastes (Management and Handling) Rules, 1989 set out 18 categories of waste with regulatory quantities. The Manufacture, Storage and Import of Hazardous Chemical Rules, 1989 contain a list of 434 Chemicals. A third set of rules issued on 5th December 1989 pertains to the manufacture, use, import, export and storage of hazardous micro-organism, etc. It is not quite clear whether all those who manufacture, process or keep any of these substances will be hit by the new enactment. On the face of it, the list notified is the only one that is binding on the owners of hazardous substances, within the meaning of the Act. The snag, however, is that besides 179 items classified in four groups, viz. toxic substances, highly reactive substances and explosive substances, highly reactive substances and explosive substances, the notification provides for flammable gases and liquids and highly flammable liquids in a separate group without specifying their names.[7]

The quantum of relief provided under this Act is inadequate to meet challenges of accident resulting from hazardous substances. A victim has to claim compensation first under this Act and later approach the forum for higher compensation. This multiplies the problems of victims. The provisions for maximum relief amount of medical treatment or disablement, are on the lines of workmen's compensation cases and evidently fixed in the light of minimum insurance coverage that a motor vehicle owner is required to take in respect of third-party properties. The relief provisions do not reflect the effect of observations of Supreme Court in *India Insurance Co. Ltd.* v. *Nirmala Devi*[8] repeated again in *Hardeo Kaur* v. *Rajasthan State Transport Corporation*[9], which said that:

> The determination of the quantum must be liberal not niggardly since the law value life and limb in the free country in generous scales.

The adjudicatory powers to award compensation have been vested in the Collector, who is a revenue official of government. The District Collector may not find enough time to spare for the settlement of claims and give immediate relief to victims. Besides, the assessment by such a person will not enthuse confidence in the mind of victims. Adjudication of claim by the Collector will not be conducive to the adminis-tration of the Act.

The Act aims at providing immediate relief to victims, but under Section 7(7) the Collector is required to dispose of an application applying for compensation as expeditiously as possible, within three months of its receipt. It may not call for inserting express conditions of ouster of civil court's jurisdiction and a bar on legal representation. Besides, the limitation period for entertaining an application for payment of compensation should take into account the diseases noticed after a long interval. To insure a claimant, a proximate nexus between the ailment and the accident must be established.

The National Environment Tribunal Act (NETA), 1995

With the emergence of environmental laws in the forefront, it was considered necessary that there should be a separate court to deal with such issues. The recent law on National Environmental Tribunal represents an effort to concentrate attention on environmental laws and to create a forum where legal issues relating to the environment may be considered in depth.

The Act had a three-fold impact on the law. In the first place, it has laid down a principle of liability for harm caused to the environment. Second, it has created a special tribunal for adjudication of litigation in which compensation is sought for such harm. Third, some of the provisions of the law are likely to have a deep impact on the Indian legal system considered desirable to highlight these three.

The NETA established strict liability for death, injury or damage sustained in accidents involving hazardous substance of everyone except the workmen of the factory, who are the first victims of any mishap. Thus the proposed enactment protects only those affected in the vicinity of the accidents, and not the workers on the spot. In such situations, the claimants affected in the vicinity might obtain better monetary compensation than the workers directly involved. It is not clear as to why a workman has been excluded. Should he not be given the choice to invoke either jurisdiction, under the Workmen Compensation Act, 1923 or under the Act, as the latter is a beneficial legislation? Workers should not be excluded as the Workmen's Compensation Act is inadequate and they would be able to get a higher compensation from the Environmental Tribunal.

There is no provision in the Act for setting up an independent technical panel to provide necessary scientific and technical assistance to the tribunal. This would indicate that the intention is to make the tribunal nothing more than claims court.

The composition of the tribunal too has been questioned. How can a vice chairperson be promoted as chairperson if he does not have required judicial background as provided under the Act. It is desirable that an independent ecological science research group should be set up so that it can assist the tribunal, rather than having experts from either side, counteracting each other and causing more confusion.

National Environment Appellate Authority Act, 1997

The Authority was established on April 9, 1997, after the enactment of the National Environment Appellate Authority Act. The most important and critical expectation was that the Authority, appointed under Section 3 (3) of the Environment (Protection) Act, 1986, would address the grievances regarding the process of environmental clearances and would implement the 'precautionary principle' and the 'polluter pays principle'.

The order in the *AP Pollution Control Board* v. *Prof M V Nayudu*[10] articulated the problems of earlier tribunals and the expectation from the National Environment Appellate authority thus:

> (T)he judicial and technical inputs in environmental appellate authorities/tribunals fell short of a combination of judicial and scientific needs. Things are not quite satisfactory and there is an urgent need to make appropriate amendments so as to ensure that at all times, the appellate authorities or tribunals consist of Judicial and also Technical personnel well versed in environmental laws. Such defects in the constitution of these bodies can certainly undermine the very purpose of those legislations. The Court opined that the Government of India should bring about appropriate amendments in the environmental statutes, Rules and Notification to ensure that in all environmental courts, Tribunals and appellate authorities there is always a Judge of

> the rank of a High Court Judge or a Supreme Court Judge—sitting or retired—and a scientist or a group of scientists of high ranking and experience so as to help a proper and fair adjudication of disputes relating to environment and pollution.

The creation of what was touted as an 'independent' authority was welcomed by a cross-section of people in the hope that there would be an improvement in environmental decision making, with transparency and accountability. However, today, over a dozen cases indicate that the Appellate is far from realising these objectives. This is mainly because the NEAA places an exaggerated emphasis on procedure, such as the time limit for filing petitions, and has a deep distrust of the intentions of NGOs, civil society groups and individuals who seek to draw the attention of the judicial process when the executive machinery, in this case the MoEF, fails to see the negative impact of a proposed project or its violation of environment clearance norms. Most of the cases have been dismissed on grounds of time delays without a single hearing given to substantive issues, such as the potential destruction to the environment or the negative impact on local people.

This emphasis on a time limit is downright unfair, as there is no requirement under the regulatory notification that mandates the publication of environmental clearance orders. In cases that have been filed and dismissed so far, on the grounds of time taken to file petitions, reasons such as the inability to access information on clearances, or the Authority being located in Delhi, or the correct procedure for filing a case have not been considered satisfactory reasons for the delay.

The second problem is the manner in which the locus standi of the appellant is determined. In several cases so far, the locus standi of the petitioner has come under microscopic examination while, in contrast, the merit of the issues raised by the petitioner has received much less time and attention.

According to the Appellate Authority Act, the Authority is to consist of a chairperson, a vice-chairperson and up to three other members. While the chairperson should have been a judge in the Supreme Court or the chief justice of a high court, the vice-chairperson must have been a secretary to the Government of India or any other equivalent post and should have expertise or experience in administrative, legal, managerial or technical aspects of problems relating to the environment. Members of the Authority require professional knowledge or practical experience in conservation, environmental management, law or planning, and development.

The Authority, when it was first established, consisted of: Justice N G Venkatachala, chairperson (former judge of the Supreme Court), Nirmala Buch, vice-chairperson (former secretary to the Government of India) and Mohinder Singh, (former principal secretary to the state government of Uttar Pradesh) and Ejaz A Malik, (former principal chief conservator of forests, Jammu and Kashmir) members. However, since 2002, the Authority has not had a chairperson or members with technical expertise in environment-related fields, thereby negating the possibility of arriving at decisions that draw on scientific reasoning for or against a project based on its potential impact. Even prior to that, the Authority's decisions gave greater importance to economic imperatives than environmental and social concerns regarding projects.

A closer look at the independence of the Authority leads to the realisation that the MoEF, whose decisions the Authority is supposed to scrutinise, is in fact the agency that regulates the functioning of the Authority. The MoEF has the power to make rules for carrying out the provisions of this Act, appoint the chairperson and members of the Authority and regulate the procedure for the investigation of misbehaviour or incapacity of the chairperson, vice-chairperson or a member. Unless the selection committee comprises non-MoEF members, it is only to be expected that the MoEF will constitute the Authority such that it accepts clearances granted by the MoEF and quashes all cases challenging these clearances. The following words of Lord Woolfe, quoted in the M V Nayadu[11] case, seem prophetic:

> Critics have objected that Judges cannot make appropriate decisions because they lack technical training, that the Jurors do not comprehend the complexity of the evidence they are supposed to analyse, and that the expert witnesses on whom the system relies are mercenaries whose biased testimony frequently produces erroneous and inconsistent determinations. If these claims go unanswered, or are not dealt with, confidence in the Judiciary will be undermined as the public becomes convinced that the Courts as now constituted are incapable of correctly resolving some of the more pressing legal issues of our day.

Environmental Impact Assessment (EIA)

Environmental Impact Assessment (EIA) involves evaluation of environmental implications and in corporation of necessary safeguards against environmentally harmful activities. Environmental Impact Assessment is essential so as to ensure that plans for development in all sectors are in harmony with the objective of main training the health of life sustaining ecosystems and other environmental resources. The National Committee of Environment planning and coordination and Department of Environment later Central Integrated Department of Environment, Forest and Wildlife of the Ministry of Environment have introduced environmental approval of projects in selected sectors such as industry, multipurpose river and mining. The project authorities are required to incorporate a chapter on environmental aspects in their feasibility reports.

The Ministry has issued a notification[12] on 27th January, 1994 making Environmental Impact Assessment statutory for 29 different activities in industries, mining, irrigation power, transport, etc. Environmental Impact Assessment reports are to be mandatory for 29 projects including industries, mining, power, transports, tourism, etc. as per the notification wherein procedure for obtaining environmental clearance, constitution of an experts committee; procedure for public hearing and a time schedule for taking a decision is provided.

The notification deals with the detailed procedure for obtaining environmental clearance constitution of an experts committee, procedure, public hearing and time schedule action. However, in a subsequent notification of dated 4th May, 1994[13] the Ministry decided that such participation and EIA can be dispensed with, where the public interest so demands. The latter notification curiously takes back in the name of public interest what former notification had granted.

On April 10, 1997[14] the Ministry of Environment issued two notifications which are believed to be the last ditch attempt to make the environment clearance procedure industry-friendly. The notifications transfer substantial discretionary powers to the state authorities, contrary to the advice of environmental experts. The power to sanction certain categories of power plants will be accorded to the state authorities even while making public hearings at the site compulsory before a final go-ahead is sanctioned. The public hearing would be carried by the State Pollution

Control Board. The State Governments can clear thermal power plants in these categories; co-generation plants, captive thermal power plants up to 500 MW capacity, coal based plants up to 500 capacity using fluidised best technology; and conventional coal-based power plants up-to 250 MW and gas are naphtha-based plants up to 500 MW capacity.[15] The second notification which makes public hearing mandatory before any project is granted environment clearance in 29 specified industrial sectors.[16] This notification set aside the earlier notification of May 1994 and restored the position of earlier notification of January 1994. However, delegation of such powers to state agencies which are severely ill equipped and lack the expertise and resources to do proper assessment is not advisable.

Besides, there is no role of independent experts with the hearing being conducted by the Board Personnel, District Collector, Local Bodies and senior citizens. However, the public interest privilege contained in the Indian Environment Assessment Notification is so wide that the Ministry of Environment and Forest (MoEF) bureaucracy may be inclined, more often than not, to use it as a tool to block any information.

The 1997 amendment seems to have cleared the dead-wood and moved towards realisation of public participation. The amendment notification makes it mandatory to have a public hearing before environmental clearance is given. It is specially provided that the finding of the Impact Assessment Authority should be based, inter alia, on the details of public enquiry. Procedure for hearing is laid down in detail. The project proponent has to submit 20 sets of documents to the State Pollution Control Board. The board has to give notice of hearing in two newspapers of wide circulation in the locality, one of which should be in the vernacular language. The notice should mention the date, time and place of public hearing. Suggestions, views, comments and objections of the public are invited within 30 days from the date of State Pollution Control Board, the State Government and the local authorities and not more than three senior citizens will hear representations.[17] Bonafide residents, environmental groups and people affected by the project of displacement can participate in the public hearing.

The provisions for public hearing now incorporated into the environment assessment process, undoubtedly, is a step ahead. The procedures laid down in the notification are to be strictly followed. However, the required procedures are only in respect of those projects noted in the schedule to the main notification. It is true that no such scheduled projects can escape the mandatory requirement of public hearing. But what about other ventures which do not come under the category and at the same time have the potential for environmental damage? Once it is found in a threshold inquiry that an unscheduled project may have significant environmental effects, public hearing must be insisted upon before a green signal is given to the project.

Recently, the Ministry of Environment has further amended the EIA notification on September 14, 2006[18]. This notification dilutes the EIA as originally notified and rectified through amendment of 1997. Comments were sent by several groups and organisations related to environment, but the final draft notification appeared without any substantial change. The major flaws in the present legislations are:

Several points of contention regarding the changes that were proposed to the EIA notification 1994 in the draft notification of 2005 remain unaddressed in the new notification. As all the repeated communication to the Ministry on the problems with the existing and proposed notifications and appeals to the Ministry to consult groups and communities (who have worked on and have been impacted by the decision making on large development projects) were ignored or rejected.

The categorisation of projects in the notification, into A and B, has been done based on 'spatial extent of potential impacts on human health and natural and man made resources'. Category A projects are to be clearance by the MoEF while Category B projects are to be cleared by the State Environment Impact Assessment Authority (SEIAA).

Handing over of the responsibility of granting clearance to a large number of projects to the state governments without any system of checks and counter-checks is not acceptable. In many instances, the state government is directly involved in seeking investments. Handing over the entire function of environment regulation into their hands means that projects will be cleared indiscriminately.

The SEIAA is a body created to grant clearance at the state level. Where will this authority be housed and who will it be accountable to? Can the decisions of the Authority be challenged in the existing Environment Appellate Authority or will it be some other body? These are not known at all. Unless this is figured out and incorporated in the notification, this body cannot be allowed to grant clearances.

Construction projects also need not go through the stages of screening or scoping because these are exempted from conducting EIA studies. They also do not need to conduct the public consultation process. So they are present in the EIA notification only in so far as having to be cleared by the SEIAA on the basis of the application form. Thus this remains a category in the notification purely for cosmetic reasons.

Several large-capacity projects have been left out of the notification altogether. All building and construction projects with less than 20,000 sq.mtr built-up area, like the Vasant Kunj Square Mall in Delhi are now exempted from the notification. (According to the June 2006 Rapid EIA report, the total built-up area is 19,021.108 sq mtr.). There are several such complexes being constructed in cities and towns today which will be totally exempted from the EIA notification.

Will thermal power projects less than 500 MW or cement plants less than 1 MTPA not require any environment clearance at all? Or, will State governments follow a separate set of rules for grant of clearance to these projects since the EIA notification does not deal with them? If indeed it is the former that is true, then this notification will in no way achieve environment impact regulation.

The new notification deals only with process of grant of environment clearance (divided into 4 stages: Screening, Scoping, Public consultation and Appraisal). And it stops there. The most critical issue of monitoring and compliance, which is an integral part of the Environment Clearance regime is dealt with in precisely three sentences. There is only a mention of the six-monthly compliance reports which are to be submitted by the project proponent. The EIA notification 1994 mandated the MoEF to maintain its independent monitoring report. This role of the MoEF finds no mention whatsoever in the new notification. This could mean several things. One, that the MoEF does not see the need to independently monitor the projects that it has cleared and that its function ends with granting clearance; two, that the project proponents will monitor themselves adequately.

The notification is also silent on the point of who would be the monitoring agency for projects cleared by the state government. Will it be the SEIAA or will these projects be self-monitored? It is absurd if the latter is what is expected to take place. There is no role for local community groups to be involved in the monitoring of projects. There are six sets of activities which have been exempted from the process of public consultation completely. There is no explanation whatsoever as to why these projects have been exempted from this extremely important step of the environment clearance process. Since this is a step to ascertain 'the concerns of locally affected persons and others', their exemption means that the Ministry is not interested in ascertaining the concerns of locally affected persons and others while clearing these projects. The Public Consultation process, as laid out in the EIA notification, 2006 is severely flawed and clearly limits public participation. Only a draft EIA report will be available to the locally affected persons at the time of the public hearing. Citizens will now not get to see the final EIA document on the basis of which the decision on the project will be made. There are enough examples in the last 12 years of the existence of the EIA notification, when project proponents have sought clearance on incomplete, and misleading data. The Ministry has not only failed to take punitive action against erring agencies but gone ahead and cleared projects based on these reports. This practice will only grow if the final EIA report is not opened to public scrutiny.

Further, the public will have no control over whether or not their inputs and concerns get incorporated in the EIA report and influence the decision-making process. The time period for which the draft EIA report will be available prior to the hearing is not mentioned in the notification. The 1994 notification mandated that it be available for a period of 30 days prior to the hearing.

The EIA notification, 2006 states that the EAC or SEAC will convey the terms of reference within 60 days of the receipt of Form 1. While the notification clearly lays down guidelines on how long it should take for each of the four stages to be completed for grant of environment clearance, there is no mention or record of how much minimum time must be spent on putting together a comprehensive EIA report.

IV. The Recent Policy of 2006 vis-à-vis Legislative Commitments: A Critique

The preceding analysis of major environmental legislations reveals that they are not framed in such a manner that they can be effectively implemented and the National Environment Policy 2006 again lays stress on these laws besides certain other laws and modifications without any detail. Various legal experts had pointed out the lacunae of these laws but their recommendations were ignored while passing the legislations. The past experience of relationship between policy and law is not very pleasant. The new policy, despite tall claims, has not addressed this problem. The EIA notification 2006, which was notified after the policy statement disappointed the environmental concerns. It may be humbly submitted that the Policy 2006 will be commented has on only on legal lines.[19] The major comments are:

According to the NEP, the regulatory mechanisms that are to be put in place to safeguard the environment are to be premised upon the Govindarajan Committee Report on Reforming Investment Approval and Implementation Procedures 2002. Thus, ignoring delays caused by widespread non-compliance with mandatory requirements under the Environment Protection Act or Forest Conservation Act, NEP condones the view that delays in environmental clearances are an impediment to the 'development' process by endorsing the Govindarajan Committee Report. The policy documents fails to make a clear commitment towards the making of EIA statements open and public. The relevant portion of policy that deals with Environment Impact Assessment (EIA) and the Coastal Regulation Zone (CRZ), and subsequent legislations based on the NEP, do not effectively protect the environment (the details of new EIA notification September 2006 may be perused).

The policy states that, it will follow the Public Trust Doctrine where the state is "not an absolute owner, but merely a trustee of all natural resources, which are by nature meant for public use and enjoyment, subject to reasonable conditions, necessary to protect the interest of a large number of people.' However, this assertion of the states not being an absolute owner may have been welcome, but for the fact that it does not limit the state power when it comes to the privatisation of natural resources. If the state is not an absolute owner, can it privatise something it does not own? But the policy is silent on how the state's power to denotify these areas for commercial or large developmental projects will also be delimited through this review of legislation.

The overall thrust of the NEP remains the same. It recognises the 'dichotomous relationship' between economic growth and environmental degradation, but justifies economics as a means to resolve the two as it can allocate resources for environmental investments. It clearly states that multilateral regimes and programmes responding to global environmental issues will not be at the cost of development opportunities in developing countries. Cliched as it may sound, the current development paradigm is being blindly promoted without a critical questioning of who the beneficiaries of such development are, and who stands to lose from this chosen developmental path.

The NEP seems to align itself with 'sustainable development', more out of an obligation to keep pace with the rest of the world without displaying the will to actually understand what it is all about. In subscribing to sustainable development by wanting to achieve a 'balance and harmony between the economic, social and environmental needs of the country', the NEP 2006 believes that it is making a reasonable compromise, when it is, in reality, a sacrifice of long-term interests to short-term goals.

The General Agreement on Tarrifs and Trade (GATT) allows nations to restrict the import of products from other members nations, so long as these restrictions do not discriminate between foreign and domestic products. For instance, under GATT the US may ban the importation of a dangerous pesticide so long as the use of the pesticide is also banned in the US. Article 20 lists the exceptions that justify a deviation from GATT's general free trade requirements. Among these exceptions are trade restrictions 'necessary to protect human, animal or plant life and health, and those' relating to the conservation of exhaustible natural resources. Most of the controversy regarding the agreement's impact on environmental protection has centred on the WTO dispute panels' interpretations of the Article 20 exceptions. GATT is one of the most powerful international agreements in the world and often supersedes environmental treaties. Why is this? Is it because GATT, unlike most environmental treaties, has an institutional dispute resolution mechanism and enforcement procedures? Or is it because the global community puts greater emphasis on commodities and trade than on the environment and conservation? The present policy has not given any adequate reply to these questions.

The MoEF at present seems set on re-engineering the country's environment clearance procedures and drafting a national policy for the environment, which would benefit investors more than any other constituency. The link between the MoEF and the NEAA seems designed to hand over the natural environment to those in the business of 'developing' dams, mines, industries, roads, ports. In fact, the bias towards developers is so strong that the best practices guidelines issued by the MoEF last year stated that no project was to be rejected only on procedural grounds. Compare this with the NEAA, where most cases have been dismissed *only* on procedural grounds!

V. Conclusion and Suggestions

Overall, the policy statement has an extremely narrow approach and does not reflect as to how Government would treat environmental legal issues in the future. The environment-related legislations are more than enough in India but the issues are still unresolved. Whether the NEP 2006 would ensure passing of good laws can only be stated, once these legislations are passed by Parliament and implemented in due course of time. The previous policies issued by the Government and accordingly the legislations enacted by Parliament were not coherent and could not meet the aspirations of the environmentalists or public at large. Some of them have not even been implemented till date. This time the Ministry spokesman claims, 'So far there were only a set of policies, but no comprehensive policy. The old approach saw the environment, as an end in itself, but now we have linked it to sustainable development.' To him this is a complete restructuring of the whole policy.

Many comments had been appeared by the Parliamentary Standing Committee as well as environmental experts after enactment of the previous legislations but they have not been carried out so far. How the Ministry revised these legislations, only the Ministry knows, as it did not react to removing the shortcomings of the previous legislations. As per the Ministry of Environment, this is a comprehensive policy in comparison to previous ones but, it is doubtful whether would serve the purpose of making legislation solve environment problems.

How was this policy approved without any substantive change in its draft notified in 2004? We need not comment on the whole process, like Ashish Kothari and other environmentalists.[20] The way MoEF has been behaving in the last two decades is really surprising. The **civil liability, polluter pays principle, precautionary principle, public trust doctrine, sustainable development** and other **international environmental principles** have now becom part of Indian legal system which are again reiterated in this policy. We have ample provisions to protect and improve the environment in existing legislations. But, we face the problem of their implementation, which is acknowledged by the implementing authorities too. To implement one legislation, there are many agencies involved and we find no coordination among them. This creates the problem of enforcement of laws. Coordination amongst various implementing agencies is a must, otherwise the tall claims of revision of the laws and enactment of new laws, given in the new policy, will remain on paper. No clear-cut guideline has been issued in this regard.

To wind up, it is astonishing to state that in the same year different policies have appeared on identical issues. For example, besides the policy under comment, the **National Tribal Development Policy 2006** (draft) as well as the **National Rehabilitation Policy 2006** (draft) have been notified by the concerned ministries. There is also a separate Forest Policy. All these issues are related to the protection of environment one way or the other. Then why does the Ministry of Environment and Forest, which is a nodal ministry, with the consultation of other ministries, not notify a consolidated policy statement? This would eliminate the overlapping and confusion in drafting legislations as well as their implementation. To ensure proper implementation of environmental laws, the policy should cover the following submissions:

(i) The right to a clean environment under Article 21 of the Constitution should be drafted in such a manner that it does not remain paperwork, in view of enormous pressure of development but at the same time it cannot be made absolute so as to bring the country's progress to a standstill. Thus, there must be a balance between right and duty (i.e. duty to protect and improve environment).

(ii) The definition of pollution of environment is very comprehensive but the general exceptions, having wider scope, have almost been given a licence to pollute the environment. It becomes necessary to seriously examine how far the exceptions have or have not tried to protect or improve the environment.

(iii) There are multiple controlling agencies under the general and environmental laws. They are at the Central, State, regional and municipal levels. First, all those authorities must work in coordination with each other, instead of creating watertight compartments. Second, the controlling agencies have to be more responsible in their duty and the negligent and inactive officers must be punished. And last, a progress report of the agencies must attract the attention of the people and in particular the members of the legislature.

(iv) The law must envisage establishment of a new set of environment courts or tribunals. The long delays and procedural technicalities have greatly hampered environmental justice. Environmental law, to be effective, must have simple procedure aiming at speedy justice.

(v) Environmental law is beneficial to the people but restricted to those who are involved in exploiting the nature. However, it has an intimate impact on each and every member of the society. Further, environmental law deals with matters with which the law-makers may not be fully conversant. In these circumstances such exercise must attract larger participation of the people inside and outside the legislature.

(vi) The environment must not be considered as just another section of national development. It should form a crucial guiding dimension for plans and programmes in each sector. This becomes clear only if the concern for environmental protection is understood in the proper context.

(vii) The governmental programmes seem to be based on the belief that concern for the environment essentially means protecting and conserving it, partly from development programmes but mainly from the people themselves. The Rajasthan Canal is a fine example of a government programme that has transformed extensive grazing lands into agricultural lands. No effort was made by the government to ensure that the *nomads,* who used these grazing lands earlier, would benefit from the canal on a priority basis. The efforts should be made to modify the development process itself in a manner that will bring it in greater harmony with the needs of the people and with the need to maintain ecological balance, while increasing the productivity of land, water and forest resources.

Endnotes

1. See for details S. Bhatt, *Environment Protection and Sustainable Development*, pp. 21–33, A.P. Publishers, New Delhi, 2004.
2. *Ibid.* at 27.
3. AIR 1996, SC 149.
4. *Vellora Citizens' Welfare Forum* v. *Union of India*, AIR 1996 SC 2715.
5. See for example, *S. Jagannatha* v. *Union of India and others* (1997) 2 SCC 87.
6. For example see the speech of Manoranjan Bhakt, Shri P. Chidambram, Shri Bal Gopal Misra in the Lok Sabha and Shri Dinesh Bhai Trivedi and others in the Rajya Sabha. Mr Chidambram spoke as under:

   ... Now you take the steel companies; SAIL is a major steel company. BOKARO; Raourkela and are Bhillai can set up a fund and then apply to the Central Government for exemption. Why should TATAs not do the same thing and say that they are setting up a fund? I think this just a kind of bureaucratic insidious interpretation which can defeat an Act. Most establishments of this country belong to the Central Government are in the Public Sector. (See *Rajyasabha Debates*, January 9, 1991).
7. The same confusion is reflected by a decision of single judge of Allahabad High Court (See *U.P. Electricity Board* v. *District Majistrate, Dehradun*, AIR 1998 Allahabad 1.
8. (1980) ACJ 55.
9. JT 1992 (2) SC 409.
10. AIR 1999 SC 812.
11. *Ibid.*
12. S.O. 60 (E) 27.1.1994 II 3 (i) Extra *Sl* 42.
13. S.O. 356 (E) 4.5.1994 II 3 (i) Extra *Sl* 225.
14. S.O. 318(E) dated 10 April 1997, The Gazette of India, Part II, Sub section (ii) 10 April 1997, Amendments to para 2 of the Main Notification [SO 60(E) dated 7 January 1994] were made. Schedule IV was added, it lays down the procedures for public inquiry.
15. S.O. 318 (E), 10.4.1997 II 3 (ii) Extra *Sl* 244.

16. S.O. 318 (E), 10.4.1997 II 3 (ii) Extra *Sl* 244.

17. *Ibid*. The panel has representatives from the State Pollution Control Board, the District Collector or his nominee, a representative of the State Government dealing with the subject, a representatives of the Department of State Government dealing with the environment, not more than three representatives of the local bodies and not more than three senior citizens nominated by the District Collector.

18. Gazette of India, Extraordinary, Part-II, and Section 3, Sub-section (ii) September 14, 2006.

19. See for details, Ministry of Environment & Forest, Government of India, National Environment Policy (2006), pp. 11–22.

20. See for details, *Down to Earth* (August 2006).

Chapter 10

Regulatory Regime Governing Biodiversity in India: A Critical Assessment

Ketan Mukhija and Yugank Goel

Introduction

Biological diversity or biodiversity is the term given to the variety of life on Earth and the natural patterns it forms. The term 'biological diversity' was coined by Thomas Lovejoy in 1980, while the word 'biodiversity' itself was coined by the entomologist E.O. Wilson in 1986, in a report for the first American Forum on biological diversity organised by the National Research Council (NRC).[1] Thus the term biodiversity may encompass both genetic diversity[2] and species diversity[3]. India is one of the twelve mega-biodiversity centres in the world.[4] Its living forms represent two of the major realms and three basic biomes of the world and its area can be divided into 10 bio-geographic regions: Trans-Himalayan, Himalayan, Indian Desert, Semi-Arid, Western Ghats, Deccan Peninsula, Gangetic Plains, North-East India, Islands and Coasts.[5] This paper attempts to study the legislative efforts undertaken in India to preserve and nurture its biodiversity.

Why is biodiversity protection an urgent need?

During the previous century, erosion of biodiversity has been increasingly observed and though estimates of extinction rates are controversial, ranging from very low to upwards of 200 species a day, scientists generally acknowledge that the rate of species loss is greater now than at any time in human history, with extinctions occurring at rates hundreds of times higher than background extinction rates.[6] Such estimates indicate that human activities are from an environmental perspective, unsustainable ecological practices and resulting in destruction of habitats and ecologies.

Coupled with such environmental concerns are the additional anxieties of developing countries. The Indian Government for example has raised and fought politico-litigative battles over the patentability of products drawn from the traditional knowledge[7] (that vary according to biodiversities of the local environments) of local populations in developing countries. Basmati, neem and turmeric are recent illustrations of the trend.[8] The phenomenon of protection being sought by predominantly multinational companies based in countries of the industrialised north for products that are based on the traditional knowledge of local populations suffers the rather dubious title—'bio-piracy'.

Biodiversity Management: International efforts to preserve biodiversity

The conservation of biological diversity has become a global concern. Although not everybody agrees on the extent and significance of current extinction, most consider biodiversity essential. There are basically two main types of conservation options, insitu[9] and exsitu[10] conservation. An example of an insitu conservation effort is the settingup of protection areas. An example of an exsitu conservation effort, by contrast, would be planting germplasts in seed banks.

At the international level, the United Nations Convention on Biological Diversity (CBD) which came into force on 29th December 2003 was a landmark step taken to reaffirm the commitment of nation-states to protection and sustainable use of their biological resources. The CBD, though only constitutive of 'soft law' in the international law framework, was the first international legal instrument for promoting conservation and sustainable use of biodiversity. On achieving this object, the CBD factored in many and often divisive concerns of nation-states regarding implementation of its principles. Primary among those divisive concerns was the concept of 'fair and equitable sharing of the benefits arising from utilisation of biological resources'. The said concept finds incorporation on account of 'the need to share cost and benefit between developed and developing countries and the ways and means to support innovation by local people'. Fair and equitable benefit sharing is a controversial concern for many reasons. Among those reasons are entwined the issues of IPR protection and biopiracy.

Local efforts at preserving biodiversity: An outline of the Biological Diversity Act, 2002 (BDA)

Among the signatories to the CBD, India is one of the first few countries to have enacted an appropriate comprehensive legislation to achieve the objectives of the convention. To fulfill its obligations under the Convention of Biological Diversity, conserve biological resources, enable their sustainable use, provide for equitable benefit sharing and check biopiracy, India enacted the Biological Diversity Act, 2002. The main intent of the legislation appears to be the protection of India's rich biodiversity and associated knowledge against exploitation by foreign individuals and organisation without sharing the benefits arising out of such use, and to check biopiracy.

The Act establishes various authorities, such as the National Biodiversity Authority, a State Biodiversity Board (SBB) and Biodiversity Management Committee (BMC) to operate at the national, state and local levels respectively. The legislation further provides for the research, commercial activities and bio-survey of biological resources or knowledge associated thereto, by non-Indians and bodies with foreign equity, subject to prior approval of the NBA. Transfer of the results of any research to such persons, also subject to prior approval of the NBA, who may impose benefit-sharing fees or royalties or both. Contravention of this provision is an offence, punishable with a fine up to $21,000 or imprisonment or both. The NBA and SBB are empowered to initiate actions for such contravention. There is also a requirement of Indians or Indian bodies to inform the SBB before undertaking the commercial utilisation and bio-survey of biological resources occurring in India. However, exemption from the above rule is provided to practitioners of indigenous medicines. Collaborative research projects between an Indian and a non-Indian, including between institutions are again subject to policy guidelines and approval of Government. Acquisition of intellectual property rights in or outside India for inventions based on research or biological resources obtained from India is again subject to the prior approval of the NBA. The NBA has also been authorised to impose benefit-sharing fees or royalties on patents obtained.[11]

2. Summary of the Principles recognised in CBD

The Convention on Biological Diversity has primarily three objectives, viz. the conservation of biological diversity, the sustainable use of its components and fair and equitable sharing of benefits arising out of the utilisation of genetic resources.[12] To elaborate, the objectives of the Convention are 'the conservation of biological diversity, the sustainable use of its components and the fair and equitable sharing of the benefits arising out of the utiliastion of genetic resources, including by appropriate access to genetic resources and by appropriate transfer of relevant technologies, taking into account all rights over those resources and to technologies, and by appropriate funding.'[13] The Convention seeks to promote the conservation of biological diversity by providing for national monitoring of biological diversity, the development of national strategies, plans and programs for conserving biological diversity, national in situ and ex situ conservation measures, environmental impact assessments of projects for adverse effects on biological diversity, and national reports from parties on measures taken to implement the convention and the effectiveness of these measures.[14] The Convention also contains a provision which gives the relatively new Global Environment Fund (GEF)[15] the responsibility for financing these national and international measures, at least until the parties to the Convention can decide on another financing organ.

In recognition of the fact that developing nations are not only burdened with the responsibility of protecting the environment but also economic development and specifically, the eradication of poverty, the Convention adopts an approach that is capable of balancing these interests. Implicit in the Convention is the notion that developing countries will undertake conservation measures more enthusiastically where economic benefits inure to them as a result. Thus, there emerges an effort to give nations a right to share in the profits of products made using the genetic materials native to their territories. In furtherance of this policy, the Convention vests in developing nations the right to exclude nationals of foreign countries from access to biological organisms found in their territory.

The CBD, through its 23 preambular paragraphs and 42 Articles, establishes commitments on conservation and sovereign rights of nation-states over their biological resources and adopts an approach which allows for access and transfer of genetic resources with prior informed consent. On a principle of benefit sharing,[16] the CBD establishes the foundation to link conservation and utilisation with sustainable development.[17] The Convention stresses on both insitu[18] and exsitu conservation[19] and the recognises the need for promoting local, regional and global cooperation.[20] The Convention further calls upon the developed countries to provide 'new and additional financial resources to enable developing countries to meet the agreed full incremental costs'. It further recognises that developing countries can only evolve conservation objectives which factor in concerns of economic and social development and eradication of poverty (Article 20). The Convention also acknowledges that there is a need to 'take all practicable measures to promote and advance priority access (regarding Handling of Biotechnology and Distribution of its benefits),[21] on a fair and equitable basis by Contracting Parties, especially developing countries, to the results and benefits arising from biotechnology based upon genetic resources provided by these countries'.[22]

3. Scheme and summary of the BDA and the Rules

Introduction

The Biological Diversity Act ('BDA' or 'the Act'), though passed by the Indian Parliament in 2002, received Presidential assent on 5th February 2003. The Act consists of XII chapters and 65 sections. The seven preambular paragraphs of the BDA make it clear that the intendment behind enactment is domestic implementation of the CBD to which India is a signatory. The BDA in acknowledgement of India's rich biodiversity and associated traditional and contemporary knowledge systems[23] mentions that the BDA is intended to ensure conservation of India's biodiversity while allowing sustainable use of its components and a fair and equitable sharing of the benefits arising out of the use of biological resources and knowledge.[24]

The BDA is primarily regulatory in nature and provides for an elaborate chain of governmental command over the management and protection of biodiversity. The conservation strategy in place therefore falls within the paradigm of the 'command and control' framework, which envisages protection of the environment through a three-fold strategy: (i) identification of a type of environmentally harmful activity; (ii) imposition of specific conditions or standards on that activity; and (iii) prohibition of forms of the activity that fail to comply with the imposed conditions or standards.[25] Towards the end of this chapter, it will be shown how these three criteria are fulfilled in the example of the BDA.

Authorities established by the BDA

The agencies figuring in the hierarchy of command under the Act may be broadly listed as the National Biodiversity Authority, State Biodiversity Boards who function respectively, under the directions and control of the Central and State Governments respectively. The Act also provides for the establishment of Biodiversity Management Committees which are to play an advisory role in the conservation strategies adopted by the above authorities.[26] The National Biodiversity Authority ('NBA') appears to be the principal agency in which is vested the primary responsibility to ensure that the mandates of the BDA are carried out. The Act provides for the establishment of the NBA as a 'body corporate' with all the arising implications.[27] Chapter II (Ss. 8 – 17) of the Act deals with the establishment of the NBA and outlines procedural matters, such as composition of members. Chapter VII (Ss. 26 – 30) deals with the finance, account and audit matters of the NBA while also establishing the National Biodiversity Fund which is to be utilised for purposes of channelling benefits to the benefit claimers,[28] conservation and promotion of biological resources and development of areas from where such biological resources or knowledge associated thereto has been accessed and socio-economic development of such areas. Chapter IV (S.18) deals with the broad functions of the NBA which may be summarised as follows:

- To regulate activities referred to in Ss.3, 4, 6 of the Act (discussed later).
- Issue guidelines by way of regulations for access to biological resources and for fair and equitable benefit sharing.[29]
- Advise the Central Government on matters relating to the conservation of biodiversity, sustainable use of its components and equitable sharing of benefits arising out of the utilisation of biological resources.
- Advise the State Governments in the selection of areas of biodiversity importance to be notified[30] as 'heritage sites' and measures for the management of such heritage sites.
- Act on behalf of the Central Government in taking any measures necessary to oppose the grant of intellectual property rights in any country outside India on any biological resource obtained from India or knowledge associated with such biological resource which is derived from India.
- Perform all other functions as may be necessary to carry out the provisions of the Act.

The Act also envisages the setting up of a State Biodiversity Boards ('SBB') for both states and union territories. Section 22 under Chapter VI deals with the establishment of SBBs as 'body corporates'[31] and related procedural matters such as composition. Section 32 under

Chapter VIII deals with the setting up of State Biodiversity Funds, which are meant to be used for purposes of the management and conservation of heritage sites, compensating or rehabilitating any section of the people economically affected by notification,[32] conservation and promotion of biological resources, socio-economic development of areas from where such biological resources or knowledge associated thereto has been accessed, subject to any order made under Section 24 and the meeting the expenses incurred for the purposes authorised by the Act. The functions of the State Biodiversity Board have been again outlined in Chapter VI under Section 23 as follows:

- Advise the State Government (subject to any guidelines issued by the Central Government) on matters relating to the conservation of biodiversity, sustainable use of its components and equitable sharing of the benefits arising out of the utilisation of biological resources.
- Regulate by granting of approvals or otherwise, requests for commercial utilisation or bio-survey and bio-utilisation of any biological resource by Indians.
- Perform such other functions as may be necessary to carry out the provisions of this Act or as may be prescribed by the State Government.

Chapter X of the Act provides that every local body[33] must constitute a Biodiversity Management Committee ('BMC') within its area for the purpose of promoting conservation, sustainable use and documentation of biological diversity including preservation of habitats, conservation of land races,[34] folk varieties[35] and cultivars,[36] domesticated stocks and breeds of animals and micro-organisms and chronicling of knowledge relating to biological diversity.[37] The BDA provides that the NBA and the SBBs must consult the BMCs while taking any decision relating to the use of biological resources and knowledge associated with such resources occurring within the territorial jurisdiction of the BMC.[38] The Act has authorised the BMCs to levy charges by way of collection of fees from any person for accessing or collecting any biological resource for commercial purposes from areas falling within its territorial jurisdiction.[39]

Chapter IX of the BDA deals with the responsibilities of the Central Government. Broadly, as outlined in Section 36(1), the role of the Central Government is envisaged to be that of a policy maker that is responsible for developing national strategies and plans for conservation of biological diversity. It has been further provided that whenever the Central Government has reason to believe that any area rich in biodiversity is being threatened by overuse etc., it must issue directions to the concerned state Government to immediately take ameliorative measures and further, offer technical and other assistance to the said state Government.[40] The Central Government is under further obligation to integrate the conservation of biodiversity with relevant sectoral programmes wherever practicable.[41] The Act obliges the Central Government to carry out an environmental impact assessment of projects that are likely to have adverse effects on biodiversity and allow for public participation in such assessments wherever possible.[42] Management of the risks associated with the use and release of living modified organisms resulting from biotechnology that is likely to have adverse impact on the conservation of biological diversity and human health is another obligation thrust upon the Central Government under the Act.[43] The BDA also mandates that it shall be the endeavour of the Central Government to protect the knowledge of local people relating to biological diversity which may include registration of such knowledge at the local, state or national levels.[44]

The Central Government has been authorised under the Act to notify, in consultation with the concerned State Government, any species which is on the verge of extinction or likely to become extinct in the near future as a threatened species and take appropriate steps to rehabilitate and preserve those species.[45] The Act has also authorised the Central Government to exempt, in consultation with the National Biodiversity Authority, biological resources normally traded as commodities from the applicability of the provisions of the Act.[46] Chapter XII of the Act has vested plenary powers in the Central Government to issue binding policy directions[47] to the National Biodiversity Authority in its discharge of functions.[48]

Chapter IX also deals with the duties of the state Government which appears to carry the primary responsibility of determining and notifying areas[49] of biodiversity importance as biodiversity heritage sites. The State Government is also required to frame schemes for compensating or rehabilitating any person or section of people economically affected by such notification.[50] Like the Central Government, the State Government has also been authorised to issue binding policy decisions to the SBBs.[51]

Regulation in the use of biodiversity resources

It may be said at the outset that the Act makes a preliminary classification on the basis of alien nationality and extra-territoriality[52] for the following purposes:

1. Access to biodiversity resources.[53]
2. Transferring the results of any research relating to any biological resources occurring in, or obtained from, India for monetary consideration or otherwise to persons falling under the groups of alien nationality or extra-territoriality.[54]
3. Obtaining any intellectual property right, by whatever name called, in or outside India for any

invention based on any research or information on a biological resource obtained from India to persons falling under the groups of alien nationality or extra-territoriality.[55]

4. Obtaining any biological resource (or knowledge associated) occurring in India for the purposes of research or for commercial utilisation or for bio-survey and bio utilisation[56] or transferring the results of any research relating to biological resources occurring in, or obtained from, India.

While persons falling under foreign and extra-territorial categories are prohibited from undertaking the activities under the first and fourth categories without approval of the NBA, all persons are prohibited from undertaking the activities under the second and third categories without permission of the NBA.[57] However, Indian citizens are allowed to carry out activities under the fourth category on intimation to the State Biodiversity Board concerned.[58]

The Biological Diversity Rules, 2003 ('BDR' or 'the Rules') notified by the Central Government under Section 62 of the BDA, however, allows the appropriate authorities mentioned above to impose restrictions on access to biodiversity resources on the following bases:[59]

- The request for access is for any endangered taxa.
- The request for access is for any endemic and rare species.
- The request for access may be likely to result in adverse effect on the livelihoods of the local people.
- The request to access may result in adverse environmental impact which may be difficult to control and mitigate.
- The request for access may cause genetic erosion or affecting the ecosystem function.

- Use of resources for purposes contrary to national interest and other related international agreements entered into by the country.

(k) Concerning the second category (i.e. the approval for transfer of results of any research), Rule 18 and Form II are to be complied with, while in matters relating to approval for applying for intellectual property protection as under the third category, Rule 19 and Form III are to be complied with. Rule 20 and Form IV of the BDR need to be complied with for approval of activities falling under the fourth category. However, it may be noted that the same are procedural obligations and do not pose substantive restrictions, as seen under the first category. Other than the above limitations on use the BDA also makes provision for fair and equitable benefit sharing of biodiversity resources.[60]

4. Conclusion

The BDA has come under criticism for many reasons. For one, it is argued that the legislation facilitates biopiracy by only providing for regulation against persons falling under the class of alien nationality or extra-territoriality and not limiting local exploitative practices. Second, it has been argued that the many layered structure of licences/ permissions only succeed in burying biodiversity under a mountain of bureaucracy that can only serve to alienate ordinary farmers from their resources while making international bio-piracy easier. Third, it has been argued that the Act is quite toothless on the issue of intellectual property rights(IPR). Given that the only stipulation is that IPR applications will have to go through the NBA, there exists no substantial restriction.[61] The Act has also been criticised for not providing citizens with the power to approach courts when they detect violations.[62] It has further been argued that the BDA is unnecessarily soft on Indian corporate and other entities, requiring only 'prior intimation' for the use of bio-resources, rather than permission as in the case of foreigners. The empowerment of local community bodies under the BDA has again been found to be somewhat incomplete.[63]

Regarding the BDR, it may be said that it was expected that the Rules would supplement the BDA in areas suffering from ambiguity or lacunae. Instead, it has been put forth that the Rules dilute the strictures provided under the Act in the following ways:

- The Act mandates the establishment of a National Biodiversity Authority (NBA), State Biodiversity Boards (SBB) and at local levels the Biodiversity Management Committees (BMC). The setting up of the BMCs could have enabled local communities to have some voice in the conservation, sustainable use, and equitable benefit-sharing of biological resources. But as per the Rules the role of the BMCs is merely limited to preparing People's Biodiversity Registers (PBRs) that document local knowledge and bioresources. This immensely undermines the rights of local communities who are the most important stakeholders when it comes to conservation of biological resources. Documentation without any legal protection is also an invitation to exploitation. The BMCs would be preparing the PBRs, but where is the power to ensure that they will not be openly accessible for theft or piracy? There is an apparent lack of faith in the competence of local groups in taking decisions, as well as an attempt to centralise natural resource management all over again. Hence the concessionary inclusion to locals, without actually vesting any power in them. Such a step, even after the 73rd and 74th Amendments to the Constitution of India have upheld the need for decision making at the village level, seems completely retrograde.[64]

- The criteria for rejecting claims of IPRs, or of revoking access/approval, are rather restricted. It is not clear whether they include, for instance, the possibility of the access or the IPR's causing serious loss of livelihood to local communities, loss of traditional knowledge, violation of the intellectual rights of communities, and so on. Form 1 (Application for access), for instance, only asks the applicant to give information on whether the collection would endanger biodiversity, but not on whether it may threaten biodiversity-based livelihoods, traditional knowledge, or related aspects.
- The Rules regarding equitable benefit-sharing do not define what is meant by 'equitable' and do not include aspects such as ensuring equity and protecting traditional knowledge. More seriously, they introduce an unnecessary layer of bureaucracy by stipulating that benefits that identified that individuals/groups are to be paid, need to be routed through the district administration. Given the difficulties that people especially in villages have, in gaining access to the corridors of power in district administrations, this provision could seriously undermine efforts to ensure that benefits reach the deserving claimants.
- The Rules contain fairly evident scientific faults and gaps. For instance, Rule 17 provides for the NBA to restrict or prohibit request for access, if it is to any 'endangered' species though by internationally accepted terminology, the word 'endangered' is a subset of the term 'threatened'. If the intent was

serious, the wording should have been 'threatened' species. However, the Rule at present leaves out of the purview of this provision, a large number of non-endangered threatened species.[65]

References

The Convention on Biological Diversity, 1993.

The Biological Diversity Act, 2002.

The Biological Diversity Rules, 2003.

Biodiversity at *http://en.wikipedia.org/wiki/biodiversity.*

Dwivedi, Kriti. Bio-Diversity and the Challenges at http:/ www.legalservicesindia.com/articles/biodiversity.html.

Indian Experience—Submission by India to WTO Committee on Trade and Environment Council for TRIPs, 14 July 2000 at *http:// iprlawindia.org/category08/1889/bio.htm*

Kothari, Ashish. Countering Biological Piracy: A Revolutionary New Bill, September 2000, at *http://www.iprlawindia.com*

Gabriel, D.C. and Hariharan, Rajeshwari, India:Legislation bids to regulate biodiversity at *http://www.mangingip.com/includes/ magazine.htm*

Indian Experience—Submission by India to WTO Committee on Trade and Environment Council for TRIPs, 14 July 2000 at *http:// iprlawindia.org/category08/1889/bio.htm*

Ghosh, Asish—Convention on Biological Diversity: A Comparative Analysis at www.cuts.org

Press Release, Conservation of Biodiversity—A Backgrounder, October 2004 at *http://pib.nic.in*

Devraj, Ranjit. Biodiversity Or Biopiracy? December 2002 at *http:// www.indiatogether.org/environment/articles02/biodiv02.htm*

A thematic guide to the Convention on Biological Diversity at *http:// www.ciesin.org/TG/PI/PI-home.html*

Kohli, Kanchi. Biodiversity ruled out!, July 2004 at *http://www. indiatogether.org/environment/biodiv.htm*

Debbarma, Sukhendu. An Assessment of the Implementation of the Indian Government's International Commitments on Traditional Forest-Related Knowledge from the Perspective of Indigenous Peoples at *http://www.international-alliance.org/documents/india_eng_full.doc*

Endnotes

1. The word 'biodiversity' was suggested to him by the staff of NRC, to replace 'biological diversity', considered to be less effective in terms of communication. See Biodiversity at *http://en.wikipedia.org/wiki/biodiversity.*
2. The term is interchangeable with 'diversity within species' or 'variation of genes within species'. The phenomenon of genetic diversity can be understood in terms of the evolutionary struggle to increase ability to adapt to disease, pollution and other changes in environment. Kriti Dwivedi, Bio-Diversity and the Challenges at *http://www.legalservicesindia.com/articles/biodiversity.html.*
3. The term is interchangeable with 'diversity between species' and refers to the variety of species within a region. *Ibid.*
4. Implementation of Article 6 of the Convention on Biological Diversity in India—National Report, MoEF (1998) at *http://www.biodiv.org/doc/world/in/in-nr-01-en.pdf*
5. *Ibid.*
6. Biodiversity at *http://en.wikipedia.org/wiki/biodiversity.*
7. Traditional knowledge associated with biological resources is an intangible component of the resource itself. Traditional knowledge has the potential of being translated into commercial benefits by providing leads for development of useful products and processes. Indian Experience—Submission by India to WTO Committee on Trade and Environment Council for TRIPs, 14 July 2000 at *http://iprlawindia.org/category08/1889/bio.htm*
8. Ashish Kothari, Countering Biological Piracy: A Revolutionary New Bill, September, 2000, at *http://www.iprlawindia.com*
9. 'In situ conservation' has been defined in the BDA to mean 'the conservation of ecosystems and natural habitats and the maintenance and recovery of viable populations of species in their natural surroundings and, in the case of domesticated or cultivated species, in the surroundings where they have developed their distinctive properties' – Explanation (b) to Section 36(5).
10. 'Ex situ conservation' has been defined to mean 'the conservation of components of biological diversity outside their natural habitats'.
11. DC Gabriel and Rajeshwari Hariharan, India:Legislation bids to regulate biodiversity at *http://www.mangingip.com/includes/magazine.htm*

12. Indian Experience—Submission by India to WTO Committee on Trade and Environment Council for TRIPs, 14 July 2000 at *http://iprlawindia.org/category08/1889/bio.htm*
13. Article 1.
14. Edith B. Weiss, Introductory Note to the Convention on Biological Diversity, 31 I.L.M. 814, 817 (1992).
15. The GEF is a financial institution set up in early 1991 and jointly run by the World Bank, the United Nations Development Programme, and the United Nations Environment Programme. All three of these organisations contribute money to the GEF, as do the countries which are participants in the fund. The GEF then acts as an outlet for the funding of environmental projects. It is currently the mechanism listed in both the Convention on Biological Diversity and the United Nations Framework Convention on Climate Change to administer financial resources needed under those treaties. It was also proposed as the funding mechanism for the Agenda 21 action plan, a broadly sweeping non-binding agreement negotiated at the Rio Summit.
16. Articles 15, 16, 19, 20, 21.
17. Articles 6, 10, 14.
18. Article 8.
19. Article 9.
20. Articles 17, 18.
21. Article 19.
22. Dr. Asish Ghosh, Convention on Biological Diversity: A Comparative Analysis at www.cuts.org
23. Preambular Paragraph 2.
24. Preambular Paragraph 1.
25. Types of Environmental Law, Encyclopædia Britannica Library, Electronic Deluxe Edition 2004.
26. The BDR however relegates the main role of the BMCs to prepare People's Biodiversity Register (PBR) in consultation with the local people, which will include comprehensive information on availability of local biological resources and traditional knowledge associated with them. Press Release, Conservation of Biodiversity—A Backgrounder, October 2004 at *http://pib.nic.in*
27. Section 8(2).
28. Defined under Section 2(a) to mean 'the conservers of biological resources, their byproducts, creators and holders of knowledge and

information relating to the use of such biological resources, innovations and practices associated with such use and application'.

29. The concept of 'fair and equitable benefit sharing' has been/will be discussed earlier.
30. Under S. 37(1).
31. S. 22(3).
32. Under S. 37(1).
33. 'Local Bodies' have been defined under S. 2(h) to mean 'Panchayats and Municipalities, by whatever name called, within the meaning of clause (1) of article 243B and clause (1) of article 243Q of the Constitution and in the absence of any Panchayats or Municipalities, institutions of selfgovernment constituted under any other provision of the Constitution or any Central Act or State Act.'
34. 'Landrace' has been defined under Explanation (c) to S.41(1) to mean 'primitive cultivar that was grown by ancient farmers and their successors'.
35. 'Folk Variety' has been defined under Explanation (b) to S.41(1) to mean 'a cultivated variety of plant that was developed, grown and exchanged informally among farmers'.
36. 'Cultivar' has been defined under Explanation (a) to S.41(1) to mean 'a variety of plant that has originated and persisted under cultivation or was specifically bred for the purpose of cultivation.'
37. S. 41(1).
38. S. 41(2).
39. S. 41(3).
40. S. 36(2).
41. S. 36(3).
42. S. 36(4)(i).
43. S. 36(iii).
44. S. 36(5).
45. S. 37.
46. S. 40.
47. The question of what qualifies as a 'policy decision' is at the discretion of the Central Govt. under S. 48(2).
48. S. 48.
49. In consultation with the concerned local bodies as provided under S. 37(1).

50. S. 37(2).
51. S. 49.
52. The said term may be understood as comprising foreign citizens, non-resident Indian citizens (as defined in S.2 (30), Income Tax Act, 1961), body corporates, associations and organisations that are not incorporated or registered in India or incorporated or registered in India (under any Indian law which has any non–Indian participation in its share capital or management) who are required to fulfil substantially different obligations than Indian citizens under the Act.
53. S. 3.
54. S. 4.
55. S. 5.
56. 'Bio-survey and bioutilization' have been defined under S. 2(d) to mean 'survey or collection of species, subspecies, genes, components and extracts of biological resource for any purpose and includes characterization, inventorisation and bioassay.'
57. Ss.3, 19.
58. It has however been provided that 'this section shall not apply to the local people and communities of the area, including growers and cultivators of biodiversity, and vaids and hakims, who have been practicing indigenous medicine' under S.7.
59. Rule 17, BDR.
60. The term being defined under S. 2(g) to mean 'sharing of benefits as determined by the National Biodiversity Authority under S.21'. Rule 21 of the BDR specifies the criteria for such sharing.
61. Ranjit Devraj. Biodiversity Or Biopiracy? December 2002 at *http://www.indiatogether.org/environment/articles02/biodiv02.htm*
62. A thematic guide to the Convention on Biological Diversity at *http://www.ciesin.org/TG/PI/PI-home.html*
63. Ashish Kothari, Countering Biological Piracy: A Revolutionary New Bill, September 2000, at *http://www.iprlawindia.com*
64. Kanchi Kohli, Biodiversity ruled out!, July 2004 at *http://www.indiatogether.org/environment/biodiv.htm*
65. Sukhendu Debbarma, An Assessment of the Implementation of the Indian Government's International Commitments on Traditional Forest-Related Knowledge from the Perspective of Indigenous Peoples at *http://www.international-alliance.org/documents/india_eng_full.doc.*

Chapter 11

Small Grants Programme makes a Big Difference: Micro-Level Evidence Across the Country

Probhjit Sodhi

The Global Environment Facility (GEF) Small Grants Programme (SGP) is implemented by the United Nations Development Programme (UNDP) in 90 countries globally. The SGP became operational in India in September 1995 and works in partnership with the Ministry of Environment and Forests (MoEF), Government of India (GoI). The Centre for Environment Education (CEE) is the National Host Institution (NHI) for the SGP India since September 2000. Presently, CEE has a presence in all the States and Union Territories of India through a local network of seven regional offices and 23 field offices.

The SGP emphasises partnership and community participation in planning, implementation and monitoring of projects in the five GEF thematic areas of biodiversity conservation, climate change mitigation, protection of international waters, prevention of land degradation and elimination of persistent organic pollutants. The SGP aims to develop community-based approaches, strategic partnerships and innovative technologies that reduce threats to the global environment through local initiatives.

SGP, India has facilitated 216 action-based and community-led initiatives across the country. The projects have successfully addressed the issues of environment, equity, poverty reduction, community empowerment and sustainable livelihoods. The grant raised and disbursed till now consist of USD 3.5 million from GEF and USD 4.2 million raised as co-financing from the communities and other partners, in cash or in kind. The bar diagram in figure 11.1 represents a synoptic view of the thematic area-wise distribution of the projects under GEF.

Figure 11.1: GEF Thematic Area-Wise Distribution of Projects

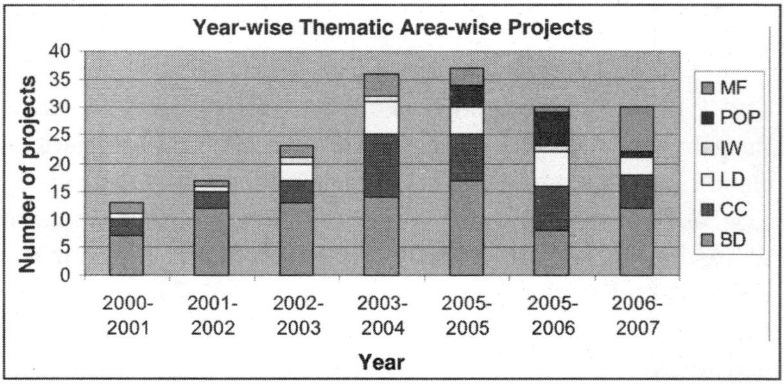

The programme lays emphasis on the following areas:

- More visibility to GEF-SGP;
- A wide range of strategic innovative project in poor areas;
- More emphasis on cross-cutting issues of community ownerships, local institution building, gender and equity and links to local governments;
- Common e-exchange set up between the Regional offices of CEE and SGP partners for more sharing;
- Responsive guidelines developed for involving the CBOs;

- New ideas of planning grants introduced to pilot, test ideas and exploring more possibilities of links;
- Decentralised, participatory Standard Operating Procedures (SOPs) developed for clarity.

Evaluation and Impact Assessment

Some of the widely reported impacts of SGP are:

- SGP is an excellent vehicle to facilitate local initiatives;
- Projects address the GEF focal areas as well as livelihood needs of local communities ;
- SGP adopts a strategy of building projects on to the efforts of the NGOs/CBOs in areas where they have a demonstrated track record, skills, knowledge and commitment;
- SGP-India organised, the South Asian Regional Consultation from 16 to 20 January 2005;
- Representatives of UNEP, Indian Chemical Industry, UNDP GEF SGP and CEE collectively drafted a paper on interventions in phasing out of PoPs in India through an action-based programme;
- A strategic partnership has been forged amongst the UNDP GEF SGP, MoEF, GTZ and CEE under which there is an action-based, participatory Solid Waste Management Program—'ECOCITIES';
- UNDP GEF SGP-India responded to the tsunami and supported CREED, SEVA, Progress and Social Welfare Trust with a grant support of USD 83,575 and leveraged co-financing of USD 760,067;
- Nearly 85 per cent of the SGP projects are with indigenous people or vulnerable/marginalised communities.

In order to elaborate upon the above-mentioned gains under the SGP, we present a synoptic view of the some of the important projects that have been undertaken under the SGP-GEF umbrella with a view to corroborate the aims and objectives with real-world experience and achievements.

Case Study I

A Community Endeavour for Bio-conservation and Employment Generation in Uttaranchal

Society for Community Involvement in Development (SFCID)

Chamoli, Village/ Post Pipalkoti, Chamoli, Uttaranchal—246 472

GEF Thematic Area: Biodiversity Conservation

Project Area: Pipalkoti, Chamoli Garhwal near Nanda Devi Biosphere Reserve

Beneficiaries: 70 community members

Total SGP Grants: USD 18,947 Total Co-financing: USD 10,405

Project Objective: The main objective of the project is to conserve and develop local biodiversity around the Nanda Devi Biosphere Reserve area and the tourist trekking routes through people's active participation and employment generation at the local level.

Project Strategy: The project involves training of community and establishment of three mother herbal gardens for conservation and cultivation of rare /threatened and extinct herbal plants and trees, and establishment of three Bio Tourism Parks for the socio-economic development of the rural community. This would provide gainful employment to rural women and youth from the nearby villages. The project will provide source of income, employment generation and awareness of biodiversity conservation for the local community through the bio-tourism park (BTP). The BTP will be managed by a team of youth selected and trained from the local community. Local resources and skill will be utilised to generate alternative livelihood options and linkages will be made with government departments, marketing agencies, etc.

Project Activities

Establishment of a Bio Tourism Park (BTP): Two bio-tourism parks will be established with communities' participation in the project area where eco tourism, adventure tourism and pilgrimage-based activities will be promoted. Sustainable rural ecotours will be promoted for income generation using local resources.

Indigenous culture/cuisines/traditional knowledge will be linked with tourism, local handicrafts and lesser known treks and destination around Bio-Tourism Park will be promoted.

Awareness generation and community participation: The BTP will be used to sensitise the local community and the tourists on climate change and Biodiversity conservation. The community will be mobilised on issues of cleaner environment, waste management in trekking routes and ownership in mountain tourism activities.

Establishment of mother herbal gardens: Two mother herbal gardens will be established in the area for conservation, cultivation and income generation through rare and endangered flora and fauna of the area. The mother herbal gardens will also help to preserve the genetic bio-diversity. New mountain-friendly and economically viable plant species will also be introduced. The community will manage the nurseries.

Alternative Income-Generation activities: Women's groups will be formed in the surrounding villages and thrift and credit activities will be promoted. The groups will be linked with the banks and activities like bamboo-based handicrafts, bio-dynamic composting and weaving.

Impacts: The emerging impacts of the project are:

- A bio-tourism park has been constructed with five tourist huts having basic amenities along with green house/Poly house, Smoking chamber for ringal (hill bamboo) craft, NADEP/ vermicompost, fish pond and nursery of fruit/fodder/fiber plants;
- Identification of potential craft clusters has taken place and training imparted to 30 craftsmen on design diversification and value addition. A marketing outlet has been established in the BTP and master craftsmen identified for training of trainers. The artisans' federation is managing the souvenir shop;
- Twenty-three SHGs of rural tribal women formed;
- Fifiteen unemployed youth trained and given gainful employment in the Bio-Tourism Park;
- Large-scale plantation of ringal bamboo has been done on the hill slopes by school children and rural artisans are being encouraged to undertake hill bamboo nurseries;
- Ffity women and youth trained and gainfully employed in nursery raising and poly house;
- Around USD 1,500 of profit generated through the BTP;
- Around USD 99,000 raised as co-financing from community, private sector and the Central and State Government;

- Waste management and environmental awareness of communities and tourists in the Bio-Tourism Park;
- Exsitu and insitu conservation of 10 rare medicinal plants done in the medicinal and herbal garden;
- Use of bio-dynamic compost introduced in 17 families;
- More than 400 trees planted along the slopes to increase green cover and reduce landslides;
- More than 200 trees planted in nearby forests and more than 340 local fruit, fodder, medicinal, ornamental plants and bamboo planted inside the BTP.

Case Study II

Tarikhet Non-conventional Energy Project

GEF Thematic Area: Climate Change

Grantee Partner **Chatrasal Seva Sansthan(CSS),** 169, Bara Bazaar Mallital, Nainital – 263 001.

Beneficiaries: 258 households in 5 villages (Khudoli, Thapla, Mori, Doba and Pali)

Total SGP Grants: USD 28,200 Total Co-financing: USD 10,301

Project Location: Tarikeht block of District Almora in Uttaranchal

The five villages of Khudoli, Thapla, Mori, Doba and Pali in the Almora district of Uttaranchal had a story similar to that of other villages in remote, hilly, infrastructure-poor areas. Increase in population, changed economic conditions and lifestyle resulted in loss of biodiversity, decrease in land fertility, erosion and over exploitation of forests for fuel and fodder and made the once self-sufficient economies increasingly unproductive.

In such a situation, the Chatrasal Seva Sansthan (CSS) intervened with the objective of conserving the biodiversity, to minimise dependence on forests and forest products by emphasising and raising awareness about non-conventional energy. CSS approached the UNDP GEF SGP and was supported with a grant amount of INR 1,260,000.

Project Goal: Conserving the biodiversity to minimise dependency on the forest and forest products by emphasising and raising awareness about non-conventional energy.

Project Activities:

- Documenting and highlighting the priority sites, species and strategies for biodiversity conservation and use of non-conventional energy;
- Ensuring community participation through formation and strengthening of Self-Help Groups;
- Initiating income-generation activities to stop migration from the villages;
- Reducing dependency on the forest for fuel, fodder and other requirements.

Project Details: The NGO, Chatrasal Seva Sansthan had identified soil erosion, reduction of land fertility and destruction of biodiversity as problem areas in the villages that they were working in. Fuel, fodder and agriculture were putting pressure on forests and land. This was also threatening the livelihood security of the communities and resulting in migration from the villages. With this understanding, the organisation decided to initiate the project.

A step-wise implementation strategy was chalked out to ensure complete participation of the community. The first step was to collect baseline data on the biodiversity resources. The survey provided data on population, dependency of the community on local biodiversity resources, fuel—fodder arrangements, livestock, forest encroachments, possible sources of non-conventional sources, utilisation of water and conditions of reservoirs, utilisation of the women's time, knowledge of medicinal plants, etc. Six Self-Help Groups (SHG) were formed in each village, with focus on the marginalised women. The SHGs were a powerful tool in motivating the community and to initiate income-generation activities. Community action plans developed through SHG meetings and all activities were carried out according to the action plan. Trainings were imparted to the SHG members on income-generation activities like food processing, detergent and incense stick making, stitching and embroidery and biogas. Twenty-five biogas plants were set up as demonstration units. This ensured that the community had full participation in all stages of development and operation of the project.

Beneficiaries: Local communities from four villages, especially women and the unemployed youth.

Observations: With the efforts put in and through involvement of the community, awareness about issues regarding environment has spread. This has shown positive results such as reduction in the use of fuel wood, acceptance of biogas as an alternative fuel, land arrangement being made for fodder and livestock management.

The six SHGs have been strengthened and are not only maintaining records of their own groups but also on sharing and utilisation of resources. The members have saved INR 90,000 in two years and have mobilised a bank loan of nearly INR 200,000. The bank has fixed a limit of INR 120,000 for one of the groups. The project has raised a co-financing of INR 459,000 from the government as well as other sources. The members are now engaged in income-generation activities such as food processing and 25 biogas plants are running successfully in the project area. The SHGs have reached a stage where they are now ready to share 30 per cent of their profits from the income-generation activities for expansion of project activities such as installation of biogas, environment protection, etc. The SHGs have been involved with the Department of Agriculture in making five biodynamic pits. In addition, the forest department is helping the communities in plantation and the dependency on forests for fodder and fuel has reduced by 50 per cent in winters and by 80 per cent in summers.

Learning from the Project

- The major component in the success of the project was complete participation of the local communities and the sense of ownership among the community.
- It was crucial to address the livelihood needs of the community as in the absence of this, it was extremely difficult to generate interest among the communities.
- The community was not ready to accept any alternative source of energy on face value of and the local government was also not proactive in this matter. In such a situation it was imperative to establish a demonstration unit and gain the confidence of the villagers before planning for scaling-up of the activities.
- The interest and participation of the local communities has helped to bring in participation from other agencies and the government and has also motivated the community to expand their thought into other areas of conservation such as pastureland development.

Global Environment Benefits

With the interventions made by CSS, the community's dependency on forests for fodder and fuel has reduced by

50 per cent in winters and by 80 per cent in summers. The introduction of biogas has been able to save 47 mts of wood per year, which has resulted in checking 5 mts of carbon emission per year. In addition, the plantation activities and livestock management has been able to increase the forest cover. The shift to organic manure by the communities has reduced the application of chemical fertilisers. The manure produced by the biodynamic heap and cowpit pats have increased the productivity by removing *Kirmul,* a type of pest.

Case Study III

Impacts emerging from the 'Securing Rural Livelihoods through Biodiversity Conservation' project with Centre for Social Research

Implementing Partner: Centre for Social Research (CSR), New Delhi

Project Duration: 1st September 2004 to 31st August 2006

SGP Grant Amount: USD 43,402.00

Project Location: Gadsa Valley, Kullu, Himachal Pradesh, India

Project Background: The Kullu district has a rural population of 92.8 per cent. A preliminary survey showed that 30 per cent of the population in the Gadsa valley lives below the poverty line. The Gadsa valley represents the scattered settlement pattern on the higher altitudes (3,000 mts plus) in the Western Himalayas. Deep-rooted social sanction against the Scheduled Castes has restricted their settlement to the periphery of villages that usually has poor, less fertile and smaller land holdings.

Harsh climatic conditions and inaccessibility increases the drudgery of poor households in meeting the livelihood needs. Even within the poor, the burden is much more on the women because of the nature of gender-defined roles and activities like collecting firewood and fodder, cutting grass, grazing cattle, fetching water along with other household duties. Heavy dependency on forest resources for livelihoods is contributing to the rapid depletion of resources and the drudgery involved in procuring the resources is increasing.

Criteria for Selection of Beneficiaries: CSR clearly defined the criteria used for selection of beneficiaries:

- Land holding of less than one acre and/or food production providing food security of less than two months for the family
- Annual income of the family less than $ 588 per annum
- Acutely dependent on forest resources
- Literacy level of all members of family less than fourth standard
- Family living in two-room non-concrete houses and animal and family staying in same house
- Family members prone to chronic illness
- Widow or deserted women-headed households

Project Purpose: To expand and diversify livelihoods of poor rural women of Gadsa Valley through local biodiversity conservation.

Project Benefits:

1. Social Benefits:

- All the groups formed under the project represent the members from marginalised and the vulnerable sections of society present in the area. In terms of social composition, 59 per cent of the members belong to the Scheduled Caste category, 25 per cent to the other backward caste and 16 per cent to the minority castes.

- 19 per cent of the women group members are illiterate and now they have been facilitated to take up gainful occupations.

2. Livelihood Benefits:

- Two quintals of apricot kernels were procured by the women, which fetched a rate of about $ 2 per kg (i.e. USD $ 400). The total cost of grading, sorting and transportation was $ 1.2 per kg.
- Further value addition by extracting oils has increased the scope for fetching better prices. An order has already been obtained from a private company for 10 quintals of kernels and 200 kg of oil per annum. The first batch of four quintals has been already delivered.
- Wild peach will also be collected for the first time in the valley and good prices have been negotiated for it.
- Two groups have undergone training in collection of high-altitude honey and in apiculture and will take it up as a source of income generation in the coming months. The honey fetches a price of around $ 7 per kg along with the bee-wax that is also expected to fetch good price.
- Market potential is being explored for aromatic plants also.
- Production of vermin compost was initiated in 12 groups. As much as $ 255 worth of compost was sold and members used $ 218 worth of compost. This has also resulted in saving around $ 9 per agricultural season on chemical fertilisers.
- A decorticator has been procured for de-shelling the apricot kernels, which has increased the efficiency to 50 kg per day per person from 3 kg per day

per person. This has also reduced the drudgery of the women.

- Daily wages worth $ 380 paid to 24 group members for nursery work.

3. Environmental Benefits:

- One nursery has been set up and visits by women to the nursery have motivated them to lease out their plot to raise a high-altitude nursery, leading to adopting conservation measures, provided the locals to have more sensitivity towards the local biodiversity
- Five species of rare medicinal plants have been sown in the nursery and the women are committed to conserve and nurture the varieties, which were getting extinct and rare. Such measures are leading to strengthen the local biodiversity conservation in the fragile zones.
- Decrease in the use of chemical fertilisers by 144 families, leading to improvement and maintain better soil fertility.

Case Study IV

Promotion of Sustainable Agricultural Practices through Demonstration of Biogasplant, Vermicompost and Nadep compost by Conservation and Optimum use of Local Natural Resources

Sarvangeen Vikas Samiti (SVS), Gorakhpur, Uttar Pradesh

Area of implementation: Dudapar, Asthaula and Pichhoura village, District Gorakhpur in Uttar Pradesh

Total SGP Support: USD 14,380

Project Objectives

- To provide an alternative energy source, based on biogas.

- To build the capacities of rural community regarding importance and use of alternative energy sources.
- To emphasise sustainable agriculture practices through efficient use and management of available organic solid wastes.
- To emphasise community participation and equality.

Project Strategy

Issues of firewood for cooking, use of chemical fertilisers in the farming process and issues of unmanaged solid waste are the bane of most rural areas in any developing country. Though there are solutions that can be used, they need to be conveyed to rural communities who can adopt them with support from an implementing agency.

The project followed a set guideline to perform the necessary steps. Self-help groups were formed and PRA techniques adopted. Promotion of renewable energy sources such as biogas and the use of alternatives to chemical fertilisers, pesticides and firewood were performed to convince the rural people to adopt the proven technology. Support services were established, which are covering the use, operation and maintenance of assets at local village level.

Drawings were made throughout the village that explained the technical details of the biogas plant. These images create awareness and are valuable assets for the support services. Selection of the right beneficiaries was one of the challenges for the organisation, for this was one of the reasons for failure in the governmental scheme. In view of these, the organisation has selected only those households who were willing to contribute for the stove and gas pipe. This contribution was made essential for the partners, so that sense of ownership could be developed.

It was observed that after the completion of the project, several villagers in the rural area adopted the technology for themselves with the help of the organisation and personal funding.

Project Impact Assessment

Poverty: The project in the villages Dudapar, Asthaula and Pichhoura, was a fine example of a poverty decreasing development. Before the installation of the biogas plants the participants were spending Rs 6,000 to Rs 7,000 annually on chemical fertilisers and pesticides. Not only was this a very expensive means, it is also a very polluting method. After the installation, they were saving Rs 5,000 to Rs 6,000 annually on fertiliser because they could use the final product of the biogas plant, the slurry, on their farming lands. The slurry is environmental friendly and is more effective than chemical fertilisers.

Producing compost to use as a fertiliser using traditional means takes a long time. They have to dig a hole in the ground in which they place the dung and waste, which then turns into fertiliser after several months. This is a long process and often the required amount of compost is not produced in time. After installing the compost pit to handle the solid waste of the house and dung, the process of producing better and more eco-friendly fertiliser only takes 45 to 60 days. Therefore, it reduces the waiting time which they can use to work on the field and apply the fertiliser which in turn increases their crop output.

When food was being prepared, the women spent hours in the woods collecting firewood. Each day there had to be at least three meals prepared, using up as much as 5 to 6 kg (1 kg per head of the family) of wood which had to be collected. Women spent two to three hours collecting firewood. This is a time-wasting effort. Using biogas, women do not need to collect wood to start a fire. This saves time, which they can use to work on their lands, thereby increasing crop output.

To manage the waste such as cowdung, large stone baskets have been placed in which the waste can be placed. Inside these stone baskets there are special kind of earthworms which feed and move on and through the waste. These worms multiply rapidly. For instance, 100,000 worms produce 1.5 million offspring within twelve months. Since there is no use for so many earthworms these can be sold to other villagers or villages. Therefore they are able to sell three to four kilograms of earthworms annually at the rate of Rs 500 per kg.

By using biogas as a cooking fuel, there is no more need to cut trees. Therefore the project saved 80 to 96 kg of wood each day, ensuring fewer natural resources were extracted. These resources are limited, so it is very important that they are preserved.

Another environment-unfriendly development that was caused by using wood and cow cake as fuel was the harmful emission of smoke into the atmosphere. For comparison: one kilogram of wood emits 305 mg of CO_2 into the atmosphere and cow cake emits up to 876 mg of CO_2 in the atmosphere. Therefore sixteen families were emitting 4,880 mg of CO_2 and even up to 14,016 mg of CO_2 into the atmosphere if they used cow cake as a fuel. The harmful emission of CO_2 is a heavy strain on nature and is one of the causes of global warming. By ensuring the use of biogas there is no more emission of smoke when preparing meals. Biogas emits only 144 mg of CO_2 into the atmosphere, therefore sixteen families are now emitting only 2,304 mg of CO_2 into the atmosphere. A reduction of more than 50 per cent when based on wood and more than 80 per cent when based on cow cake! Furthermore, because biogas is fully burned, there is only a very limited amount of CO entering the atmosphere.

Social: There are large social impacts when using a biogas plant. Earlier, the women of the village were involved in the task to collect firewood to use as a fuel for making fire. According to the women, this task was a drudgery. Using biogas, there is no need to collect wood to start a fire. This made the women of the villages very happy.

Second, because the women and children do not need to collect firewood to start a fire, they can spend their time on other fruitful activities.

Health: Using wood as a fuel also caused fumes of unhealthy smoke developing in the kitchen, which is harmful to the women and to the infants into spend a lot of time inside the house with their mother. For comparison: burning one kilogram of wood releases 128 PPM (Parts Per Million) of harmful CO—exceeding the health standard 15 times—while the safe concentration is 8.6 PPM. Furthermore burning wood releases 3300 ì g/m^3 of harmful particles exceeding the health standard 33 times which were not burned in the fire while the safe concentration is only 100 ì g/m^3. These particles are the cause of many deceases in rural areas such as tuberculosis, eye diseases and even cancer. Using biogas there is no need to use wood to start a fire, thus removing the existence of high concentrations of harmful fumes in the kitchen. The concentrations that do exist are far below the health harming standard.

The biogas plants allow waste to be managed in a controlled matter by putting dung and waste material in a stone container. Before this system was installed, waste would be put at places were it could be retrieved easily for later use. During the rainy season the rain would wash the waste on the streets and all the way to the houses. This increased the growth of many harmful bacteria which caused illnesses. Therefore by putting the waste inside the stone container, it does not spread during rainfall and so does not cause any illnesses.

Awareness: The project enabled massive awareness among the villagers. By giving many demonstrations and explaining the benefits of using a biogas plant there was a change in attitude. The beneficiaries were trained for maintenance and processing of the biogas plants, which slowly removed their doubts on this efficiency and sustainability and was seen as an improvement in terms of health, social and economic factors. Using wall paintings which could be found throughout the village, the people were constantly aware of the system as a valuable asset.

The women who were using the biogas plants slowly made other non-participants aware of the usefulness of the plant. Because of this, several new participants have installed biogas plants using their own funds.

Training sessions were organised with the local community leaders. The project has increased living conditions for many villagers. Not only did the project have a positive impact on their livelihood, they are also realising these benefits themselves. They are more open towards future plans and place more interest in attracting development projects to the village.

Successful completion of the project is being used as an example for the government. Using this project as an example enables the government to replicate it in other villages. Thus awareness is spreading not only at a village level, but also on a global scale.

Financial: For the implementation of this project, funding was required, therefore UNDP was approached. A request was made for the SGP (Small Grants Programme). After reviewing the project plan, the response was very positive. They were given a funding of Rs. 7,00,000 for finalising the project.

All participants of the project have spent nearly Rs 1500 to pay for the tube and metal piping that leads to their houses from the biogas plant. After they were convinced of its usefulness and realised it would increase their livelihoods, they all decided to help co-fund the project. A total of Rs 2,4000 has been put into the project by sixteen families.

Case Study V

Conservation of biodiversity with the help of traditional knowledge and skills of the Pardhi community

Dalit Sangh

Flat no. 13, Lata Marg, Mitra kunj, Sohagpur, District Hoshingabad, Madhya Pradesh-461771

Project area: 5 villages of block Sohagpur, Dist. Hoshangabad, Madhya Pradesh

Project Goals:

- To explore sustainable livelihood options based on the traditional knowledge and skills of the Pardhi community
- To enhance community-based conservation of biodiversity through mitigating negative impacts of traditional livelihood activities

Project Objectives:

- Capacity building of 107 families (out of 1000 families in the areas) of the Pardhi Community from five villages for sustainable livelihood options, using their own traditional knowledge and skills and linkages with BD conservation efforts.

- Facilitating the socialisation process for the Pardhi community

Project Rationale:

A large part of the Hoshangabad district is within the Satpuda Tiger Reserve, Bori Wildlife Sanctuary and Pachmarhi Biosphere Reserve. These parts are rich in biodiversity. The mammalian wildlife is threatened by illegal hunting especially tigers. The Pardhi community is often blamed for this, and shunned as a 'criminal tribe' by other village communities. The Pardhi Community has traditional skills in relation with wildlife and animal behaviour which could be utilised positively especially in the context of the recent Tiger Crisis in the country. The NGO has been working with Pardhis for the last few years and has insights into their attitudes, knowledge and skill base, as well as their ability to adapt to other livelihood options. Specific components of the project aim to enhance the socialisation and integration processes of the Pardhi families into the village community, and activities have been designed to enhance this. The livelihood options being explored for the Pardhi community build upon their existing knowledge and skill base, and this is likely to be a key factor that will encourage the participation of the Pardhis.

Project Activities:

Eco-tourism: Fifteen youths will be trained as eco-tourism guides and will be linked to the tourism management authorities of the:

- Satpuda Tiger Reserve;
- Bori Sanctuary;
- Pachmarhi Biosphere Reserve;

- Madhai Ecotourism Centre ;

Medicinal plants and medicine

1. Documentation of traditional knowledge of medicinal plants and medicine preparation with the help of local community
2. Training for value addition in herbal medicinal products
3. Establishing marketing linkages for medicinal products

Non-forest based activities: Training 107 families for poultry farming and a 1 day camp for veterinary check up of poultry

Apiculture training: Training 25 families in apiculture techniques and providing apiculture kits. SHGs will be formed and linked to the activity for dissemination of technique. Linkages with Khadi Gram Udyog will be made for marketing.

Art and craft: Two Training programs will be conducted for enhancing handicraft skills. In initial stage raw material will be made available by the NGO, and local sources will be explored for supply over long-term

Linkages with Government: Awareness campaigns will be done for mixed groups of local people and local government officials through meetings, film shows, street plays and exhibitions. Meetings with officials of human rights department will be done and linkages with officials from police, forest and other state departments through workshops

Education: Strengthening education of Paradhi children through innovative activities in enhancing life skills through infrastructure development, teaching-learning material (TLM) development and trainings

1. Set up two class rooms for school in 2 villages
2. TLM and other resource development
3. Employment of teachers for schools

Positive Connections with the local community/ society: Pardhis will be enrolled in voters list and facilitated for obtaining caste certificate and ration cards. Participation in gramsabha/ Janpad meeting will be facilitated.

Linkages with local government schemes: Workshops on health, water management issues will be conducted to make Pardhis aware about various schemes in these areas. Facilitation for linkages with other relevant local schemes will be done. State level/ national level summit of Pardhi community will be arranged for sharing experiences.

Project Outcome:

1. Pardhi families will be engaged in livelihood options such as:
   - Ecotourism activities in Satpuda Tiger Reserve, Bori WLS and Pachmarhi Biosphere reserve areas through 15 trained Pardhi youths
   - Documentation of TK and impact of traditional practice on BD, Medicinal plants identification, conservation, processing and marketing with involvement of *vaidus*.
   - Non forest based activities like:
     - Poultry with 107 Pardhi families
     - Apiculture with 25 Pardhi families
     - Two Art and craft related skill enhancement workshops
2. Awareness about rights and improved status in the society

3. Innovative educational activities for Paradhi children at schools enhancing life skills and their positive use.

Enabling Pardhi community to better access of local government schemes, linkages between TK and local resource conservation and establishing positive stakes in conservation especially in context of tiger conservation.

Appendix

INTRODUCTION OF THE *PARDHI* COMMUNITY*

Dr Gopal Authey and Ratan Umre
Dalit Sangh, Madhya Pradesh

The *Pardhis* have been a part of the Indian society for long. Hunting in the forests and selling the hunt to the non-vegetarian members of the society has been a long stood ill-famed source of livelihood for the community. The occupation is so well known that not just Kabir but other Saints too have made references to the same, in their spiritual addresses.

Intellectuals too have enlightened the society through innumerous educative tales based on *Pardhis.* The *Pardhis,* known by names like Shikari, Cheeta Pardhi, Bail Pardhi, Phaans Pardhi, Nona, Chamari, Chidimaar, Baheliey, Phandiey, Babariey, Daphair etc. in the various parts of the country, are organised into numerous caste-groups displaying varying customs and traditions. While each of these groups believes it is different from the other, in totality they are all hunter-communities who have survived on hunting down the wild for centuries. Though it is true that many of them have adopted other occupations, the society in these fiercely competitive times, has pushed them to their old occupation time and again.

Pardhis are nomads and it is difficult to identify them in ordinary attire. They have their own language and usually set up camps on the outskirts of villages that are close to forests. Looking for prey, they do not stay at a place for more than five to seven days. While the women go around selling small items of day-to-day use, the men go hunting along

* Thanks to Ms Simin Akther MA Final Economics, JMI for translating the Hindi version of this article into English.

with the young boys. The *Pardhis* usually hunt down wild pigs and animals of the deer family which destroy crops and such hunting enjoys the silent consent of the villagers whose crops face the threat of harm from these animals. The same villagers provide a consumer base for the low-priced meat so procured.

The most unfortunate consequence of the community's nomadic lifestyle is that their children, who grow up roaming around in the wild, remain totally illiterate. Thousands of *Pardhi* kids neither know the alphabet nor about medical aid. Nonetheless, they acquire the knowledge of hunting early in their lives. For hunting lions and tigers the community prepares irontraps that are hidden under grass or planted on the riverside. When the animal gets trapped, the *Pardhis* beat it up with sticks sine die. The skin is then pulled out securely and sold to traders. They know that the deer can be killed only by hiding behind other animals, in the direction opposite to the wind. They are also aware that the wild pig rally one behind the other, oblivious of the direction, and so it is best to plant traps in their known paths in such a manner that at least three to four of them get trapped in one go. While the *Pardhi* kids may not know the arithmetic of 'two twos are four', they are well aware that half the prey belongs to the owner of the trap while the rest is equally divided among others who ally in the hunt.

These people also hunt down birds like *Teetar* (Black Francolin) and *Batair* (Quail), for which traps with ten loops are used. Two of which are sacred loops, which the bird generally escapes. The female Francolin is released if caught. These illegal birds are then sold at local hotels and road-side eateries. The *Pardhis* generally know that it is best to chop off the ears and tails of their prey-dogs, for these parts often get bitten by the prey in combat. *Pardhis* also know well, which animal parts, like pig's tooth, owl's claws, deer skin, etc., are put to which use by practitioners of black magic. They also know who'll pay them the highest and they are so well versed in this that even though the *Kasturi Mrig* (Scented Deer) may be found in the Himalayas, they learn the art of preparing and selling fake Kasturie's from their parents right since childhood.

For thousands of *Pardhi* children, the forest is the school, prey-dogs their trusted companions, camps their houses and traps and nets their books, copies and pens.

In the utter insecurity of the forest, superstition gets deeply ingrained in the minds of these children. They believe that *Matabai* and *Pipli Mata* are their sole saviours in the frightful dark nights. To satisfy *Matabai,* each family must sacrifice at least two he-goats each year. The children believe that if this is not done, *Matabai* might gobble up the family as punishment. They also believe that if *Matabai* is with them, all problems will get solved in seconds. They are taught to keep away half of their first earning for next year's sacrificial ritual. A *Pardhi* family, on an average spends almost half of its annual income on the ritual. As part of the ritual the *Pardhis* are supposed to offer Puries right out of boiling oil to the village priests. 'Whatever Pardhi earns, is lost to *Matabai*' is a popular Pardhi proverb.

Under the assumption of a welfare state, even after 57 years of independence, the *Pardhis* continue to struggle with hunger and insecurity. Nationwide, numerous *Pardhis* continue to live on hunting as they used to in the times of Emperor Ashoka, King Harshvardhana or Emperor Akbar. The villagers dislike them and maintain a safe distance. Brought up in this environment of insecurity, the *Pardhi* children know no skills apart from hunting.

When the husband returns from the hunt, the wife must prove her chastity by what is an old and cruel tradition. To establish the wife's chastity an iron axe, weighing a kilo and a fourth is heated till red-hot. Tying nine *Peepal* leaves to the woman's hand, the axe is kept atop. The woman must walk nine steps before she throws it in the grass, which being dry catches fire. If blisters surface on her hands she's made to pull out water from the well using one hand and is deemed infidel. A slice of her ear is then chopped off in front of the deity so that her infidelity becomes known and visible to one and all in the community. If no blisters appear, the woman is deemed chaste. A *Pardhi* woman has to undergo this test many times in her life.

There is a provision for punishing the menfolk too. If a man marries an outcaste or is seen indulging in illegitimacy, he's forced into a pit covered with dry grass tin sheets, stones, etc. The grass is then set afire and the man is pulled out after five minutes. If he receives burns or injuries, he's pronounced guilty and is taken for the ear-chopping ceremony to tell all that he dared challenge the community's rules.

Apart from these, there are other customs too which are not as widely practised these days. A woman walking with five to seven stones on her head, man diving to fetch a stick thrown a hundred hands away without raising head out of water, etc., are also tests of chastity believed in by a few tribes.

The *Pardhis* have no specific language and thus no script of their own. They use Gujarati, Marathi and Hindi words. There pronunciation is more or less like the Haryanavis and the Rajasthanis and they usually speak fast, which makes it difficult to comprehend their language.

Illiteracy causes superstition to rule wide in the community. *Piplaj Devta, Pardan Devi, Dabi Mata, Khudiar Mata* and *Dhaniraja Devta* are the worshipped deities, worshipped according to Hindu traditions. Wine is used in these prayers.

In the old times, *Cheeta Pardhis* used to be deft hunters. In the times of the Mughal emperors, their services were used to catch and tame the fastest of wild leopards, hence the name *Cheeta Pardhi.* For their entertainment, the Mughal emperors and kings used leopards to hunt deer because, being the fastest, the leopards are the only ones capable of chasing down the deer. *Phans Pardhis* were those who used loop-traps to catch the deer.

According to available information, *Phans Pardhis* know how to train deer calves. They tie loops to the antlers of the trained one and leave it in the herd. Chasing away the stranger, members of the herd lock their horns with the latter and in turn end up getting caught in the loop-traps.

The modus operandi of *Bail Pardhis* is also unique. As soon as they begin learning the ways of the world, *Bail* kids learn to imitate the *Teetar* (Francolin's) call. While some do it by turning their lips in, others use an instrument made of bamboo or plastic pipes to produce the sound. This call is so real that even the faraway bird responds to it in its shrill and resounding voice.

Nationalisation of forests and wildlife protection has posed an employment problem for the community. Lack of awareness about alternative employments causes their indulgence in anti-social activities. While other sections of the society use them to hunt down crop-destroying animals, the *Pardhis* are the ones who are punished if caught.

The *Pardhi* community too wants its children to study in schools but their nomadic lifestyle has not allowed the same so far.

Sectoral Perspectives

Chapter 12

Sustainable Development Aspects of the Doha Round: Implications for Developing Countries

Naushad Ali Azad

Introduction

The World Trade Organization (WTO) in its Doha Development Agenda (DDA) of 2001 stipulated that trade, especially trade liberalisation policies, should be seen not as an end in itself but as a means to development. In other words, policies of trade liberalisation resulting from multilateral trade negotiations (MTNs) should not aim just at maintaining and furthering the growth objectives of the advanced countries. In addition, they should be able to take care of the development needs of the developing and the least-developed countries by providing enough policy space for protecting the interests of the vulnerable and the marginalised groups in these countries. On the other hand, the World Summit on Sustainable Development (WSSD) held at Johannesburg in 2002 emphasised the social and environmental dimensions of sustainable development (SD). This implies that a successful implementation of the DDA requires that MTNs should incorporate both the social and environmental aspects as the key factors for sustainability of the existing growth patterns across the globe. Keeping in view the fact that the world economy is globalising at an unprecedented fast rate, and that the effects of trade and trade policies on the developed and the developing countries continue to be (seen and felt) asymmetric, it is imperative that the ongoing and the future negotiations at the WTO do not remain confined merely to what trade can do in terms of productivity and efficiency gains, but also what it does to the social and environmental conditions across the globe. The main objective of this paper is to explore the links of trade, growth, environment and development with reference to their implications for the developing countries.

Generally, the cost of trade liberalisation to developing and LDCs is discussed in terms of preference erosion, tariff revenue loss, balance of payments problems, trade-related adjustment costs arising from various packages of policy reforms, impact of increased food prices for net food-importing developing countries (NFIDC), etc. In particular, the socio-economic and environmental impact of trade liberalisation is more important in the context of the vulnerable and the marginalised groups of people living in the developing and the least-developing countries. In the world of today, almost one-third of its population still lives in poverty. The poor, especially the rural poor, are dependent on the natural earth-resources to meet their basic needs of food, water and shelter. As the natural environment is also frequently the source of their small incomes, the poor are the most immediately affected lot by environmental degradation. Further, exports of most of the developing countries are also predominantly based on natural resources. It is but natural, therefore, that the developing and the LDCs are sensitive to interactions between trade and SD and that the progress in the multilateral trading system should be inextricably linked to progress on account of social and environmental dimensions of development.

Paragraph 51 of the Doha Ministerial Declaration (on identifying and debating developmental and environmental aspects of negotiations, in order to help achieve the objective of having SD appropriately reflected) is of special importance in this regard that, among other things, suggests that the successful outcome of the negotiations will require particular focus on issues of key developmental concern to developing countries, such as environmental requirements and market access, trading opportunities for environmentally preferable products and services, the protection and sustainable use of biodiversity and traditional knowledge, and the effective implementation of packages of supportive measures for developing countries in multilateral environmental agreements with trade measures to meet their objectives.

Thus, with DDA as the main plank of multilateral trade negotiations at the WTO, there is a strong case for a further exploration of the nexus between trade and sustainable development, rather than between trade and environment only. In Section 2 we explain the concept of sustainable development and highlight its social and environmental dimensions in the context of DDA. In Section 3, we look at the linkages between trade, development and environment. Section 4 deals with implications for SD in developing countries while in Section 5 we present conclusions and suggestions.

About Sustainable Development

Sustainable Development (SD) is perhaps the single most important challenge facing the contemporary world. For the purpose of conceptual clarity, we discuss here the analytical and operational definitions of SD. Let us take the analytical aspect first. The notion of SD is derived from two words—sustainable and development. SUSTAIN means 'to maintain' or 'to uphold' something (e.g. a life, a system) by providing a minimum necessary support and/or a conducive environment. Under a given set of conditions, *sustainable* then implies 'capable of maintaining' the life-supporting systems and, therefore, has the objective and subjective dimensions of viability and desirability respectively. Analytically, it looks like a constrained maximisation problem where the subjectively defined targets are to be achieved under the objectively defined constraints. In this way, *sustainability* refers to the capacity of systems for achieving the viable and desirable results from an activity under consideration. It can be seen that the desirability aspect of SD is easily understood if we differentiate between the notions of *development* and *growth* respectively. As we understand, the latter is a concept that implies vertical change in the economy requiring an efficient use of resources (based on the criterion of cost minimization or profit maximisation). A synonym of economic progress, it is limited to an increase in the material production of goods and services produced in a year and can be measured quantitatively in terms of per capita income of a country. On the other hand, development is a 'multi-faceted' phenomenon of horizontal change demanding a judicious use of resources and it is based on the notion of welfare maximisation for the society as a whole. It may be noted that growth is a necessary condition for development but not the sufficient one, i.e. growth is necessary for development but growth need not always lead to development. Next, the viability aspect of SD requires that we compare the costs and benefits of all activities required for economic growth and development in terms of their economic, environmental and social repercussions. It is here that distinction between explicit and implicit forms of benefits and costs becomes important. The explicit costs and benefits so involved are normally physical in nature and, therefore, easy to calculate. On the other hand, the implicit benefits and costs arise due to positive and negative externalities respectively. Unlike the explicit ones, they are usually non-material in nature and, therefore, difficult to compute. Once computed, the comparison of aggregate (explicit plus implicit) costs and benefits of the activity in question should give us an idea of its sustainability. Any activity whose aggregate benefits exceed aggregate costs will be sustainable in the viability sense of the term.

Traditionally, the concept of SD has been limited to environmental issues only, but it has now evolved to encompass many other socio-economic aspects such as poverty, income inequality, human rights, etc. Development levels achieved by different countries (or states within a country) are nowadays measured by human development index (HDI) that, in turn, is based on four parameters—per capita income (material well-being), literacy (education), life expectancy (health), and gender disparity (social or community health). SD now relates to overall fabric of life of present and future generations. The need of such a holistic approach to SD was demonstrated at the World Summit of Sustainable Development (WSSD) held at Johannesburg during 2002. As a result, many other aspects of social concern such as poverty levels, gaps in income-distribution, etc. are also the desired variables of HDI but not included in its calculations so far. Of late, the process of globalisation and modernisation has added some more aspects of desirability, e.g. financial, cultural, spiritual, human rights so that peace, stability and conflict resolution are also being realised as essential requirements of SD at the global level. Depending on new contexts in future, one may expect to have some more interpretations and operational definitions of SD.

Amongst the various issues, however, the focus is on 'eradication of poverty and hunger' and 'preservation of environmental sustainability' in all countries. Due to their prime importance and widespread recognition, 'Eradication of Extreme Poverty and Hunger' and 'Ensuring Environmental Sustainability' were also incorporated into the eight Millennium Development Goals (MDGs) initiated by the United Nations that are targeted to be achieved by 2015. In this context, 'eradication of poverty' and 'environmental sustainability' may be considered as two most important instrument variables for the purpose of achieving SD. Finally, it is important to realise that the viability and desirability aspects of sustainability may also vary with respect to space and time.

Contextualising Sustainability—The Space and Time Dimensions of SD

It may be noted that the two types of costs and benefits discussed here may vary with respect to space and time. In other words, the concept of SD is basically contextual. It depends on whether our concerns are local or global in nature and whether they pertain to the present or future generations. Thus, for a meaningful discussion and for purposeful policy implications, SD needs to be contextualised in terms of space and time. Although the conceptual framework mentioned here provides a reasonable basis for an analytical definition, for practical purposes we require some operational definitions that, in turn, require a proper contextualisation of the environmental and social concerns. In the next section we move on to a discussion on the linkages of trade, growth, development and environment and then explore some major concerns of developing countries with special reference to DDA.

Environmental Concerns and Environmental Sustainability

Given the limited nature of the life-supporting resources of Mother Earth, environmental sustainability can be defined and studied in terms of ecological imbalance and environmental pollution. The environment, left to itself, can continue to support life for millions of years. Unfortunately, however, human beings with modern technology have the capacity to bring about, intentionally or unintentionally, far-reaching and irreversible changes in the environment. In the context of environmental and ecological concerns, a number of operational definitions of SD or 'environmental sustainability' available. The first operational definition of SD is perhaps the one given by Brundtland Commission also known as the World Commission on Environment and Development (WCED, 1987): '(economic) progress that meets the needs of the present generations without compromising the ability of future generations to meet their own needs'. This approach of WCED for defining SD basically reflects our temporal concern about the utilisation rates of natural resources with respect to three parameters—resource base, technology and preference structure. Raskin and others later gave a similar definition in 1996. They defined 'environmental sustainability' as the capacity of unimpaired persistence and maintenance of the ecosystems and the natural resource base into the future. These definitions incorporate the well known 'save the earth' spirit voiced at the summits of Stockholm (1972) and Rio (1992).

Environmental Pollution and Global Climate Change

Environmental concerns relate to what human beings have done to the earth's reservoirs of natural resources (land, water and air) and the consequent effects on its climate. Unfortunately, the quality and quantity of these resources are fast deteriorating and depleting. *Environmental pollution* implies contamination of air, land and water caused by human products and activities. The sources of pollution include chemicals released by industrial processes, exhaust from gasoline-powered vehicles like automobiles, refuse and gases emitted by factories, sewage and garbage disposed of by cities, and pesticides and herbicides used in agriculture. *Global climate change* implies the long-term fluctuations in temperature, precipitation, wind, and all other aspects of global atmosphere. These changes are in addition to natural climate variations observed over comparable time periods. *Global warming* (the heating of the earth's atmosphere as a result of the greenhouse effect) is a unique feature of global climate change. It is caused by an increase in the atmospheric concentration of greenhouse gases that inhibit the transmission of some of the sun's energy from the earth's surface to outer space. The gases that trap sun's radiation within the troposphere include carbon dioxide, water vapour, methane, chlorofluorocarbons (CFCs) and other chemicals. In addition, depletion of the ozone layer and ultra-violet rays may also be involved in this process. The long-term effects of global climate change are not yet precisely known, but it is widely believed by scientists that global warming is a threat to most forms of life on the planet. The shantytown dwellers of Latin America and Southern Asia are already experiencing its effects in terms of lethal storms and floods. In Europe, they are seen in disappearing glaciers, forest fires and fatal heat waves. Recently (August–September 2005), the occurrence of category 5 hurricanes Katrina and Rita that hit New Orleans and other cities of the industrial belt bordering the Gulf of Mexico of the USA (in Texas, Louisiana and Mississippi) are considered a 'wake-up call' to the world about the dangers of global warming.

Table 12.1: Top 20 Carbon dioxide-emitting Countries

	Countries	Total CO_2 (Metric Tonne 000s)	World Proportion (%age)	Per Capita (Metric tonne)	Per Capita (Rank)
1.	US	1486801	23.8	5.43	1
2.	China	848266	13.6	0.68	18
3.	Russian Fed.	391535	6.3	2.66	6
4.	Japan	309353	5.0	2.45	9
5.	India	289587	4.6	0.29	20
6.	Germany	225208	3.6	2.75	5
7.	UK	148011	2.4	2.51	8
8.	Canada	127517	2.0	4.17	3
9.	Italy	113238	1.8	1.97	12
10.	Mexico	102072	1.6	1.07	17
11.	France	100951	1.6	1.72	14
12.	Korea (Rep.)	99260	1.6	2.64	7
13.	Ukraine	96510	1.5	1.9	13
14.	S. Africa	93808	1.5	2.38	10
15.	Australia	90470	1.4	4.88	2
16.	Poland	87807	1.4	2.27	11
17.	Brazil	81758	1.3	0.49	19
18.	Iran	79119	1.3	1.2	16
19.	Saudi Arabia	77237	1.2	3.83	4
20.	Spain	67468	1.1	1.7	15
	All 20	4915976	78.7	2.35	
	World	6243592	100	1.13	

Source: Marland et al. (2001).

In the present context, the implication of the increasing environmental pollution (see table 12.1) and the devastating potential of global warming and various environmental hazards caused by them have created pressure for the enactment and enforcement of and stringent environmental standards (laws that control the amount of released pollutants), properly structured plans and concerted efforts for the conservation of natural resources (such as recycling), and last but not the least, innovative development and use of greener technologies in the production of goods and services.

Social Concerns and Social Sustainability

The last two decades of the 20th century witnessed unprecedented growth in the world economy, which was also accompanied by alarming rates of poverty and income inequality. A high incidence of unemployment and rapid increase in absolute poverty increased human sufferings in many countries of Africa and Asia. It was also realised that international (or spatial) patterns of growth and development had been uneven. With time, the poor–rich divide deepened and widened. The striking difference in the use or overuse of these resources by the developed and the developing countries has also been a moot issue at every global meet. In a way, the process of economic growth seemed to have belied human expectations at the local and global levels. It was natural, therefore, that after the turn of 21st century, the focus shifted to what may be termed as *social sustainability*—a concept that includes all *major issues of human concern* e.g. poverty, illiteracy, gender discrimination, education, health, and so on. It was no wonder that the Johannesburg Summit, also known as World Summit on Sustainable Development, (WSSD) took a comprehensive view of SD by defining it in terms of three pillars— environment, economy and society (SEE*). In this way, SEE became a more comprehensive operational definition of SD*. It was argued that, unless transformation of society and the management of the environment are addressed integrally along with economic growth, growth itself will be jeopardised in the long run.

Global Initiatives for SD

During the previous three decades, there has been a number of important events for creating awareness and evolving consensus for global initiatives, such as:

The Stockholm Conference on Human Environment (1972),

The Brundtland Commission, also known as World Commission on Environment and Development (WCED) (1987),

The Rio Earth Summit (1992) (In this summit, Advanced Countries (ACs) undertook voluntary commitments to reduce their emission levels in 2000 to the level of 1990. Through a resolution, ACs defined emission levels in terms of past histories, but was not agreed by LDCs as it looked unjust.)

The Kyoto Conference (1997) (This conference defined 'Clean Development Mission' and resulted in the creation of Global Environment Facility (GEF) by ACs for helping LDCs in lowering their emission levels. ACs also wanted to put binding commitments on LDCs but was resisted by the latter.)

The Johannesburg Summit (2002), also known as The World Summit on Sustainable Development, advocated for a 'holistic' approach to SD by including environmental, social, economic and other aspects of human concern.)

In October 2005, a symposium on trade and sustainable development within the framework of paragraph 51 of the Doha Ministerial Declaration was organised by the WTO at Geneva to 'identify and debate developmental and environmental aspects of the negotiations, within the framework of Paragraph 51 of the Doha Ministerial Declaration, in order to help achieve the objective of having sustainable development appropriately reflected'. The terms of reference of the symposium included: (a) concept of sustainable development and its relevance in the context of the Doha Work Programme, (b) potential contribution of trade towards achieving the objective of sustainable development, (c) substantive issues of the Doha negotiations, namely Agriculture, Fisheries Subsidies and Environmental Goods and Services, (d) other selected issues of the Doha Work Program of particular interest to developing countries: (i) relationship between the Convention on Biological Diversity and the TRIPS Agreement, and (ii) the role of intellectual property rights in facilitating Transfer of Technology.

Linkages: Trade, Development and Environment

In this section, we attempt to explore the various channels of international trade affecting economic, social and environmental aspects of SD. Trade affects environment directly as well as through the channels of growth and development. In other words, we discuss some theoretical explanation and empirical evidence of the linkages between trade, development and environment. To begin with, we briefly discuss the 'Kuznets' Curve Hypothesis' and the 'Pollution Haven Hypothesis'.

Development and Environment–Kuznets' Curve Hypothesis The Kuznets' curve (an inverted-U) was first used as a graphical representation by Simon Kuznets to explain his theory about the relationship between growth and development. Known as 'Kuznets' hypothesis', it states that economic inequality first increases over time (the well-known trade-off between growth and equity) but begins to decrease after a critical point. Another but similar application of this hypothesis is found in the relationship is of development and environment. It is claimed that many environmental health indicators, such as water and air pollution, show the inverted U-shape relationship: in the beginning of the process of growth and development, little weight is given to environmental concerns, raising pollution along with industrialisation. After a threshold, when basic physical needs are met, interest in a clean environment rises, reversing the trend. Now society has the funds, as well as willingness to spend to reduce pollution.

This relation holds most clearly true for a few pollutants, such as sulphur dioxide and nitrogen oxide, but there is little evidence that the relationship holds true for other pollutants, especially those with non-local effects (e.g. greenhouse gases) or for the environment in general. For example, energy, land and resource use do not fall with rising income. While the ratio of energy per real GDP has fallen, total energy use has been rising across the globe and proportionately more in developed countries. In addition, the status of many key 'ecosystem services', such as provision and regulation of fresh air and water, soil fertility, and fisheries, also continues to decline. It can be argued that the fact that the Kuznets curve has been found for some environmental health concerns (such as air pollution) but not for others (such as biodiversity) does not necessarily invalidate the theory. It is possible that we may still be on the 'upward' leg of energy use Kuznets' curve and have to get even richer still before we see a decline. It is also important to note that much of the environmental damage associated with economic growth, such as extinct species and loss of wilderness, is irreversible.

Trade and Environment—Pollution Haven Hypothesis (PHH) versus Factor Endowment Hypothesis (FEH):

Just like the linkage between development and environment, we have only a limited understanding of how international trade affects the environment and that too is not very clear. A number of hypotheses has been put forward linking 'openness to trade' and 'environmental quality'. Noticeable amongst these are: (i) 'Pollution Haven Hypothesis' (PHH) and its natural counterpart (ii) 'Factor Endowment Hypothesis' (FEH). PHH states that relatively low-income developing countries will be made dirtier with trade. In this hypothesis, the movement of pollution-intensive industries to LDCs is explained in terms of low-cost advantage as well as a lax environment regulation climate. On the other hand, FEH suggests that dirty capital-intensive processes will be located in the relatively capital-abundant developed countries and is based on a number of assumptions of the well-known neo classical model of international trade.

Implications of PHH for Developing Countries

As stated above, PHH follows from an assumption that differences in pollution regulation across countries are a significant determinant of trade patterns, and firms in pollution intensive industries respond to these differences by relocating in lax regulation locations. Free trade then offers firms the same access to markets they had before, but with the now lower costs attainable in the lax regulation country. The immediate implication of this hypothesis is that with trade liberalisation, firms engaged in pollution-intensive industries would move from their current location in the developed world to new low-cost locations in the developing world. As a consequence, environmental quality may be lowered in developing countries, overall pollution in the world may rise, and governments in the developed world may respond to this migration of industry by refusing to adopt tighter regulations. In this way, PHH gives rise to technical barriers to trade (TBTs) on account of environmental arguments and supports calls for the restriction of imports produced by methods that are less environmentally friendly than those in the developed world, because this competition is unfair and raises world pollution. In terms of its application to developing countries, PHH suggests that non-signatories to Multilateral Environmental Agreements (MEAs) must be penalised via reduced market access to developed country markets because they are obtaining large (and unfair) cost advantages by not adopting tighter environmental standards. In other words, the view that free trade leads to unfair competition and a worsened environment is a strong motivation for many so-called environmentally-friendly policies that restrict trade (such as eco-labelling products according to their production method). Even if many of these trade barriers are not WTO consistent and will not withstand challenge, they are a potential threat to the economic well-being of developing countries. Therefore, it is not clear whether PHH is an adequate or even useful description of world trade in dirty goods. As a matter of fact, environmentalists criticise PHH as an intellectual foundation for imposing additional TBTs against developing country exports.

Under the alternative hypothesis, FEH, which has strong theoretical underpinnings, international trade is unlikely to lead pollution intensive production to move to less developed countries because regulatory costs are only a small part of total costs, and many of these conventional determinants of relative costs favor production of pollution intensive goods in the developed world. If this view is correct, then international trade would lead to the relocation of pollution-intensive production from the human and physical capital-scarce and relatively poor developing countries, towards the rich and capital-abundant developed economies. Pollution levels should fall in developing countries but may rise in the developed world. However, world pollution may fall with trade due to tighter environmental regulations in the developed countries. Dirty industries do not migrate to countries with lax regulation and the environment in less developed countries may improve with greater access to international markets.

Therefore, the factor endowments view of world trade in dirty goods predicts a very different environmental impact of trade liberalisation. Most of the policy prescriptions suggested by the pollution haven hypothesis are now no longer supported. The question whether free trade is good or bad for the environment, remains unanswered.

Empirical Evidence on Trade and Environment Linkage

We have seen that PHH and FEH provide diametrically opposite explanations of the impact of trade on environment. The controversy about their implications for developing countries coupled with lack of sufficient empirical support has prompted researchers to further investigate the trade environment linkage by using alternative techniques of analysis such as decomposition models etc.

Empirical studies by Tobey (1990), Copeland and Taylor (1994) and Jaffe et al. (1995) cast serious doubt on the strength of PHH. They find that trade flows are primarily determined by factor endowment considerations and apparently not by differences in pollution abatement costs. Grossman and Krueger's (1993) study of NAFTA used a decomposition model to examine if free trade had any effect on environment and put forward an argument suggesting that NAFTA improved Mexico's environment. Clearly, the results of these works do not support PPH and they at the best imply that free trade had no impact on environment.

In this context, another important study is due to Antweiler et al. (2001) who have also used a decomposition model to investigate the impact of trade liberalisation of international goods markets on pollution concentrations. Treating pollution policy as an endogenous variable, the model divides trade's impact on pollution into three components: scale, technique and composition effects. The model is based on a number of assumptions that help to explain the expected directions of these effects. For example, it is hypothesised that trade liberalisation may spur economic growth and thereby alter the scale of national output as measured by real GDP. Holding the composition of national output constant (i.e. the mix of industries the country produces) and the pollution intensity of its production techniques, this increase in the scale of economic activity must raise pollution. Therefore, the scale effect of trade liberalisation necessarily lowers environmental quality. Next, it is assumed that trade liberalisation almost always affects the composition of industrial output leading countries to specialise in those industries where their relative costs are lowest. If trade leads a country to specialise in the production of pollution-intensive (dirty) goods, then ceteris paribus, this composition effect will supplement the scale effect in raising pollution levels and, hence, worsening the environment. Alternatively, if trade leads a country to specialise in the production of relatively less pollution-intensive (clean) goods, then ceteris paribus, this composition effect lowers pollution and tends to improve its environment. Thus, the composition effect of trade liberalisation is assumed to either offset or supplement the scale effect. Finally, trade liberalisation can raise incomes per capita. If these income gains either create demands for better environmental protection or provide funds for investments in environmental protection, then changes in the techniques of production may be forthcoming. Therefore, rising incomes brought about by trade may lead to lower pollution intensities and lower pollution via a technique effect.

Antweiler et al. (2001) used data on sulphur dioxide concentrations from the Global Environment Monitoring

Project[1] to examine the effect of trade liberalisation on pollution concentrations. The study produced three key findings. The first is a very strong technique effect: a one per cent increase in national income per capita lowers pollution concentrations by over one per cent. The second is that factor endowment motivations for dirty goods trade appear to be more important than pollution haven motives. This implies that the developed world's cost advantages created by its abundant human and physical capital and superior technology at present more than outweigh the developing world's advantage of less stringent regulation. Thus, the study refutes PHH. The third is that economic growth, fuelled by capital accumulation, is likely to raise pollution levels while growth fuelled by technological progress will lower it. It suggests that growth in heavy industry may be worsening environmental outcomes in developing countries while eroding the natural comparative advantage of the developed world in these same, relatively dirty, industries. In other words, the study of Antweiler et al. have tried to establish that international trade creates relatively small changes in pollution concentrations when it alters the composition, and hence the pollution intensity, of national output. The estimates of the associated technique and scale effects created by trade imply a net reduction in pollution from these sources. Combining together the estimates of the effects of scale, composition and technique, the study yielded a somewhat surprising conclusion that freer trade was good for the environment.

Trade and Social Sustainability—Theory and Evidence

As already discussed, poverty and income inequality are the two most important parameters of social sustainability. Trade and trade-liberalising policies are supposed to affect these parameters through the process of economic growth.

Trade as an Engine of Growth: Trade theory provides us a fairly good understanding of the role international trade plays in fostering economic growth and its impact on the inter- and intra-country distribution of income. Classical and neo classical models of trade are unambiguous about free trade leading to quantitative and qualitative gains to all the trading partners, in aggregate terms. In other words, under normal conditions (assumptions), free trade is a positive sum game. The quantitative gains in production, consumption, technology and capital resources make it work as the 'engine of growth'. Besides, there are qualitative gains such as more varieties of goods and services. Together, they lead to welfare gains measured in terms of increased levels of per capita income. This is then the theoretical basis for arguments in support of free trade and the multilateral trading system.

In practice, however, world trade is anything but free trade. Moreover, the increase in the aggregate welfare does not determine the political economy of international trade. In fact, it is the distribution of income within and across countries that is instrumental in deciding the present and future course of trade policy.

Effect of Trade on Income-Distribution: Interestingly, trade theory also provides a good account of why the distribution of gains from trade within and between countries is asymmetric. In particular, the neo classical model of trade explains that the structure of tariff and non-tariff barriers entrenched in trade policy of a particular country or a group of countries is largely the result of the considerations of the policy making authorities to protect the interests of persons engaged in production or consumption of the exported and imported goods. The basic argument is that trade as well as the related commercial policies affect the producers and consumers of the exporting and importing goods in different ways leading to asymmetry in the distribution of gains from trade and hence, in the distribution of income within a country. Further, the income distribution effects of trade on the exporting and importing countries also depend on a number of country characteristics including the nature of the traded goods. As a result, the distribution of gains from trade between developed and developing countries also turn out to be asymmetric. In other words, a substantial amount of international trade theory is devoted to explain the asymmetries in inter- and intra-country distributions of income and, hence, provides a good account of why most countries follow restrictive trade policies in a selective and determined manner.

Environment and Poverty: As already stated, environment and poverty are two major concerns of SD. By and large, research and policy in this area have tended to focus on the relationship between poverty and environmental degradation in terms of pointing out that the poor are both victims and agents of environmental degradation—victims in that they are more likely to live in ecologically vulnerable areas and agents in that they may have no option but deplete environmental resources thus contributing to environmental degradation (SIDA, 1996; UNEP, 1995). At the same time, it is also acknowledged that the poor often have practices that conserve the environment.

As regards empirical evidence on these inter-relationships, again there exists huge literature but with only a limited clarity on the issues, as most of them end up in providing uncertain and inconclusive results. Similarly, there is a number of studies that reflect on the impact of trade on development including poverty and income inequality.

Effect of Trade Liberalisation on Poverty: Like environment, poverty is another big challenge of the fast globalising and modernising world. About 1.2 billion people in the world earn half a dollar a day. About 2.5–3 billion people of the global population are 'poor' (living below the consumption of $ 2 per day) while nearly half of them are 'very poor' (living below the consumption of $ 1 per day). Most poor of the world are concentrated in the continents of Asia and Africa. The global conditions about the distribution of income and wealth are also quite disturbing: In 1998, just 20 per cent of the population owned 86 per cent of the wealth while the poorest 20 per cent received just 1 per cent of global income. No wonder then, fighting poverty and reducing inequalities must be priorities of the agenda of SD.

The long-run poverty estimates for India, though controversial, have shown a declining trend. The latest of these are the ones recently announced by the Planning Commission for the year 2004–05.[2] Based on NSS 61st Round of consumer expenditure survey and using URP (uniform '30-day' recall period) data, the official figure for the headcount ratio (H) is 27.5 per cent (with a break-up of 28.3 percent for rural and 25.7 for urban areas respectively). Compared with the figures of 1993-94 (36%) and 1999-2000 (26%), these poverty estimates are not only controversial but also somewhat confusing. The current estimates on poverty have resulted in two controversial and embarrassing observations: (i) Poverty in terms of H seems to have gone up by 1.5 percentage points with respect to 1999-2000 (from 26 per cent to 27.5 per cent between in 1999-2000 to 26 per cent in 2004-05), (ii) H has declined about 9 percentage points when compared with 1993-94. The rate of decline in H since the mid-1990s (between 1993-94 and 2004-05) has been only 0.8 per cent per annum which is much less than the annual rate of decline of over 1 per cent in the period from 1977-78 to 1993-94. However, the official announcement also added a clarification saying that the estimate of 21.8 per cent for 2004-05 announced in earlier month was not comparable as it was based on MRP (mixed recall period) data obtained in the same 61st Round of NSS implying that, notwithstanding the observation in (i) above, the incidence of poverty has declined since the mid-1990s, but the rate of poverty reduction has slowed down in this period, which has also been the period of reform and higher GDP growth. It is, therefore, safe to conclude that the national and international estimates of poverty for India are indicative of a declining trend since early 1990s.

Yet, the absolute number of poor is very large! Notwithstanding the controversy about the data and the methodology, the size of 'very poor' group of India (about 350 million) is more than one-third of its own population and about one-fourth of the global 'very poor' population. Poverty of this order (often accompanied by high incidence of unemployment) not only hampers the economic growth of the economy but may also shatter the social fabric of the nation.

Implications for SD in Developing Countries

The discussion in the previous sections points towards the fact that, by now, all the channels of the relationship of trade, development and environment are not properly understood and a number of questions and the issues remain unresolved. We have also seen that there exists a huge amount of theoretical literature suggesting a trade-off between positive and negative effects of trade and trade liberalisation. The existence of such a trade-off makes the process of SD a challenging task.[3] International trade may play a positive role in this effort, but to assess its implications we need to measure trade liberalisation's impact on the environment (or on resource use). As trade expands and transforms, its environmental and social ramifications seem to become more important requiring urgent attention. Due to uncertain and inconclusive linkages between trade and the environment, environmentalists and trade protagonists very often square off (take a fighting stance) over the environmental consequences of liberalised trade. Further, due to asymmetry in the benefits and costs of ongoing process of globalisation (economic integration), both the developed and the developing countries have their own shares of apprehensions about the ultimate effects of trade on their respective economies. In the following section we focus on the key issues pertaining to the Doha Development Agenda that essentially have a bearing on the sustainability of growth and development processes of the developing countries.

Doha Concerns and the Challenge of SD

The most important challenge of SD is to provide a balanced programme of fostering growth in real incomes while maintaining or improving environmental quality. These concerns, with special reference to DCs and LDCs were reflected in the Doha Declaration. The DDA includes the negotiating objectives of clarifying on the WTO rules, specific trade obligations set out in the multilateral environmental agreements (MEAs) and reducing or eliminating tariff and non-tariff barriers to environmental goods and services. Why is DDA important to DCs and LDCs? It is so because secure and reliable market access to developed country markets is critical to their growth prospects. Access to developed country markets can also bring new technology, knowledge and capital via foreign direct investment; it can foster a more competitive domestic manufacturing base, raise the returns to education, and offer a ready market for many of the natural resource-based exports of the LDCs.

However, the access to developed country markets is not easy and automatic. In fact, it is argued that the economic integration through trade liberalisation may also pose risks to the process of SD in DCs and LDCs. If international trade alters the composition and scale of output in favour of relatively pollution or resource-intensive industries, as suggested by the PHH, and then increased market access may imply greater environmental degradation. Other risks arise not from the environmental impact of trade per se, but from limits (or denial) to market access brought about by developed country trade restrictions or by consumer boycotts of certain products.

Thus, many of the industries that are key sectors of DCs and LDCs, such as textiles, agriculture, forestry, fish, food, beverages, tobacco, are also highly protected and face high walls of non-tariff barriers (NTBs). For example, while the US and European Union have, on an average, quite low tariffs (1.9 and 2.7 per cent respectively) on these goods, NTBs are extremely significant. Moreover, since many of the exports of developing countries are natural resource based, their exports may be vulnerable to import restrictions on environmental grounds. For example, the European Union is on record for taking a position to restrict trade in agriculture on environmental grounds, i.e. to protect agriculture's role as a protector of the environment, and a guarantor of animal welfare, food safety and food security. The US and other member governments have denied taking up such positions, but the EU has made its position makes clear its interests in employing environmental grounds for import protection. In the past, concern with the environment in the developing world has also led to calls for a ban on tropical timber imports, for restrictions on the technologies used in capture fisheries and for the use of green countervail measures to level the playing field across countries, etc.

In addition to these risks, developing countries have now been asked to play a role in conserving global public goods by entering into Multilateral Environmental Agreements (MEAs). While the Kyoto Protocol contains no firm obligations for developing countries in reducing carbon emissions, it was of course the absence of these commitments that led the US to withdraw from the treaty. Similar concerns arise with discussions over conserving the world's biodiversity. Much of this diversity exists in tropical rainforests located in developing countries. The value of this resource is great from a global perspective, but low from the perspective of the developing economy owning the resource. Economic theory tells us that a globally efficient level of conservation or a globally efficient programme of carbon reductions can be obtained, but it most surely requires large monetary transfers from those in the developed world with the highest valuation for these global public goods to those in the developing world that are in the best position to either ensure conservation or lower carbon emissions. Large cross-country transfers are, however, difficult to implement, which means countries will seek other solutions that are necessarily less efficient and perhaps more costly to developing countries.

In total, developing countries face a significant threat from environmental concerns—both from the potential for liberalised trade to worsen environmental outcomes at home, but also from rising environmental awareness in the rest of the world that may translate into reduced or more costly access to rich developed country markets. With these facts as a background, it is worrying then that the Doha found includes the discussion of the relationship between MEAs and the WTO. Many of the over 2000 MEAs in force contain provisions for members to apply trade restrictions on countries both within the agreement and without. As well, some of the MEAs adopt a strong form of the precautionary principle and allow countries to limit imports under far weaker conditions than do WTO rules. It is therefore a concern of developing countries that these discussions are ongoing. If the discussions lead to negotiations that provide for the precedence of MEA rules over WTO rules, the scope for protection against developed country imports will increase dramatically.

The Role of WTO in Assisting Developing Countries

There are three main areas where technical and capacity building assistance may help developing countries. The first form of assistance is in building and maintaining systems of environmental monitoring. In many countries, monitoring of air, water and soil quality is at best rudimentary and in many cases non-existent. In addition, monitoring of the state of national forests and fisheries is difficult if not impossible for some of the poorest nations. Assistance in this regard would be important in several ways. Information concerning environmental quality is a necessary input into any evaluation of the impact of trade liberalisation. Moreover, the existence of data on environmental outcomes may provide some insurance against claims made that developing countries are excessively depleting their natural resources or producing ruinous conditions of air quality. In some cases, NGOs have joined with developing country governments in an effort to help with environmental monitoring but far more assistance is necessary if we are to accurately measure environmental quality in developing countries.

A second form of assistance is in the capacity-building area. Developing countries should start a process of environmental review of major policy changes and major industrial projects. Such reviews are commonplace in developed countries. To start, these reviews could be ex-post analyses of the environmental costs and market benefits from past policy changes or major industrial projects. With time and experience, these reviews could play an important role in assessing the likely environmental impacts of prospective policy changes. The United Nations Environment Program has already commissioned a group of studies with a view to creating the required expertise in developing countries. These studies examine the environmental consequences of trade policy and trade-related projects and established guidelines for environmental reviews. Further funding along these lines is necessary if developing countries are to build the human capital needed to undertake these studies and generate the data needed to make this type of work useful for policy analysis.

Finally, developing countries need to strengthen their capacity to engage in trade negotiations at the WTO. It is well known that the burden of administering existing trade agreements is already very large in relation to the regulatory capacity of many developing countries. Apart from the capacity to maintain and administer their current trade obligations, developing countries must develop a capacity to defend their own interests in the trade and environment area. Part of this defence could come from their better monitoring of environmental quality, and the proposed environmental assessments of major projects and policy changes. But even with this information at hand, trade negotiators in developing countries are likely to face ongoing calls from the developed world to restrict market access on environmental grounds. This may be the single largest threat to the successful integration of developing countries into the world economy.

The Integrated Framework Initiative: After the formation of the WTO, an initiative known as *Integrated Framework* (IF) was started in 1997 for building LDCs' trade capacities. Later, at the Hong Kong Ministerial Meet, it was further strengthened and incorporated as a formal clause in Article 57 as *Aid for Trade* (AFT). The main objective of this programme is to promote growth and development in DCs and LDCs with provision of financial assistance for capacity building measures such as (i) Institutional Development and Trade Policy Regulation, (ii) Development of Trade and Trade-Related Infrastructure, (iii) Building productive capacity to produce goods and services in competitive ways, (iv) Meeting Trade-Related Adjustment Costs arising from policy reforms, preference erosion, tariff revenue loss and balance of payments problems, and (v) Trade-Related Technical Assistance and Other Needs. In addition, AFT also sectors identified should belong to the national development agenda of the countries and that due attention is paid for the poverty reduction strategies.

Concluding Remarks

- For a holistic approach to SD, we need to combine the social, economic and environmental aspects of development. In order to achieve these objectives of SD, especially in the developing countries, a substantial rate of economic growth is necessary. Because not all types of economic growth can support SD, a changed pattern of investment becomes the sufficient condition.
- International trade and foreign investment are important drivers of the process of economic growth. They can also work as instruments of economic integration, regional and global cooperation, peace and stability.

- However, growth driven by trade can sometimes generate harmful effects in terms of environmental degradation or erosion of livelihoods. In fact, disregard of the trade policy makers of such harmful effects is one of the sources of tension with the environmental and development communities. The same can be said for foreign direct investment. Appropriate investment can spur sustainable development, but much investment in developing countries has been environmentally, socially and often economically questionable.
- Trade liberalisation is good for enhancing aggregate output and welfare but it can also result in various types of costs. In the context of DCs and LDCs, major costs of trade liberalisation involve preference erosion (erosion of preferential market access), adjustment costs (due to policy reforms and reallocation of resources), increased food prices for net food importing developing countries (NFIDCs), etc.
- The existing empirical evidence on linkages amongst trade, development and environment (including evidence on the PHH) is incomplete and inconclusive. Thus, there is little evidence that dirty industries migrate to low regulation developing countries. On the other hand, economic development and capital accumulation may be the most likely cause of the rising shares of pollution intensive goods in the GDP and exports of many developing countries.
- Environmental requirements in key export markets are becoming more frequent, stringent and multidimensional. Reason! Developed countries, which are still the primary producers of pollution

intensive goods, are losing their competitive position in the heavy industries. This, coupled with increasing environmental awareness, has resulted in demand for protection from developing countries' exports on environmental grounds in the form of NTBs.

- The environmental requirements or NTBs are more prevalent in sectors of export interest to developing countries, such as food production, electrical and electronic equipment, textiles and clothing, leather and footwear, timber and chemicals.
- Most of the requirements are imposed through mandatory governmental regulations. Even the voluntary requirements that fall outside WTO's disciplines become de facto mandatory through market power.
- A foregone conclusion is that sustainability of the trade-induced development process in both DCs and LDCs remains at great risk unless efforts are made at global, regional and local levels to mitigate them.
- However, trade liberalisation and sustainable development are not unavoidably incompatible. Trade liberalisation can advance sustainable development goals, just as it can retard their achievement. The net effect depends on how policies in the respective areas are crafted and negotiated.
- While integrating with the world economy to foster their development, DCs must ensure continued access to developed country markets with the help of various measures such as monitoring environmental quality, making environmental assessments for major projects and policies, and develop capacity needed to maintain and defend their interests in WTO negotiations.

DCs need to strengthen their capacity to meet higher standards in their key export goods and markets. They need to move away from the usual fire-fighting approach to a proactive and strategic adjustment approach. Ex-ante impact assessments of new standards and active participation in WTO negotiations for standard setting can be effective means of reducing undesirable effects. It is time now to move beyond rhetoric and address the real issues underlying trade and sustainable development.

It is a mistake to regard developmental and environmental issues as separate and largely unrelated. The issue for the WTO is sustainable development. Environmental goals cannot be reached without equity for the developing countries, whereas developmental goals cannot be pursued in a way that further undermines the environment.

References

Antweiler, W., Copeland, B.R. and M.S. Taylor (2001), "Is Free Trade Good for the Environment," *American Economic Review*, 91, 877–908.

Copeland, Brian R. and M. Scott Taylor (2004), "Trade, Growth and the Environment", *Journal of Economic Literature*.

Grossman, G.M. and Krueger, A.B. (1993), "Environmental Impacts of a North American Free Trade Agreement", In P. Garber (ed). *The Mexico-U.S. Free Trade Agreement*, Mass.: MIT Press, Cambridge.

Jaffe, A., S. Peterson, P. Portney and R. Stavius, (1995), "Environmental Regulation and the Competitivenens of U.S. Manufacturing: What does the evidence tells us?" *Journal of Economic Literature* 33 (March 1995): pp. 132–163.

Raskin, R., Chadwick, M., Jackson, T. and Leach, G. (1996), The Sustainability Transition: Beyond Conventional Development, Stockholm Environment Institute, Stockholm.

Swedish International Development Cooperation Agency (SIDA) (1995), "Promoting Sustainable Livelihoods: A Report from the Task Force on Poversty". SIDA, Stockholm.

Taylor, S. (2004), Trade, Development and the Environment,

Tobey, J.A. (1990), "The Effects of Domestic Environmental Policies on Pattern of World Trade" Kyklos, 43(2), pp 191-209

UNEP(1995)," Poverty and the Environment: Reconciling Short term Needs with Long term Sustainability Goals, UNEP Nairobi.

WECD (1987), Our Common Future, Oxford University Press, Oxford.

Endnotes

1. Lack of data on regulations in most developing countries forced *Antweiler* et al. to examine the impact with the help of a reduced form model. A major drawback of the reduced form estimation is that structural parameters remain hidden.
2. *Economic and Political Weekly*, Mar 31–Apr 06, 2007.
3. The debate about the trade-off between costs and benefits from trade and trade liberalisation was initially fuelled by negotiations during the North American Free Trade Agreement (NAFTA) and the Uruguay round of GATT. It is interesting to note that such standoff in the negotiations occurred at a time when social and environmental concerns such as widespread poverty, increasing income gaps, global warming, species extinction and industrial pollution, etc. were rising. For example, see Taylor (2004).

Chapter 13

Corporate Governance, Environmental Record and Market Valuation of Companies

Shahid Ashraf

Introduction

'...the results of human activity are putting such a strain on the natural functions of Earth that the ability of the planet's ecosystems to sustain future generations can no longer be taken for granted'. *(Millennium Ecosystem Assessment, 2005).* One of the important objectives of the National Environment Policy of 2006 is to integrate environmental concerns into policies, plans, programmes and projects for economic and social development. The emergence of sustainable development gave the world the famous three pillars model of economy, society and environment, first described by Barbier (1987). While subsequent thinkers have added pillars such as technical, political, or institutional (Hill and Bowen, 1997), the three-pillar idea stuck. It was put forward by the World Business Council on Sustainable Development into the triple bottom line of 'people, planet and profit', and continues to form the indicators of sustainability assessment initiatives, both in business and in buildings.

It may be possible to create a future where the damage done to the biosphere and to our social systems has been restored, and people can live in mutually supportive symbiosis with their social and biophysical environment (their whole ecological system)—the one nurturing the other. McDonough and Braungart (2003) develop this thinking when they ask us to imagine 'buildings that make oxygen, sequester carbon, fix nitrogen, distil water, provide habitat for thousands of species, accrue solar energy as fuel, build soil, create microclimate, change with the seasons and are beautiful—just like a tree'.

However, as Eisenberg and Reed (2003) warn, it would be wise to exercise some humility about our ability to understand and manage natural systems, and understand that they cannot be managed as though they are machines or businesses that can be dealt with in a way that will make them act in a uniform, predictable manner. We should also guard against an interpretation of the ecology that attempts a superficial reproduction of nature to solve one problem in an otherwise conventional product or process, or a rearrangement of current harmful processes. Changing to an ecological paradigm does not just mean a shift in technology and materials, but also in the overall perspective.

Ethical Investment and Corporate Governance

Concern about the environment has led many investors to restrict their investments to corporations that are perceived to be ethical in their social behaviour (Jaggi and Freedman, 1982). The emergence of ethical investing can be inferred from the appearance of ethical or green mutual funds in US and Europe. Moreover, many institutional investors, such as state pension funds, are sensitive to political influence and are thus reluctant to allocate financial resources to investments that may later prove to be controversial or non respectful of social values.

According to N. Vittal, former Central Vigilance Commissioner, in his 5th JRD Tata Memorial Lecture, corporate governance calls for three factors:

(a) Transparency in decision-making

(b) Accountability, which follows from transparency because responsibilities could be fixed easily for actions taken or not taken, and

(c) The accountability is for the safeguarding the interests of the stakeholders and the investors in the organisation.

Implementation of corporate governance has depended upon laying down explicit codes, which enterprises and organisations are supposed to observe. The Cadbury's code in United Kingdom was the starting point, which led to a number of other codes. In India itself, we have the Kumaramangalam Birla code as a result of the committee headed by him at the behest of the SEBI.

The idea that ecological and social sustainability is compatible with a free market economy was officially embraced in industrial circles in 1989. This was when the Coalition for Environmentally Responsible Economies (CERES), a US network of investors, major environmental organisations, and public interest groups, pooled these to advance environmental stewardship in business organisation. This coalition was formed as a reaction to the Exxon Valdez oil spill in Alaska, USA. The negative environmental impact of this accident, encouraged environmental and investor communities to cooperate with one another. This accompanied a move towards higher standards of corporate environmental performance, responsibility and disclosure. CERES set up a pioneering 10-point list of principles, or codes, of corporate environmental conduct.

In 2001, Elkington published '*The Chrysalis Economy: How Citizen CEOs and Corporations can Fuse Values and Value Creation*'. This book argues that sustainability will be the premier business challenge of the 21st century. It draws insight into the most important development of our times; in particular, how values can generate valuable companies, and to call for firms' CEOs to be good citizens, to generate values that reflect citizen aspirations; and to integrate these values as part of the firms' strategic thinking. This, according to Elkington, is critical, if companies are to survive the pressures and demands of a changing 21st-century society. The ideas in these books, among many others, have made sustainability the leading force behind corporate strategic change. Since its establishment, it has advised companies on environmental auditing in Europe; and has assumed a key role in advising companies globally. The 'Triple Bottom Line' model was soon and widely adopted by no less than 150 international organisations, ranging from corporations, government organisations to NGOs.

Investment and Environment Concerns

A rapidly evolving change in corporate management and strategy has been the recognition of and response to environmental concerns by industries. Often called 'greening', this response is a process by which human activity is made compatible with biospheric capacity. Industries have begun to develop and implement policies, programmes and tools to meet their environmental opportunities and constraints. This process has occurred partly in response to a desire to reduce potential liabilities. However, it has also been fuelled by opportunities in new markets and demand by diverse constituents including the public, shareholders, customers, and employees.

Society has become increasingly concerned with the health of the natural environment and the role of companies in impacting ecosystems and human health. This has led to asking the question, 'Which company is greener?' Though no clear or agreed upon definition of 'greenness' exists, judgments are frequently made as to which companies are green and which are not.

Expectations are that companies with a good (bad) environmental record should be valued at a premium (discount) by the stock market. Such a relation results from the emergence of 'ethical' (or Green) investing and from an increased awareness by investors of the potential negative consequences from corporate environmental damages. Two factors will influence company's stock market valuation. Companies with a poor pollution record will generate less cash flows because of investment in anti-pollution equipment. Second, they will face sanctions from governments. This will affect future market valuations from ethical investors.

Two mutually reinforcing trends can influence a company's stock market valuation. First, companies with a bad pollution record will generate less free cash flows in the future since they will be forced to invest in additional anti-pollution equipment. Also, companies with a bad pollution record will most likely face increasingly costly sanctions and penalties imposed by governments upon corporate polluters. Firms with a bad pollution record thus face a potential liability that should reduce their stock market valuations. Also, it is assumed that ethical investors have influence in the stock market and they bid up (down) share values of firms with good (bad) pollution records. Their behaviour will be reflected in the premium (discount) at which a firms stock is selling.

From a survey of 115 institutional investors, Longstreth and Rosenbloom (1973) indicate that 57 per cent of respondents base their investment decisions both on economic and social considerations; 28% of institutions mentioned that they avoided investments considered to be socially undesirable; 34 per cent of respondents, mostly banks and mutual funds, also argued that there is a correlation between socially responsible business enterprises and good monetary returns. Surveys of mutual fund managers by Buzby and Falk (1978) also find a need for social information, especially with respect to environment performance. Thus there seems to exists a sizeable number of investors that actively looks and is thus willing to pay a premium for ethical investments.

Using an accounting identity framework, Cormier et al. (1993), investigated the relationship between corporate pollution indices and a firm's market valuation. The results weakly suggest that a firm's pollution performance negatively affects its market valuations. Some support is thus provided for the ethical investor hypothesis. However, it can also be argued that investors perceive that firms that are not meeting current environmental standards may only get in worse financial shape in the future. Both explanations do signal that timely non-financial quantitative information about environmental performance could be useful to market participants.

Investment, Environment Compliance and Indian Industries

The Indian economy and financial market has received a boost by the introduction of ABN Amro's mutual fund in India for the first time. Dutch banking major ABN Amro's Indian asset management arm has opened its Sustainable Development Fund, a three-year close-ended scheme. The fund will invest in Indian companies that rank high on number of attributes, like environment compliance, compliance to corporate governance rules, etc. indicating a high level of social responsibility. Ratings major CRISIL will do the ranking based on a set of parameters and review the list annually. Thereafter, ABN Amro's fund team will do the financial analysis of the companies that make the final cut for final investment decision.

Globally, these types of funds are categorised as socially responsible investment (SRI) funds and currently have over $ 3 trillion in assets. In the US, about $ 2 trillion (or 9 per cent of total assets under management is in SRI funds while in Europe the corresponding figure is about $ 1 trillion. In about 24 funds, ABN Amro manages about $ 2 billion in SRI funds. Although this is a completely new area as far as mutual fund products in India are concerned, the ABN Amro team is excited that a fund like this will be a successful product. In India, environmental and social problems are not as abstract as in some other parts, of the world, so 'I think India is ready for this kind of funds', said David Morrow, global product specialist, SRI. Now, awareness about global warming and climate change is starting to migrate from government and big businesses down to individual consumers.

A wide range of issues that a decade ago were considered non-financial such as climate change, environment, bio-diversity are now coming to the fore as factors that can have significant impact on investment value. In Table 13.1 around 10 different funds from across the world with investments themes are shown. These funds now have a major impact on the investment behavior of other mutual funds. Some of them, like CaLPERS, have also registered as Foreign Financial Institutions in India also and have started investing in the Indian stock market.

According to the *Economic Times* of 10/4/07, taking into account environment and social factors and linking them with financial progress, is catching up with Indian companies.

Over 30 companies, including Tata Steel, Reliance Industries, ITC, etc. have bought the triple bottom line

Table 13.1: Investing Responsibly

Funds	**Total Assets ($ bn.)**	**Country**	**Themes/Investment Rationale**
ABP Pension Fund	265	Netherlands	Climate changes, environment, bio-diversity
CIA	5	Switzerland	Affordable housing, best practice architecture, energy saving qualities
Environment Agency Pension	2.7	UK	Climate change, emission, contaminated land, natural habitat and wildlife
ERAFP	4	France	Good labour policies, bio-diversity, carbon emission
Fonds de Reserve	42	France	Environmental responsibility, human rights, fair trade practices
Govt. Pension Fund Global	280	Norway	Environment, labour welfare, corporate social responsibility
Govt. Pension Fund	9	Thailand	Pollution and environment, good morals and customs
PGGM	97	Netherlands	Clean technologies, social housing, greenhouse gas emission projects
PREVI	50	Brazil	Environmental preservation, social development
CaLPERS	230	USA	Environmental technology solutions, real estate, green building technologies

Source: United Nations Environment Programme, as quoted in *Economic Times*, 6/6/07.

concept coined by John Elkington, which uses the tools of People, Planet and Profit (triple Ps) for assessing the success of a business and its sustainability. With adherence to the sustainability norm, companies gain financially by discovering and realising ways to conserve energy and resources, the like water. They also gain by making employees and other stakeholders happier about their association. And these corporates gain automatic esteem with investors who put a high premium on being environment friendly and socially aware.

The application of Elkington's theory takes the form of alternative reporting, separate from the standard financial reporting. Ideally such a report links financial progress to social and environmental aspects like the health of the employee, energy efficiency, industrial processes and economic use of resources, like water. The guiding principles are laid down under the Global Reporting Initiative. By measuring elements like packaging, transportation costs, health of employees among many factors, the need for efficiency is realised and many cost-saving opportunities are created. Industries with scattered supply chains, the like cement and chemicals, benefit the most as there is much scope in such industries for better utilisation of resources.

While preparing its sustainability report in 2004, ITC had measured cost savings that followed reduction in consumption of water per unit of production. It was reported that ITC's hotels business had reduced energy consumption by 20 percent which translates into a direct saving of 20 per cent on energy costs. Jubilant Organyses has said that during the process of reporting, performance in various fields get measured and thus come into focus of management. It has said that while incorporating GRI guidelines, actions are identified for improvement and again get reviewed and measured in the process of further reporting. Tata Steel is preparing its sixth sustainability report and they say that a direct benefit is seen in the case of fund raising from national and international financial institutions. These institutions demand such reports for vetting the reliability of their prospective clients.

While a majority of Indian companies stick to financial reporting alone, of the 80 that are engaged in alternative reporting, 30 per cent conform to GRI guidelines. Even as sustainability reporting is in its early stages in India, with the Indian economy opening up and positioning itself to be competitive in the global economy, more companies will be taking it up. Therefore, the introduction of the mutual fund by ABN Amro in the Indian investing arena and the acceptance of the global reporting initiative by listed Indian companies augers well for the beginning of a possible increase in environment awareness and compliance.

References

Barbier, E.B. (1987), 'The Concept of Sustainable Economic Development.' *Environmental Conservation*, 14/2, Summer.

Buzby, S. and Falk, H. (1978), 'A Survey of the Interest in Social Responsibility Information by Mutual Funds', Account Organisation Society, March.

Cormier, D., Michel Magnan and Bernard Morad (1993), 'The Impact of Corporate Pollution on Market Valuation: Some Empirical Evidence'. *Ecological Economics*, 8.

Eisenberg, D. and Reed, W. (2003) 'Regenerative Design: Toward the Re-Integration of Human Systems within Nature.' Article from the *Pittsburgh Papers*, Selected Presentations from the Greenbuild Conference 2003. Online: *http://www.integrativedesign.net/articles/pdfs/Regenerative_Reintegration.pdf*

Elkington, J. (2001), *The Chrysalis Economy: How Citizen CEOs and Corporations Can Fuse Values and Value Creation*, Capstone Publishers, UK.

Hill, R.C. and Bowen, P.A. (1997) 'Sustainable Construction: Principles and a Framework for Attainment.' In *Construction Management and Economics*, 15(3), 1997.

Jaggi,B. and Freedman, M.(1982), "An Analysis of the Information Content of Pollution Disclosures", *Finance Review*, September.

Longstreth, B. and Rosenbloom, D.(1973), *Corporate Social Responsibility and the Institutional Investor*, Praeger Publisher, New York.

McDonough, W. and Braungart, M. (2002), *Cradle to cradle*, North Point Press, New York.

McDonough, W. and Braungart, M. (2003) 'Towards a Sustaining Architecture for the 21st Century: The Promise of Cradle-to-Cradle Design'. *UNEP Industry and Environment*, April to September.

Millennium Ecosystem Assessment. (2005), *Living Beyond Our Means: Natural Assets and Human Wellbeing. Statement of the Board*. Available Online at: *http://www.millenniumassessment.org/*

Chapter 14

The Emerging Issue of Electronic Waste in India

Asheref Illiyan

1. Introduction

The modern lifestyle of people, population growth, rising standard of life, coupled with urbanisation, industrialisation, and increased reliance on information technology in different walks of life has enhanced the consumption of electronic products. This has resulted in the generation of greater amounts of solid wastes in industrialising countries, including India, which in turn has made electronic waste management an issue of environment and health concern. It is a point of great concern as many components of such equipment (e-waste) are considered toxic and non bio- degradable. 'The share of electronics in generation in overall industrial waste may not be very high at this stage but it is necessary for us to take preventive steps to contain this before it reaches unmanageable proportions'.[1] In this backdrop, the paper attempts to analyse various dimensions of e-waste management and recycling in India.

The paper is organised into nine sections. While Section 1 is introductory, Section 2 gives the definition of e-waste. Section 3 describes why is e-waste is a threat to human health and environment, and Section 4 explains the extent of e-waste generated in the world, with special reference to India. While Section 5 describes factors responsible for piling up of e-waste in India, Section 6 narrates measures taken in other countries for minimisation and disposal of e-waste. Section 7 highlights absence of a comprehensive legislation in India for minimisation and disposal of e-waste and stresses the need for promulgation of a national e-waste recycling and management policy. The elements of such legislation are also discussed along with various measures taken in India. Finally, possible ways out, conclusions, policy recommendations and suggestions are included in Sections 8 and 9 respectively.

II. The Definition of Electronic Waste

There is no generally accepted definition of electronic waste. It is commonly known as 'e-waste', and consists of used computers and technology equipments such as used TVs, refrigerators, mobile phones, etc. that are deemed obsolete or unwanted, broken, un repairable and needing to be disposed of.

According to the European Union Waste Electrical and Electronic Equipment, (WEEE) directive (one of the most comprehensive legislations for management of eletronic waste in Europe), electronic waste includes the following:

- Large household appliances (ovens, refrigerators etc.)
- Small household appliances (toasters, vacuum cleaners, etc.)
- Office and communication (PCs, printers, phones, faxes, etc.)
- Entertainment electronics (TVs, HiFis, portable CD players, etc.)
- Lighting equipment (mainly fluorescent tubes)
- E-tools (drilling machines, electric lawnmowers, etc.)

- Sports and leisure equipment (electronic toys, training machines, etc.)
- Medical appliances and instruments
- Surveillance equipment
- Automatic issuing systems (ticket-issuing machines, etc.)

III. Why is E-waste a Threat to Environment and Human Health?

E-waste is a threat to environment and human health because it contains several different toxic substances and chemicals, such as lead, cadmium, mercury, copper, zinc, chromium, PVC plastics, brominated flame-retardants, which likely to create adverse impact on environment and health. E-waste is a hazardous waste, as per the Basel convention (1989).[2] The worrisome point is that much of the current e-waste is being improperly disposed of or not scientifically recycled at all.

Chemical elements contained in electronic waste are:

- lead, zinc, chromium, cadmium, mercury, copper

Elements in trace amounts are:

- germanium, gallium, barium, nickel, tantalum, indium, vanadium, terbium, beryllium, gold, europium, titanium, ruthenium, cobalt, palladium, manganese, silver, antimony, bismuth, selenium, niobium, yttrium, rhodium, platinum, arsenic, lithium, boron, americium

Other

- silicon, carbon, iron, aluminium, tin, copper

Toxic constituents	Constituents components
• Lead and cadmium	Printed circuit boards
• Lead oxide and Cd	Cathode ay tubes (CRTs)
• Mercury	Switches and flat screen monitors
• Cadmium	Computer batteries
• PCB	Capacitors and transformers
• Brominated flame retardant	Printed circuit boards, plastic casings cable
• PVC	Cable insulation / coating

Table 14.1 gives the specific health and environmental hazards due to e-waste.

IV. The Extent of E-waste

Having discussed the definition and environmental and health hazards of e-waste, it is imperative to see the extent of e-waste generated all over the world, with special reference to India.

E-waste World Scenario

According to an estimate by UNEP 2005:

- More than 500 million computers will become obsolete in the USA alone between the years 1997 and 2007.
- 130 million cellular phones will be discarded in the USA by the year 2005, resulting in 65,000 tonnes of phone waste (BAN, 2004).
- 610 million mobile phones are to be disposed of in Japan by 2010 (Uryu et al., 2003).
- Every year, an EU citizen leaves behind 25 kg of e-waste (SECO & EMPA, 2003).
- 20 to 50 million tonnes of e-waste are generated per year worldwide.

Today's e-waste is mainly generated by industrialised countries, which already have a high number of Personal Computers (PCs) and were the first to automate their economies. But, in the near future, a large quantity of waste will be originating from countries in economic transition, such as Zimbabwe, China, Sri Lanka, India, and Eastern European countries along with the large-scale dumping of e-waste by developed countries. A close look at the extent of e-waste in India will remind us of the gravity of the problem.

E-waste—Indian Scenario

In India there is no reliable estimate of e-waste; however, a rough estimate provided by DATA QUEST and Toxic Link, which is presented below.

- It is estimated that 1.38 million PCs have become obsolete in India by the end of 2005.
- Annually 1.46 lakh tonnes of e-waste is produced in India.
- In a single month in Ahmedabad Airport 30 million tonnes of e-waste is imported.
- Every year, in and around Bangalore, 6,000 tonnes of e-waste gets stored.
- The minimum number of computers procured by an average scale scrap dealer per month is 20–25 .
- The approximate number of scrap dealers specialising in electronics, in and around Delhi, including large-scale dealers handling thousands of PCs per month, is 40 plus.

Source: Quoted from DATA QUEST, June 2005 issue.

In addition, India at present has 15 million computers and this figure is expected to grow five-fold to 75 million by the year 2010. This will add large volumes of electronic

Table 14.1: Environment and Health Hazards of E-waste

Computer/E-waste components	Process	Potential occupational hazard	Potential environmental hazard
Cathode-ray tubes (CRTs)	Breaking removal of copper yoke and dumping	• Solicosis • Cuts from CRT glass in case of implosion • Inhalation or contract with phosphor containing cadmium or other metals	Lead, barium and other heavy metals leaching into groundwater, release of toxic phosphor
Printed circuit boards	Desoldering and removing computer chips	• Tin and lead inhalation • Possible brominated dioxin, beryllium, cadmium, mercury inhalation	Air emission of same substances
Dismantled printed circuit board processing	Open burning of waste boards that have had chips removed to remove final metals	• Toxicity to workers and nearby residents from tin, lead, brominated dioxin, beryllium, cadmium and mercury inhalation • Respiratory irriation	Tin and lead contamination of immediate environment including surface and ground waters. Brominated dioxins beryllium, cadmium and mecury emissions
Chips and other gold plated component	Chemical stripping using nitric and hydrochloric acid along riverbanks	• Acid contact with eyes, skin may result in permanent injury	Hydrocarbons, heavy metals, brominated substances, etc., discharged directly into river and

		cause respiratory irritation to severe effects including pulmonary oedema, circulatory failure and death	
Plastics from computer and peripherals e.g. printers keyboards	Shredding and low temperature melting to be reutilised in poor grade plastics	Probable hydrocarbon, brominated dioxin and heavy metal exposure	Emissions of brominated dioxins and heavy metals and hydrocarbons
Computer wires	Open burning to recover copper	Brominated and chlorinated dioxin, polycyclinc aromatic hydrocarbons (PAH) (Carcinogenic) exposure to workers living in the buring works area	Hydrocarbon ashes including PAHs discharged to air, water and soil
Miscellaneous computer parts encased in rubber or plastic, e.g. steel rollers	Open burning to recover steel and other metals	Hydrocarbon including PAHs and potential dioxin exposure	Hydrocarbon ashes including PAHs discharged to air, water and soil
Toner cartridges	Use of paint-brushes to recover toner without any protection	• Respiratory tract irritaiton • Carbon black possible human carcinogen • Cyan, yellow and magenta toners unknown toxicocity	Cyan, Yellow and Magenta toners unknown toxicity
Secondary steel or copper and precious metal smelting	Furnace recovers steel or copper from waste including organics	• Exposure to dioxins and heavy metals	Emission of dioxins and heavy metals

Source: Boralkar, 2004.

waste to the waste stream and the environment. According to one estimate, two million PCs are nearing disposal, which include 286, 386 and 486 vintages being rendered obsolete. New upgrades are appearing in the market at small gaps increasing obsolescence rate of the existing models (Toxic Link, 2003).

Added to this is the problem of replacement of personal computers by laptops and notebooks. For instance, as per MAIT estimate, laptop sales have recorded a 76 per cent compound annual growth during 2000-2006, sales having increased from 41,670 units in 2000-01 to 43,1834 units in 2005-06 (see Figure 14.1). All these have the potential to add up to the waste stream after some time. Here the issue is that the laptop itself may become e-waste after some time and also that many of the personal computers may become redundant once replaced by laptops, consequently adding to the waste stream.

Figure 14.1: Notebook Shipment 2000–06

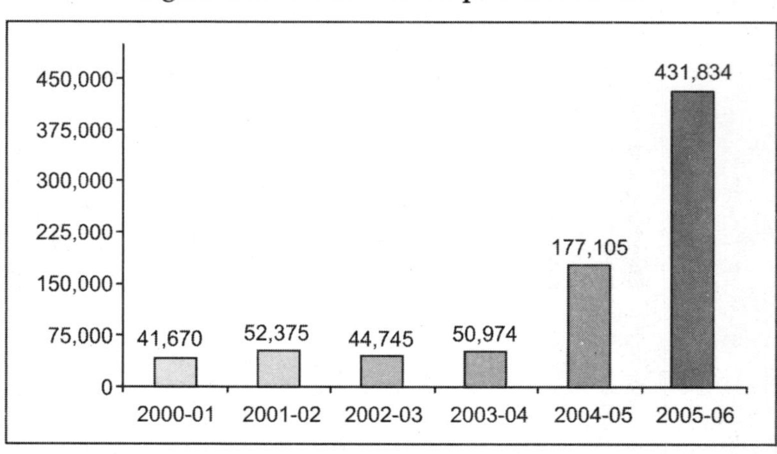

Source: Manufacture's Association of Information Technology (MAIT) website www.mait.org

In addition, if we take the case of mobile phone use, the picture is alarming. India today has about 145 million cell phone users. This figure is expected to touch 200 million by the end of 2007. The Average lifespan of cell phone use is coming down rapidly, which is expected to aggravate the existing waste stream in India.

As noted above, the e-waste generated through computer waste poses significant environmental and health hazards. Hence, the focus of this paper is computer e-waste. Rapid advances in information technology, with new and varied innovation in computers, have lead to parallel product obsolescence—adding to the toxic waste stream. 'A single computer can contain over 50 highly toxic metals and compounds, in the over 1000 materials it uses. All these get released soon as the computer is disposed and lands up in the hands of a recycler in a country like India' (Ravi Aggarwal, 2006). These materials are not a problem when the consumers use the products, but have the potential to cause serious harm when they eventually enter the environment after disposal (through burning, or crushing and subsequent leaching at landfills).

V. Factors responsible for piling up of E-waste

Where does computer e-waste come from? First, the progress of the IT sector by leaps and bounds has created a great volume of old technology-based computers. Every day new development take place in computer development making the older one obsolete. Second, fast obsolescence of computers is another major reason for the piling up of e-waste. Previously, we used to throw away a computer after seven years, but now this has decreased to three to five years. Third, the fast development of the IT industry and faster penetration of computers into various fields have increased the volume of e-waste. The computer penetration ratio has increased from 1 computer per 1000 people in 1990-91 to 14 computers per 1,000 people at present and is slated to go to

20 per 1,000 by 2010, amounting to about 20 million computers or 20 million devices with many kilos of toxirs in each. We have also been actively seeking donations of computers, which are up to ten years old. Finally, large-scale dumping of old computers by developed courtiers to developing countries aggravates the problem. This has been confirmed by various independent research studies. For instance, as per Green Peace (www.greenpeace.com) as much as 48 per cent of Western Europe's waste goes to the developing countries. Again, there is large-scale dumping of second-hand computers in India by developed countries in the name of mixed scrap or under some other guise. This has been confirmed by a Toxic Link study in 2004, which has found that more than 70 per cent of the recycled computers in Delhi are dumped by the developed countries especially USA. The British Environment Agency (BEA) in December 2004 has also confirmed large-scale dumping of e-waste from UK to India, Pakistan and China. This is in violation of the WEEE Directive.

VI. Measures taken by other countries for minimisation and disposal of e-waste

Prior to 1991, landfill was the most widely used technique by most of the countries to dispose of e-waste. However, in the 1990s, some European countries banned the disposal of electronic waste in landfills (due to its negative environmental effects). This led to the establishment of an e-waste processing industry in Europe.

Credit goes to Switzerland which, for the first time, constituted in 1991 the electronic waste recycling system, beginning with the collection of refrigerators. Over the years, all other electric and electronic devices were gradually added to the system. In Switzerland, legislation followed in 1998 and since January 2005 it has been possible to return all electronic waste to the sales points and other collection points free of charge. There are two established PROs (Producer Responsibility Organisations): SWICO, mainly handling electronic waste and SENS, mainly responsible for electrical appliances.

The European Union has implemented the Waste Electrical and Electronic Equipment Directive (WEEE 2002/96/EC) which came into existence since 13 August 2005. As per the Act, manufacturers are responsible for the recycling of the products they produce. In order to prevent the generation of hazardous waste, Directive 2002/95/EC (RoHS) requires the substitution of various heavy metals (lead, mercury, cadmium, and hexavalent chromium) and brominated flame retardants, polybrominated biphenyls (PBB) or polybrominated biphenyl ethers (PBDE)) in new electrical and electronic equipment put on the market from 1 July 2006.

The United States Congress is considering a number of electronic waste bills, including the National Computer Recycling Act introduced by Congressman Mike Thompson. This bill has continually been stalled, However, in the meantime, several states have passed their own laws regarding electronic waste management. California was the first state to enact such legislation, followed by Maryland, Maine, and Washington. From 2004, the state of California introduced an Electronic Waste Recycling Fee on all new monitors and televisions sold to cover the cost of recycling. The amount of the fee depends on the size of the monitor. Similarly, South Korea, Japan and Taiwan have already demanded that sellers and manufacturers of electronics be responsible for recycling 75 per cent of them.

VII. E-waste and Legislation in India

What is our national response to this grave problem of environmental and health threat? In India we do not have any comprehensive legisltion to deal with the problem of e-waste. The existing law that is Hazardous Waste (Management and Handling) Rules, 2003 is not comprehensive enough to containg e-waste problem. Even the latest comprehensive policy i.e. National Environment Policy 2006 does not speak much about this looming problem of e-waste. Therefore, there is an urgent need to promulgate a national e-waste management and recycling policy just as the European Union has done. The elements of a model e-waste policy are as follows.

VII.1. Elements of a Model E-Waste Legislation[3]

1. Definition. Effective legislation must define 'electronic equipment' sufficiently broadly to embrace legacy waste (old TVs, computers, etc.) and anticipate new gadgetry likely to come on the market; a definition of electronic equipment should include anything with a circuit board, complex circuitry, signal processing, or that contains one or more hazardous substances.
2. Producer Responsibility. Effective legislation must require development and implementation a system of brand owner/producer financial responsibility for equipment currently entering the marketplace. Legislation should state a non-specific requirement that brand owners, producers and distributors, or a consortium of brand owners, develop an approved system for financing the environmentally superior collection and recycling of discarded electronic equipment, with applicable rates and dates, and leave the specific details to be developed by affected companies.
3. Performance Measures. Effective legislation must set performance measures and timetables for meeting

these performance goals. Performance could be measured in one of several ways, including:

- Collection, recovery and recycling of a percentage of the brand owners' products;
- Collection, recovery and recycling of an amount per person based on the population of the state in question (e.g. 4 kg per person per year);
- A level of service and convenience, measured by a required number of drop-off or collection locations per unit of population.

4. Comprehensive Scope. Effective legislation would frame a system for e-waste collection and recycling that applies to all brand owners, regardless of sales channels, and to all end users.
5. Legacy Waste. Effective legislation must also create and finance a system of brand owner/producer responsibility for our stockpiles of so-called 'legacy waste,' electronic equipment sold and discarded prior to the effective date of the legislation. Financing for such a system should be based on market share or other means of allocation across the industry.
6. No Taxpayer Liability. Effective legislation must ensure that government and taxpayers are held harmless from all costs associated with collection, handling, transportation, storage, recycling, and disposal of discarded electronics, as well as oversight and enforcement of systems established to handle these products.
7. Disposal Bans. Effective legislation must ban electronic equipment from landfills and incinerators.
8. Toxin Reduction. Effective legislation must phase out specific hazardous materials from the manufacture of electronic equipment, including but not limited

to lead, mercury, polyvinyl chloride, and brominated flame retardants.

9. Labelling. Effective legislation must require labelling of electronic equipment containing hazardous materials. Legislation should also require labelling or information provided to consumers about the system for managing discarded products.
10. Responsible Recycling. Effective legislation should establish verifiable performance standards for electronics recyclers, including reporting and penalties for violations, worker health and safety and other criteria, to ensure that materials are managed in an environmentally superior manner.
11. Procurement. Effective legislation should establish procurement requirements for public agencies' information technology purchases, relating to product specifications and end of life product management.
12. No Waste Export. Effective legislation should, to the extent possible, prohibit export of non-working CRTs and CRT glass waste for any reason.
13. Governance and Enforcement. Effective legislation must include means for ensuring compliance and enforcement. Legislation should require specific periodic reporting by producers selling in the state, as well as public availability of all such reports. Legislation could require a multi-stakeholder advisory board to review these reports and make additional recommendations. Legislation could prohibit sales in the state, or sales-to-state agencies and units of government, for failure to abide by the terms of the legislation.

Additional elements

14. Economic Development. Effective legislation could harness the economic power of recycling and reuse industries by establishing preferences/incentives for local economic development and job creation through electronics recycling.
15. Recycled Content. Effective legislation should close the electronics recycling loop by requiring recycled content standards for materials used in electronic equipment

VII.2. Measures taken in India

The absence of a comprehensive legislation does not mean we have not made any attempt to mange and recycle e-waste in India. A few and isolated attempts have been made at organised level and there exists a large number of recycling units at unorganised sector. It is true that the subject of recycling itself is quite new'in India. We do not have any formal industry which is doing this task. However, only recently TERI has kickstarted a programme that lays out organisational procedures for e-waste recycling. The Central Pollution Control Board has constituted a national working group to address the issue. The Bangalore pilot project in 2005 under CPCB and Max Muller Bhawan is the first formal recycling unit in India. Of late, the Confederation of Indian Industry (CII) has set up a recycling plant at Chennai. The 11th Plan approach paper also lays emphasis on capacity building and infrastructure development for recycling of electronics products in India.

Recently, some hardware companies have also taken the initiative to reduce the use of hazardous substances such as lead, cadmium, and mercury and hence comply with RoHS (Reduction of Hazardous Substance) directive. For instance, 'HCL Info systems on Tuesday October 9, 2007 launched a new range of environment' friendly computers known as 'HCL eSafe', which are compliant with restriction of hazardous substance directive and having negligible or zero quantities of hazardous materials and are easily recyclable. HCL has also started various environment protection initiatives under the comprehensiveness 'HCL e Safe' programme. HCL is also the pioneering company in India to have comprehensive e-waste management programme for its manufactured products through a tie-up with an authorised recycler. They have tied up with the Chennai–based government-approved recycling plant, Trishiraya, where all collected products are recycled. People can submit their used mobiles, electronic gadgets and computers in their 200 collection centres across the country' (*Hindu*, October 10, 2007).

11th Plan approach paper also lays emphasis on capacity building and infrastructure for recycling of electronic products in India.

VIII. Possible ways for management and minimisation of e-waste

The present approach for management of e-waste can be summarised in 3 R's, viz. Reduce, Reuse and Recycle. But some countries have also adopted shortcuts, like export of e-waste to the developing countries.

(a) Reduce

The best technique to avoid the problem of e-waste is to reduce the use of electronic products, that is use them at the minimum. It is heartening to note that the new convergence technology wherein a mobile, a computer, a multimedia and internet browser everything under one roof will automatically reduce the requirement of having multiplicity of electronic products.

(b) Storage and landfill disposal

Storage and landfill disposal were widely used earlier but are not at all a viable option because of the environmental impact. Landfill may lead to leaching and contamination of water, mercury vaporisation to the air and inflammation.

(c) Re-use

Another possible way out is to lengthen the life of the computers by reusing. Corporate can donate their two to three years PCs to schools and other institutions. But care must be taken to ensure that very old computers are not donated. A certain minimum time-period has to be fixed for computers beyond which it cannot be donated.

(d) Recycling

An important way out is recycling. Electronics 'recycling' is a misleading characterisation of many disparate practices, including de-manufacturing, dismantling, shredding, burning or exporting. Recycling is mostly unregulated and often creates additional hazards itself. Although the amount of e-waste rises steadily, the industry has not yet developed very sophisticated or automated recycling procedures. In 2001, only 11 per cent of personal computers retired in the USA were recycled. Nevertheless, modern recycling plants can recover 80 per cent of the material and use another 15 per cent for burning. Only 5 per cent finishes as waste. However, the situation is worst in the developing countries, including India. The recycling and disposal of computer waste becomes a serious problem since their treatment methods remain rudimentary. Such activities pose grave environmental and health hazards; for example, the deterioration of local drinking water, which can result in serious illnesses. A riverwater sample, from the Lianjiang river near a Chinese 'recycling village' revealed lead levels that were 2,400 times higher than the World Health

Organisation's Drinking Water Guidelines. Often, workers in e-waste recycling operations in developing countries face dangerous working conditions, as they may be without protection (no masks or gloves, for example). Released gases, acid solutions, toxic smoke and contaminated ashes are some of the most dangerous threats for such people, and the local environment (UNEP, 2005).

(e) Export

Export to developing countries is a dangerous but cost-effective, and sometimes illegal waste management option chosen by some companies in industrialised countries. Sometimes illegal export is phrased as or hidden under the umbrella of charity ('computers for the poor') or as recycling. This comes from the fact that environmental and occupational regulations are lax or not well-enforced in some developing countries. Disposing of a computer in the US can cost up to USD 20, while an Indian trader can buy it for USD 10 to 15, a net gain of 30 to 35 USD for the exporter in the US. The computer will then be auctioned off in India, as part of a large consignment, and after being rerouted through an address in the Middle East, to hide the country of origin. While the recycling worker will make a pittance of less than a dollar or two a day, the traders make a killing. A PC of 32 kg can yield gold up to 0.0016kg, copper up to 6.93 kg, plastics up to 23kg and glass (silica) up to 24.9 kg, along with the toxins (UNEP, 2005).

IX. Conclusion and Suggestions

Thus, electronics waste has emerged as an important issue of health and environment in the country. Although the quantum of electronic waste generated in overall industrial waste may not be very high at present, it is imperative for us to take all possible and preventive steps to contain this before it reaches uncontrollable proportions. The policy recommendations and suggestions emanating from the study are:

First, strictly implement the Basel Convention. *Second,* implement a model legislation in line with EU directives on Waste Electrical and Electronic Equipment (WEEE) and Reduction of Hazardous Substances (RoHS). *Third,* implement Extended Producer Responsibility (EPR): Some countries are implementing policies and programme to prevent pollution and promote waste minimisation. Key among these approaches is the 'Extended Producer Responsibility' (EPR). Its objective is to make the manufacturer (financially) responsible for the entire life-cycle of their products, especially when they become obsolete. The underlying assumption is the company's interest in easier recycling and decomposition, and as such resource use limitation, pollution prevention and waste avoidance through ecological ('green') design, re-use, re-manufacturing and efficient recycling (UNEP, 2005).

Fourth, the informal sector, which is currently involved, must be incorporated as an actor in the whole scheme through a system of incentives and capacity building. A new legislation is needed, which outlines the responsibilities of each sector clearly. Also create a manual for recycling units so that they use appropriate technologies. *Fifth,* redesigning of computers for longevity and with clean materials and processes is very much needed. *Sixth,* a special fund has to be created wherein contribution from manufactures, consumers and government are collected for recycling. *Seventh,* create awareness among all stakeholders such as domestic industries, recyclers, schools and government agencies. *Eighth,* develop environmentally-sound technologies for recycling and recovery. *Ninth,* promote the use of Green Products. Recently, Greenpeace has released 'A Guide to Greener Electronics', which ranks major Indian computer manufacturers on their green performance. The ranking guide presents a snapshot of company policy on harmful substances and e-waste (Greenpeace 2007 a). More such ranking of electronic products are needed. *Tenth,* promulgate a national e-waste policy which is consistent with our socio-legal environment. *Last but not least,* international cooperation for sharing of knowledge and technology among nations is very much needed.

Bibliography

Aggarwal Ravi (2006) www.toxiclink.org.

Boralkar, D.B. (2004), 'Perspectives of Electronics Waste Management', Paper presented at National Workshop on 'Electronic Waste Management' organised by CPCB/MoEF/ EMPA / GTZ in New Delhi during March 2004.

Data Quest (2005), 'The Toxic Tech' June 27, 2005.

Economic and Political Weekly (2006), "Electronic Waste Need for Comprehensive Solution", June 17, 2006. p. 2400.

Greenpeace (2007a), "A Guide to Greener Electronics", *The Hindu,* Wednesday, August 15, 2007 p. 4.

Greenpeace (2007 b) 'India Indifferent to E waste Contamination', *The Hindu,* Monday, February, 19.

The Hindu (2007), 'HCL launches Eco-Friendly PCs', *Hindu* Delhi edition, October 10, p. 15.

Whankade Kishore (2004a),"E-Waste System Failure Imminent", *www.toxiclink.org* published on 01/03/2004.

Whankade Kishore (2004b), "Is India Becoming a Dumping ground for British E-Waste" available at *www.toxiclink.org* published on 24-09-04.

Manufacture's Association of Information Technology website *www.mait.org.*

National Environment Policy 2006, Ministry of Environment and Forests, Government of India, New Delhi.

Toxic Link (2004), 'E-waste in Chennai—Time is Running Out' published by Toxic Link, 01/01/2004.

Toxic Link (2006). 'E-waste Time to Act Now', by Satish Sinha, www. toxic link issue 01 December 2006.

Toxic Link (2003), 'Scrapping the Hi Tech Myth-Computer Waste in India', *www.toxclink.org* published on 01/01/2003.

UNEP(2005), "E waste- the Hidden Side of IT Equipment's Manufacturing and Use" Environment Alert Bulletin, UNEP, January.

Endnotes

1. 'Maran's prescription for tackling e-waste' by K.S. Sudhakar Published in The Hindu, 11/08/2004.
2. The Basel Convention is an international convention on the Control of the Trans-boundary Movement of Hazardous Waste and their Disposal and was adopted in 1989 and entered into force in 1992. It was created to prevent the economically motivated dumping of hazardous wastes from richer to poorer countries. The Basel Ban Amendment, adopted in 1995, prohibits all exports of hazardous wastes from Parties that are member states of the EU, OECD and to all other Parties to the Convention. The United States is the only OECD country not having ratified the original Basel Convention, nor the Basel Ban Amendment. Thus, the export of e-waste as has been witnessed to China, India and Pakistan is in violation of the Basel Convention and the Basel Ban Amendment (UNEP 2005).
3. Adopted from *www.computertakeback.com/legislation_and_policy/essentials.cfm*

Micro Perspectives

Chapter 15

Industrial Effluent and Rivers: Lessons from Noyyal River Basin in Tamil Nadu in the Light of National Environment Policy, 2006

K. Govindarajalu

Introduction

Industrial pollution has been and continues to be a major factor causing the degradation of the environment around us, affecting the water we use, the air we breathe and the soil we live on. But, of these, pollution of water is arguably the most serious threat to current human welfare. Water is polluted not only by industries but also by households. Both industries and household waste water contains chemicals and biological matter that impose high demands on the oxygen present in water. Polluted water thus contains low levels of dissolved oxygen as a result of the heavy biological oxygen demand (BOD) and chemical oxygen demand (COD) placed by industrial and household waste materials discharged into water bodies and water systems, both above and below the earth's surface. In addition to low levels of dissolved oxygen in water, industrial wastes (effluents) also contain chemicals and metals that are directly harmful to human health and to the ecosystem.

The supply of water through river valley projects and groundwater extraction thus has repercussions for the health and safety of people. Apart from health effects, which indirectly affect human productivity, polluted water also affects land productivity. Crop production suffers from using contaminated irrigation water from both surface sources and from groundwater aquifers. The definition of pollution in economics is based not only on the physical effect on the environment but crucially on the human response to the physical effect. The physical effect can be biological (e.g. change in biodiversity, or ill-health) or chemical (e.g. effect of harmful chemicals from industrial waste on aquatic life in a river). The human response is as a result of a loss of welfare (Seneca and Taussig, 1984; Pearce and Turner, 1990). Thus, while physical scientists base their perception of pollution solely on the physical effect on the environment, economists rely on perceived utility losses to recognise and define pollution.

Previous Studies

The following are a few important studies related to the impact of water pollution on agriculture, livestock and human health. Murthy, James and Smita's (1999) study 'Economics of Water Pollution' has mainly focused on the impact of water pollution on different sectors. This study has traced the theoretical and applied approaches to pollution control and its accountability, in particular the study by Smita Misra. A case study of the Nandesari Industrial Area has been considered a significant work in this field. The urban survey conducted for Vadodara city and the rural survey conducted for six villages surrounding Nandesari Industrial Estate are attempts to estimate these villages. The findings indicate that large potential welfare gains could be generated by water pollution abatement practices at Nandesari and underscore the need to undertake the necessary abatement measures.

Another very relevant work by Xia Guang (2000), 'An Estimate of the Economic Consequences of Environmental

Pollution in China', has empirically and quantitatively assessed the impact of environmental pollution. Bhagirath and Reddy's (2000) study on 'Environment and Accountability—Impact of Industrial Pollution on Rural Communities' has attempted to study the environmental impact of water pollution on rural communities in general and on agricultural production, human health and livestock in particular.

Some of the important issues analysed in this study are:

(a) Linkages between industrial development and changes in the micro (local) environment,

(b) Damage to crops and animal husbandry due to industrial pollution,

(c) Impact on health and sanitation in rural communities.

To conduct this study, the authors have selected a village under the Patancheru industrial belt in the Metak district of Andhra Pradesh. It is one of the oldest and most environmentally degraded areas by industrial pollution. The entire village has been suffering from various diseases arising out of water pollution. From this study it was observed that most of the diseases, such as skin infection, defective vision, fever, teeth corrosion, joint pains, loss of appetite, abdominal pain, respiratory diseases, and diarrhoea are water-borne.

Another significant study, 'Environmental Impact of Industrial Effluents in Noyyal River Basin' conducted by the Madras School of Economics (2002) has analysed the impact of polluted water on agriculture sector, drinking water and fisheries activities. The study has been conducted using primary data collected from the basin farmers and respondents in Tirupur area and also secondary data collected from the concerned officials. Data related to water quality, area under cultivation, cropping pattern, sources of drinking water, workforce participation, etc. have been used to assess the net economic loss in the Noyyal river basin. For this purpose, the values are estimated both for affected area and unaffected area, as well as before pollution and after pollution. In this background, the present study has also been undertaken to assess the impact of polluted water on the health status of the villages in Noyyal river basin.

Noyyal River Basin

The major and minor rivers flowing in Tamil Nadu used to be the major source for surface and groundwater. One of such rivers flowing in the Coimbatore, Erode and Karur districts is Noyyal River. It is a seasonal river, originating from the Vellingiri hills in the Western Ghats of Coimbatore district. It flows through Coimbatore, Erode and Karur districts and finally joins Cauvery River near the Noyyal village. It flows over a distance of 175 km. The catchments area of the river is 3.49 lakh hectares. Throughout its distance on both sides of the banks of River Noyyal, more than 100 villages are situated. Noyyal was the major source for irrigation, drinking water and other activities of the people living on both sides of the river and even for people living beyond 3 km from the river.

The average rainfall in the basin is about 700 mm. This river is the only source for 31 tanks and many minor canals. All the supply canals and tanks have a command area of 6,550 hectares. All these activities and systems were working well up to 1989 and the farmers were also growing commercial and food crops. What happened afterwards is the major issue of this study. In this context, the industrial effluent released by dyeing and bleaching factories in Tirupur, a major hosiery centre in South India, has become a serious issue because it has a severe impact on water bodies. The effluents released after semi-treatment or without treatment are let into the Noyyal river. At present there about 750 dyeing and bleaching industries in Tirupur. The effluents released by these units are stored in Orathupalayam Dam, which was constructed during 1991 at a cost of Rs. 1,646 lakh. The water-spread area is 1049 acres and it was expected to irrigate around 20,000 acres of land in three districts.

The present study was undertaken to assess the impact of polluted water (industrial effluent) on agricultural sector, health, livestock and fisheries activities. Thirty-one villages were selected for the study and household survey was conducted using pre-tested questionnaires during October, November and December 2002. Around 600 households have been selected by getting assistance from the concerned Village Administrative Officer (VAO) and then using a random sampling technique to get first-hand information about pre- and post-pollution agricultural and allied economic activities of the farmers, health status, water source and availability, livestock details and participation in fisheries activities.

Impact on Agricultural Production

Before pollution

In this section of analysis, the average agricultural production per acre in sample villages (before pollution) for different crops is presented in the Table 15.1. Before pollution (1985-86), almost all the crops including paddy, sugarcane and turmeric were cultivated in 80 per cent of the land in the sample villages. They gave very good yield during this period. The sample villages got tank and well irrigation. The average production, area under cultivation, total production and gross income is presented in Table. 15.1. The average annual gross income from the cultivated lands was worked out from the village survey data and VAOs record and it is estimated as Rs. 23.89 crore for the 31 sample villages.

Table 15.1: Average production per acre in sample villages—Crop wise (Before pollution – 1985-86)

Crops	Average production in kg/tonnes	Area cultivated (in acres)	Total production kg/tonnes	Gross income (In lakhs)
Paddy	2500x2 kg	956	47,80,000	23.90
Sugarcane	60 tonnes	392	23,520	112.89
Cotton	1500 kg	2854	42,81,000	1070.25
Tobacco	1500 kg	1723	25,84,500	387.67
Turmeric	2500 kg	484	12,10,000	302.50
Coconut	15,000 kg	217	32,55,000	130.20
Banana	1200 units	220	2,64,000	105.60
Cholam	1000 kg	2958	29,58,000	147.90
Vegetables	375 kg	1427	535125	21.40
Oilseeds	500 kg	1584	792000	87.12
Total	-	-	-	**2389.43 (23.89 crores)**

Source: Village primary survey and VAOs record.

After pollution

The average production per acre in the sample villages for different crops during 2002 (after pollution) is presented in Table 15.2. It is evident that the area under cultivation has remarkably declined during the past 10 years. The accumulating water pollution had a severe negative impact on water quality, soil condition and consequently on agricultural production. Paddy has almost disappeared from the cropping pattern of the sample villages. Recently, a few farmers had tried paddy crop using their well water, in a small area. But due to the salinity of water they were unable to be fruitful in their attempt. The same was the result in the case of sugarcane, cotton, tobacco, banana, turmeric and oilseeds. Only a few farmers were growing cotton, tobacco and oilseeds in a smaller portion of the land. Occasional rain in that area only had helped them to raise a meagre production of these crops.

Table 15.2: Average production per acre in sample villages—Crop wise (After pollution –2002)

Crops	Average production in kg/tonnes	Area cultivated (in acres)	Total production kg/tonnes	Gross income (In lakhs)
Paddy	Nil	Nil	Nil	Nil
Sugarcane	35 tonnes	80	2800	15.40
Cotton	250 kg	723	180750	36.15
Tobacco	900 kg	389	350100	52.51
Turmeric	1200 kg	62	74400	46.12
Coconut	3750 units	64	240000	4.80
Banana	1200	41	49200	7.38
Cholam	It is cultivated for cattle feed only. No yield	1098	-	65.88
Vegetables	245 kg	74	18130	1.08
Oilseeds	375 kg	186	69750	10.46
Total	-	-	-	**239.78 (2.39 crore)**

Source: Village primary survey and VAOs record.

The average production of the study area during this period was only Rs. 2.39 crore. Most of the lands in this area were sown with cholam. It is cultivated mainly for feeding the cattle and selling it in the market as cattle feed. The approximate and average net loss in the agricultural sector is estimated as Rs. 21.50 crore per annum.

Depreciation of land values in sample villages

Another significant observation made in the study is the depreciation of land values during this period. Land value per acre both for wet and dry has been estimated using the market value and VAO's record. Due to water pollution and deterioration of soil quality, the market value of land in the sample villages has considerably reduced. Having poor agricultural activity on the one side and decrease in the market value of their land on the other, the farmers' livelihood is worst affected. Before pollution, the lowest land value of wet land was at Rs.55, 000 per acre in Kullayoor village and the highest value was at Rs.2 lakh in Edakkadu village. The average land value per acre was Rs. 1.45 lakh in the study area. But, after pollution the average per acre values was estimated at Rs.25,000 per acre.

Percentage decrease in livestock population

In Table 15.3, the percentage decrease in livestock population is furnished. For this purpose, details about cattle population were collected for 1987, 1992, 1997 and 2002. During this period, the cattle population has gradually decreased. The total population was 25,231 during 1987 and it reduced to 12,169 during 2002. The category-wise percentage decrease was 62.27 per cent in cows, 60.70 per cent in the case of buffaloes and 48.50 per cent in the category of sheep and goat. On an average, there was 51.80 per cent decline in the cattle population in the study area during the past 15 years.

Table 15.3: Percentage decrease in livestock population

Cattle	Population				Percentage Decrease
	1987	1992	1997	2002	
Cow	1760	1226	884	657	62.67
Buffalo	5066	3941	3097	2032	60.70
Sheep and Goat	18405	14661	11745	9480	48.50
Total	25231	19828	15726	12169	51.80

Source: Village primary survey and village records (2002).

Cost of maintenance and return from livestock

In Table 15.4, the details about the average cost of maintenance and return from livestock are presented both for the period before pollution and after. Before pollution, the net income from buffalo was Rs. 2,100, Rs.1,150 from cows and Rs.1,450 from sheep and goat. But, it was only Rs.120, Rs.330, Rs. 750 for the three respectively after pollution.

Table 15.4: Cost of maintenance-Return from livestock (in Rs.)

Category		Before pollution (1987)			After pollution (2002)			Total
	Average value	Cost of maintenance	Return	Net Income p.m	Cost of maintenance	Return	Net Income p.m	Loss per cattle per month
Buffalo	12,000	700	3,600	2,100	2,040	2,160	120	1,980
Cow	9,000	1,050	2,200	1,150	1,890	2,220	330	820
Sheep and Goat	2,000	-	1,450	1,450	-	750	750	700

Source: Computed from village primary survey and village records (2002).

Net loss in livestock

On the basis of the average cost and return from livestock rearing, the net loss is approximately estimated and presented in Table 15.5. Due to significant reduction in return and depreciation in the market value of the animal and increasing maintenance cost, the average net loss per animal was Rs.23,760 for buffaloes, Rs. 9,840 for cows and Rs. 8,400 for sheep and goats. The total loss approximately estimated was Rs. 6.52 crore between 2000 and 2002.

Table 15.5: Net loss in Livestock rearing

Category	Number of animals	Average net loss per annum/ per animal	Average loss (Rs. in crore (2000-2002)
Buffalo	657	23,760	1.56
Cow	2,037	9,840	2.00
Sheep and goat	9,480	8,400	2.96
Total	-	-	6.52

Source: Computed using village primary survey data and village records (2002).

Apart from this estimation, data regarding deaths and sickness of livestock during 2000 to 2002 was also collected through village survey. It was evident from the medical camp and doctor's observation that there was death of sick cattles during 2000–2002. It is presented in the Table 15.6. There were about 1,183 cattles found sick and 809 cattles found dead during the period 2000–2002. The average total loss due to death of animals is estimated as Rs.48.10 lakh.

Assessment of Health Status

The health status of the villagers of the study area was assessed through health camps exclusively conducted for the present study. For this purpose, three health camps were conducted, covering 21 villages. In the first medical camp, conducted at Arugampalayam centre around 250 villagers attended. In the second camp around 445 villagers attended. In the third medical camp, around 425 villagers attended. It is observed from all the three camps that there were symptoms of skin allergy, gastritis and respiratory problems among the villagers. The details are presented in the Tables 15.6, 15.7 and 15.8.

Table 15.6: Health status - First medical camp

Name of the villages covered	Date of the camp	No. of doctors participated	No. of patients attended	Specific diseases identified
Anaipalayam Anaipalayam pudur Suppanoor Arugampalayam Kullayoor Edakkadu Pallapalayam	15.12.02	4	250	Skin allergy, Gastritis, Hypertension, Joint pain, Respiratory problem

Source: Medical camp 2002.

Table 15.7: Health Status—Second Medical Camp

Name of the villages covered	Date of the camp	No. of doctors participated	No. of patients attended	Specific diseases identified
Kathanganni, Reddipalayam, Kannimarkovil pudur, Karaipudur, Pallanaickanpalayam, Vayakkatupudur, Thottipalayam,	29.12.02	4	445	Skin allergy, Gastritis, Hypertension, Joint pain, Respiratory problem

Source: Medical camp 2002.

Table 15.8: Health Status—Third Medical Camp

Name of the villages covered	Date of the camp	No. of doctors participated	No. of patients attended	Specific diseases identified
Orathupalayam, Ramalingapuram, Kodumanal, Siviarpalayam, Pallakkatupudur, Sokkanathampudur,	5.01.2003	4	425	Skin allergy, Gastritis, Joint pain, Respiratory problem, Ulcer

Source: Medical camp 2003.

Findings

1. The majority of villages in Noyyal River Basin below Tirupur was once (up to 1990) a fertile region producing paddy, sugarcane, cotton, turmeric, banana, oilseeds etc.
2. Consequent to the development of dyeing and bleaching industries in Tirupur after 1991, Noyyal River has become a river of effluente.
3. There was clear evidence that the partly treated and untreated industrial effluent has spoiled the surface water and well water in the Noyyal River basin.

4. The available water in the wells, tanks and reservoirs was not suitable for human and domestic use.
5. Due to high TDS, the available water in the wells and tanks were more saline in nature and in turn it has spoiled the soil quality.
6. The study has identified those 31 villages in the Noyyal River basin which were directly affected by the polluted water.
7. In the agricultural sector, around 4325 acres of land were directly affected and around 10,000 acres of land had been indirectly affected.

Present Status

1. The Madras High Court Verdict (2006) on Industrial Effluent Discharge in Noyyal River imposed a fine of 6 paise per litre of effluent.
2. In addition to this, dyeing units have been asked to install Reverse Osmosis (RO) plants to ensure zero effluent discharge before 31.12.2006.
3. Most of the units failed to abide by the High Court Verdict, quoting the reason as heavy financial commitment on the part of dyers.
4. Due to this problem, the dyers halted production.
5. In the meantime, Centre for Ecology Loss of Anna University studied the issue and recommended a compensation of Rs. 118 crore to the affected parties. But the Justice Baskaran Commission awarded only Rs. 24 crore as compensation.
6. Again farmers have appealed to the High Court.
7. In this issue, the Tirupur Exporters' Association has suggested discharge of effluent into the sea.
8. Another suggestion floated by a legislature was that Government should pay the fee or it can install

effluent recycling plant and collect user charges from dyers.

9. Now, the dyers' responsibility is on two issues—on the payment of compensation to the affected farmers, and on meeting the cost of safe disposal of effluent. The problem has yet to be resolved.

Provisions in National Environment Policy 2006

Lessons from Noyyal River Basin are a clear example of the impact of water pollution on different sectors. It cannot be considered as a negligible issue. In fact, it is a case of negligence and failure to protect natural resources and streams when we plan for industrial development and export promotion. The National Environment Policy 2006 has made clear objectives and principles related to water pollution, which need strict adherence.

Key Environmental Challenges

- Environmental degradation is a major causal factor in enhancing and perpetuating poverty, particularly among the rural poor, when such degradation impacts soil fertility, quantity and quality of water, air quality, forests, wildlife and fisheries.
- Policy failures can emerge from various sources, including the use of fiscal instruments, such as explicit and implicit subsidies for the use of various resources, which provide incentives for excessive use of natural resources. Inappropriate policy can also lead to changes in commonly managed systems, with adverse environmental outcomes.

Objectives

- Minimise adverse environmental impacts
- Management and regulation of use of environmental resources

- To integrate environmental concerns into policies, plans, programmes and projects for economic and social development.

Principles

1. **The Precautionary Approach**

   Where there are credible threats of serious or irreversible damage to key environmental resources, lack of full scientific certainty shall not be used as reason for postponing cost-effective measures to present environmental degradation.

2. **Polluter Pays**

   'If the costs (or benefits) of the externalities are not re-visited on the party responsible for the original act, the resulting level of the entire sequence of production or consumption and externality is inefficient'.

3. **Entities with 'Incomparable' Values**

   Significant risks to human health, life, and environmental life-support systems, besides certain other unique natural and manmade entities, which may impact the well-being, broadly conceived, of large numbers of persons, may be considered as 'Incomparable' in that individuals or societies would not accept these risks for compensation in money or conventional goods and services. A conventional economic cost – benefit calculus would not, accordingly, apply in their case, and such entities would have priority in allocation of societal resources for their conservation, without consideration of direct or immediate economic benefit.

4. **Legal liability**

   The present environmental redressal mechanism is predominantly based on doctrines of criminal liability, which have not proved sufficiently effective, and need to be supplemented. Civil liability for environmental damage would deter environmentally harmful actions, and compensate the victims of environmental damage. Conceptually, the principle of legal liability may be viewed as an embodiment in legal doctrine of the 'polluter pays' approach, itself deriving from the principle of economic efficiency.

5. **Public Trust Doctrine**

   The State is not an absolute owner, but a trustee of all natural resources, which are by nature meant for public use and enjoyment, subject to reasonable conditions, necessary to protect the legitimate interest of a large number of people, or for matters of strategic national interest.

6. **Preventive Action**

   It is preferable to prevent environmental damage from occurring in the first place, rather than attempting to restore degraded environmental resources after the fact.

Conclusion

In the light of the National Environment Policy 2006, environmental issues, like water pollution due to industrial effluent and sewage, have gained utmost importance and prompted serious discussion. It is painful to note that degradation of river streams and consequent impacts on urban and rural community are irreversible. Hence, the role and mechanism of processing industries, chemical and pharmaceutical industries can be observed seriously and the cost of pollution abatement should not be an excuse by the concerned. It is time to act in the right direction, at least for the sake of future generations.

References

Appasamy P.Paul. et al. (2000), "Economic Assessment of Environmental Damage-A Case Study of Industrial Water Pollution In Tirupur", Madras School of Economics, Chennai.

Azeez P.A (2001), "Environmental Implications of Untreated Effluents from Bleaching and Dyeing", in Senthilnathan.S.(Ed) *Ecofriendly Technology for Waste Minimisation in Textile Industry* , CEE, Tirupur and Environment Cell Division, PWD, WRO, Coimbatore.

Bhagirath Behera, Reddy, V. Rathna (2002), 'Environment and Accountability—Impact of Industrial Pollution on Rural Communities', *Economic and Political Weekly*, January 19.

Bhattacharya. R.N.(Ed) (2001),"Environmental Economics-An Indian Experience, Oxford University Press.

Chang Yonguan, er.al. (2001), "The Environmental Cost of Water Pollution in Chongquing, China", Environment and Development Economics, Vol.6, Part 3. July.

Charles Ford Runge (1987), "Induced Agricultural Innovation and Environmental Quality: The Case of Groundwater Regulation", Land Economics, vol 63. No.3 August.

Charlisle W. et al. (1992), "Valuing Environmental Quality Changes using Averting Expenditures: An Application to Groundwater Contamination", Land Economics, vol. 68 (2), May.

Ecological Economist Unit – Institute for Social and Economic Change, (1999), 'Environment in Karnataka' A Status Report, Economic and Political Weekly, Sep.18.

Govindarajalu.K. (2003), " A study on the Assessmnet of Economic Losses and Environmental Damages in and around Tirupur and Orathupalayam Dam in Noyyal River Basin", Report submitted to Environmental Cell Division, Public Works Departemnt, Government of Tamilnadu.

Gunnar S. Eskeland and Emmanuel Jimenez (1992), "Policy Instruments for Pollution Control in Developing Countries", The World Bank Research Observer, vol.7, No.2.

Jodha.N.S. (1998), "Poverty and Environmental Resource Degradation- An Alternative Explanation and Possible Solutions", Economic and Political Weekly, Sep.5-12.

Kadekodi, Gopal, (2004), "Environmental Economics in Practice", Oxford University Press.

Karl-Goran Malor, (1997), "Environment, Poverty and Economic Growth", World Bank Annual Conference on Development Economics. Page 251-270.

Living Standards Measurement Survey Unit, (2002), "Survey of Living Conditions- Village Questionnaire (Uttar Pradesh and Bihar) ", World Bank.

Living Standards Measurement Survey Unit, (2002), "Survey of Living Conditions- Manual", World Bank.

Madras School of Economics(2002)" Economic Assessment of Environmental Damages- A Case Study of Water Polllution in Tirupur", Research Report

Murthy, M.N., A.J. James and Smita Misra (1999), Economics of Water Pollution—The Indian experience, oxford University Press, Paperbachs., New Delhi

M.N., et al. (2000), '*Economics of Water Pollution—The Indian Experience*', Oxford University Press, New Delhi.

National Environment Policy-2006, Ministry and Environment and Forests, Government of India.

Nemat Shafik (1994), "Economic Development and Environmental Quality-An Econometric Analysis", Oxford Economic Papers, 46.

Partha Dasgupta and Korl-Goran Malor,(1990), " The Environment and Emerging Development Issues", Proceedings of the World Bank Annual Conference on Development Economics, Page 101-131.

Pearce, D.W. and R.K. Turner (1990), '*Economics of Natural Resource and Environment*' Harvester Wheatsheaf, New York.

Purnamita, Dasgupta (2006), "Valuing Health Damages from Water Pollution in Urban Delhi, India: A Health Production Function Approach", Institute of Economic Growth, New Delhi.

Reddy.V. Ratna (1995), "Environment and Sustainable Agricultural Development", Economic and Political Weekly, March 25.

Ronald.H. Coase. (1960), "The Problem of Social Cost", Journal of Law and Economics, No.3.

Shankar.U. (2001), "Environmental Economics", Oxford University Press.

Seneca, Joseph and Michael. K. Taussig (1984) *Environmental Economics.* Prentice Hall, New Jersey.

Venkatachalam.L. (2001), "Environmental Economics for Efficient Water Resource Management- Manual," Water Research Organis ation, PWD, Coimbatore. vol.1, 2, 3.

Vyas. V.S. and Reddy, V. Ratna (1998), "Assessment of Environmental Policies and Policy Implementation in India", Economic and Political Weekly.Jan.10.

Water Resource Organisation, (2001), "Noyyal River System- Hydraulic Particulars- Report", PWD, Coimbatore

Water Resource Organisation, (2001), "Report on Noyyal Orathupalayam Dam Schemes", PWD, Coimbatore.

Xia Guang (2000), 'An Estimate of the Economic Consequences of Environmental Pollution in China', Policy Research Center of the National Environmental Protection Agency, Beijing, China.

Chapter 16

Export Earning Industries vs Environmental Sustainability: The Case of Tirupur Knitwear Industries in Tamil Nadu, 1980-2005

Velayutham Saravanan

I. Introduction

Environment and sustainable development has been accorded great emphasis since the last quarter of the 20th century. Sustainable development is defined as meeting 'the [human] needs of the present without compromising the ability of future generations to meet their own needs' (World Commission on Environment and Development 1987: 8). This concept implies that there is a limit to environmental resources and the ability of the biosphere to absorb human activities. These limits are seen to have roots in technological inadequacies and inequitable social organisation. Thus, sustainable development must entail: a process of change, in which the exploitation of resources, the direction of investments, the orientation of technological development, and institutional changes to be made consistent to meet future as well as the present needs (World Commission on Environment and Development 1987: 9).

In recent years, sustainable resource use has been emphasised invariably both by developed and developing countries alike. It is believed that sustainability can be ensured through the market mechanism. The role of market is the exchange of commodities (goods, services, or resources) between producers/sellers and consumers/buyers within a specific geographic area during a given period of time. But the market mechanism has failed to ensure the sustainability of the resources. It is because the cost of goods or benefit of the production or consumption is not included in the market price, neither in the supply price nor in the demand price. In other words, market does not take into account the transaction cost while determining the cost or benefit of the goods. This spill-over effect produces the externalities in the economic system. The entire process is known as market failure. For example, the dyeing and bleaching industries are using a large quantity of water and discharge almost the same amount into the Noyyal river and other public places. This effluent discharge affects the ecology and environment in these geographical regions. Neither the producers nor the consumers meet the cost of the damages created to the ecology and environment.

The basic theory of environmental policy says, '...pollution as a public "bad" that results from "waste discharges" associated with the production of private goods' (Cropper and Oates, 1992: 678). To control the pollution and protect the scarce resources, theory suggests imposing taxes. Based on the theories, the state has imposed several restrictions by several laws and taxes. However, state intervention proved to be ineffective. The failure of market led to great threat for the ecology and environment. Given this background, here an attempt has been made to show how the export-earning industries challenge environmental sustainability, with focus on the Tirupur knitwear industries and the role of state, judiciary and civil society in Tamil Nadu, in a historical perspective.

The Tirupur knitwear industry has emerged as one of the largest foreign exchange-earning industries in the past two decades. It accounts for about 90 per cent of India's cotton knitwear export, worth an amount of Rs. 5000–6000 crore. It also provides employment opportunities for more than three lakh persons. But the grave menace caused by these industries to environment and ecology is immense. Though the state has enacted several laws, its intervention has proved to be ineffective. Imperatively, the state has encouraged the knitwear industries to export so as to accrue foreign exchange at the expense of pollution caused to ecology and environment of this region owing to its ineffective implementation of the pollution control acts and rules. Neither the producers nor the consumers meet the cost of the damages created to the ecology and environment. Hence, the civil societies and non-governmental organisations brought the issues to the notice of the judiciary. Even the judicial interventions were ineffective in curtailing the pollution.

The industrialists are causing great damage to the natural resources in two ways. One is that they are diverting the water from the neighbouring regions and river basins, creating water scarcity in these basins. The Second is that, after using the water, effluents are being discharged to the neighbouring regions, which pollute the groundwater, health and the total environment of the region. The state is facilitating the industrialist by providing the necessary inputs and it has overlooked the damages caused to the entire region and its sustainability. In other words, the profit is going to the industrialists but the damages are pertinently distributed equally among the life and environment of Noyyal River Basin. Given this scenario, the obvious question to be raised is: do the export earnings of some industrialists compensate for the havoc done to the environment of the entire basin areas?

This paper consists of nine sections. The second section analyses the growth of Tirupur city and its economy. The third section sketches the origin and growth of industries in Tirupur. The fourth section gives the trends of export from Tirupur. The fifth section discusses the existing environmental laws and rules. The sixth section analyses the ecology and environmental damages. The seventh section narrates the role of civil societies and judiciary intervention. The political economy of knitwear industries is discussed in the eighth section and the last section ends with the concluding observations.

II. Growth of Tirupur City and Economy

Tirupur city is popularly known as 'Dollar City', 'Knit City', 'Cotton City' and mainly 'Hosiery Centre'. The city is located in the south-western parts of Tamil Nadu on the banks of Noyyal River, about 50 km to the east of Coimbatore city, lying between the 11°7' Northern Latitude and 77° 5′ Eastern Longitude. The geographical extension of Tirupur city was a very recent phenomenon. Till Independence, and even for some years later, Tirupur was a characteristic small town; it became a town with the inclusion of Thennampalayam, Karuvampalayam, and Valipalayam villages on 1st December 1947 (Tirupur city Corporate Plan 1999:23). In 1971, the total geographical area of Tirupur town was only 11.92 square km; but by 1991 it had increased to 43.52 square km. Similarly, the Tirupur urban agglomeration area, which was 54.86 square km in 1961, had spread to 90.98 square km in 1991.

The population growth rate of Tirupur city was moderate till 1921. In fact, the population growth rate had declined between 1911 and 1921 due to an endemic epidemic in 1917-18. After that, population growth rate has witnessed a sharp rise. The decadal population growth rate was more than 50 per cent between 1931 and 1961. It was very high, i.e. 83 per cent in 1941 (see Table 16.1 and Figure 16.1). Though the population growth rate has declined since 1961, the decadal growth rate was higher when compared with Coimbatore city. For example, the population growth rate of Coimbatore city was 15.45 per cent in 1981–1991 whereas it was 42.63 per cent for Tirupur city. A large number of people from the villages of Coimbatore district and other districts of Tamil Nadu migrated towards Tirupur mainly in search of unskilled labour in the dyeing and bleaching industries.

Table 16.1: Population growth in Tirupur city and Agglomeration: 1881–2001

Year	Town			Urban Agglomeration		
	Area (sq.km)	Population	Decadel Growth Rate (%)	Area (sq.km)	Population	Decadel Growth Rate (%)
1881		3,681				
1891		5,235				
1901		6,056			6,056	
1911		9,429	55.70		9,056	55.70
1921		10,851	15.08		10,851	15.08
1931		18,059	66.43		18,059	66.43
1941		33,099	83.28		39,195	117.04
1951		52,479	58.55		60,465	54.27
1961	27.20	79,773	52.01	54.86	97,965	62.02
1971	31.92	113,302	42.03	73.65	151,127	54.27
1981	43.52	165,223	45.83	90.98	215,859	42.83
1991	43.52	235,661	42.63	90.98	306,237	41.87
2001		346,551			542,787	

Sources: Census of India 1981, Series 1, *Primary Census Abstract, General Population,* Part IIB(i), pp.435; Census of India 1991, *Primary Census Abstract, General Population,* Part IIB(i).

Census of India 2001, Paper 2 of 2001.

A large number of population has migrated to the Tirupur city. However, there is no immigration data available for Tirupur city till 1961. The total population of

Figure 16.1 Population trends in Tirupur City and UA

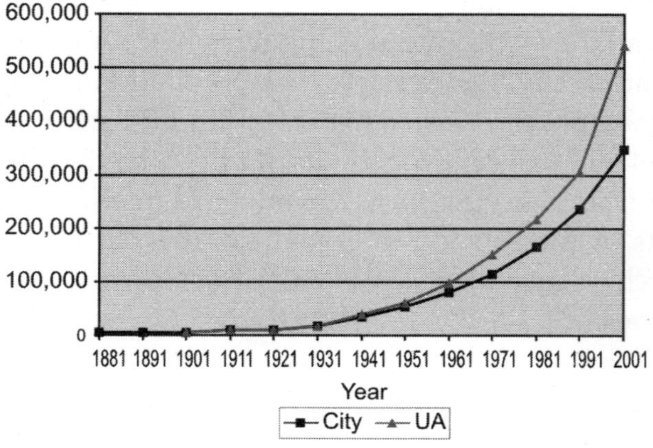

Tirupur city in 1961 was 151,127, of which 63,820 persons or 42.23 per cent were migrants. Of the immigrants, 79 per cent were from Coimbatore district alone; 13 per cent from the other districts of the state and 7 per cent from the other states of the country. In 1971, inmigrants were largely drawn from Coimbatore district itself, following a spurt in the number of housing industries, which demands heavy manual labour. The 1991 Census shows little details of immigration of Tirupur city and for the agglomeration area.

The immigration trend was high not only in the Tirupur city but also for agglomeration area in recent decades. In 1971, the total population of the Tirupur agglomeration area was 151,127 people, of which 63,820 (42.23 per cent) was migrants. Of these, about 80 per cent were from the Coimbatore district; 13 per cent from the other districts of Tamil Nadu and about 7 per cent from the other states of the country. The rate of migrants also declined and the pattern also had changed in the Tirupur agglomeration area in the subsequent decades. In 1991, half of the migrants were from within the district whereas 41 per cent were from the other districts of the state. Over a period, the proportion of immigrants from Coimbatore district has come down remarkably. A recent survey also suggested that a large proportion of migrant workers were from the neighbouring districts like Erode, Salem and Madurai as well as from other parts of the State (Cawthorne, 1993:23, Neetha, 2002). Initially, the Backward Community people were mainly engaged in these activities and later other weaker sections also got engaged in the works. Neetha states: 'Traditionally, the weaving community belonged to the Backward Caste (BC), 'gounders', and initially the workers in all the units were mostly from this caste. However, with the growth of the industry and with the flow of the migrants the workers caste composition has changed. At present Scheduled Castes, who were traditionally agricultural labourers, are seen joining the industry in large numbers though it is still dominated by the backward castes' (Neetha, 2002). It categorically shows that industrialisation and the economic transformation process had attracted immigrants largely from the other districts of the State in recent years. The declining rate of inmigration may be the reason for the low growth rate of population in Tirupur city as well as its agglomeration area in recent years.

Until the first quarter of the 20th century, the Tirupur economy was predominantly agrarian in nature. Almost all the workers were involved in the primary sector activities. This trend changed during the second quarter of the 20th century. From 1930 onwards, there was a shift towards the secondary and tertiary sectors. In 1961, a little more than half of the workforce in Tirupur city was engaged in the secondary sector activities. It got increased to two-third (65 per cent) of the workforce in 1991. As against this, the proportion of the workforce in the primary sector had declined from 4.11 per cent in 1961 to 0.68 per cent in 1991. However, in the service sector the proportion of the workforce has declined from 45.34 per cent to 34.75 per cent for the same period. In short, within a period of six decades, the Tirupur economy has changed completely from the primary to secondary and tertiary sector activities as the hosiery industries are more labour intensive in comparison to other industries. Not only Tirupur city, the agglomeration area had also transformed remarkably. The percentage of the workforce involved in the primary sector remained around 4 per cent between 1961 and 1991. It indicates that the economy of the agglomeration area also had transformed into the centre of non-agricultural activities. About two-third of the workforce in Tirupur was engaged in the industrial sector due to the large number of dyeing and bleaching industries, and only one-third in the service sector.

III. Origin and Growth of Industries in Tirupur

Since the late 19th century, Tirupur was a centre of textile business dominated by yarn and cotton (Chari, 2000: 589; Neetha, 2002). However, Tirupur Municipality established the Cotton Market Committee only in 1921 (Tirupur City Corporate Plan). Till the early 20thcentury, Tirupur's economy was predominantly or completely dependent on agriculture and its allied activities. The first ginning mill was established in 1904. There were no major industries in Tirupur city till 1911, and even till the 1930s, there were only a few knitting and ginning mills. The first hosiery factory with hand-operated machines was set up in Tirupur in 1935. Since the 1930s, infrastructure facilities were developed in the Tirupur city. With the commissioning of Pykara electricity system in 1933, Tirupur city was electrified between 1936 and 1945. The Pykara electricity system proved to be a boon for the phenomenal growth of Tirupur. However, the industrial growth was very slow till Independence, in fact till the late 1960s (Krishnaswami, 1989: 1354). After Independence, the hosiery industries engaged in the manufacture of banian products was launched and in the subsequent decades their numbers also increased. The industrialists from other regions have also established their industry due to availability of factors of production. For instance, about 150 industrialists transferred their business centres from Calcutta (now Kolkata) to Tirupur between 1966 and 1975 (Shanmugam, 1994: 2).

In 1991, about 3,823 different kinds of hosiery industries viz. knitting mills, ginning mills, dyeing and bleaching mills, printing mills, etc. existed in Tirupur city. It has increased to 4150 in 2001 (see Table 16.2). It is traced that the figures are not clear as it covers only registered firms and it does not include other related subsidiary units—manufacturing of cartonnes, tapes, polythene bags and others (Neetha, 2002). In other words, the official data have not brought out much of the informal and unregulated activities giving a wide variation (Chari, 2000: 582; Steele, 2002). In addition to this, a number of small units has increased over the period. A large proportion of the knitwear firms employ 10 to 50 workers (Chari, 2000:583). The number of registered small-scale units of cotton textile and textile product (hosiery) units in Tirupur was only 1,143 in 1980, and it has increased to 9,319 units in 1997 (Appasamy 2000: 22). Due to the increase in number of hosiery industries, a large number of labourers has migrated to Tirupur in search of employment opportunities. Unlike the industries in Coimbatore city, which required some technical knowledge, industries in Tirupur are largely manual labour oriented. Consequently, Tirupur absorbed a large number of migrants. Another facet of industrial growth in Tirupur was the setting up of water-consuming industries. Although there is a large number of water-consuming units in Tirupur, only 184 units were registered under SSI as water consuming industries till 1997 (Appasamy, 2000: 23).

Table 16.2: Industrial Growth in Tirupur City: 1911–2001

Year/Type of industry	1911	1921	1931	1941	1951	1961	1971	1981	1991	2001
Knitting Mills	-	-	5	22	35	250	415	1300	2800	2500
Dyeing and Bleaching mills	-	-	-	2	15	42	67	68	450	750
Printing on cloth	-	-	-	2	8	15	20	22	125	350
Label Looms	-	-	-	-	1	4	6	9	100	150
Cardboard making	-	-	-	-	2	9	18	48	253	200
Calendering	-	-	-	-	4	12	23	41	95	200

Notes: For 2001, compiled from the different sources.
Source: Municipal Office, Tirupur.

The numbers of industries also increased since 1925. In 1995, there were 8,437 industrial units in Tirupur city. Of these, 713 were water intensive units. The requirement of water for industrial use was 90 million litres per day. A large quantity of water is required mainly for the 526 dyeing units (Kalaimani, 1995: 8). As on January 1, 1997 there were 851 water-consuming bleaching and dyeing units in Tirupur (Madras School of Economics, 1998:33). Between 1997 and 2001, the numbers of dyeing and bleaching industries declined because of the closure of 164 units ordered by the High Court as they failed to control pollution. In the year 2000, as many as 2,500 knitting and stitching units, 750 dyeing and bleaching units, and another 735 supporting units existed in Tirupur. A large quantity of salt is used in the dyeing process, hence the wastewater (90 million litres per day) is highly saline in content and is contaminated with a variety of chemicals.[1] The increasing number of water-consuming bleaching and dyeing industries in Tirupur made the city one of most demanding for water over a or b the period (Saravanan 2007; Saravanan [a]; Saravanan 1998; Saravanan 2001; Saravanan 2004).

IV. Trends of Export Earnings in Tirupur

Tirupur knitwear products were started in 1973, mainly for the domestic market and the business was continued through the exporters in Mumbai. In 1978-79, Tirupur knitwear producers directly exported the product to other countries. Now, Tirupur knitwear products were exported to more than 35 different countries over the world during the last two decades.

Table 16.3: Exports of hosiery garments from Tirupur: 1984–2004

Year	No.of Pieces (Lakh)	Value in Rupees (in lakh)	Year	No.of Pieces (Lakh)	Value in Rupees (in lakh)
1984	104	969	1995	2,171	1,59,183
1985	172	1,869	1996	2,651	2,07,684
1986	289	3,748	1997	3,122	2,22,571
1987	334	7,448	1998	3,472	2,63,100
1988	459	10,424	1999	3,784	3,01,700
1989	614	16,739	2000	4,255	3,57,400
1990	889	28,985	2001	4,343	3,91,300
1991	905	42,948	2002	4,210	3,91,000
1992	1399	77,493	2003	3,812	4,53,400
1993	1839	11,6,243	2004	4,114	
1994	1964	13,1,800	2005		

Source: Computed with help of AEPC and TEA data – 2000. Up to 1999, data taken from Paul Appasamy and the rest of years from AEPC website.

The prominent countries are the European Union, the United States of America and Canada. About two-third of the knitwear was exported to European Union countries, about 10 per cent to United States of America, 7 per cent to Finland, Austria and Canada and the remaining percentages to several other countries, viz., Russia, Switzerland, Japan, Sweden, New Zealand, Singapore and numerous other countries. The export value has been increased from Rs.969 lakh in 1984 to Rs. 4,53,400 lakh in 2003. In 1984, 104 lakh of pieces were exported which increased to 4,114 lakh pieces in 2004 (Table 16.3). Recently, the Central Government Export Import Policy for 2003-04 has further encouraged the supportive schemes for export clusters, resulting in the increase of the knitwear industry in Tirupur. In addition, the quota-based curbs for textile exports to the United States and European nations were lifted on January 1, 2005. According to Sakthivel, President of Tirupur Exporters' Association, the export will increase to Rs.10,000 crore (50 billion US dollars) by 2010. The increasing trend of export products aggravates the problem forced by environment and ecology on Noyyal River Basin.

V. Existing Environmental Laws and Rules

The environment-related laws were enacted by the Parliament under Articles 252 and 253 of the Constitution of India. Due to the emergence of environmental movements in different parts of the world around the 1970s, the Indian government also launched environmental protection measures, viz. pollution control agency and pollution control acts. In other words, there was no pollution control agency either at the Central level or at the State level until the 1970s. At the Central level, the Ministry of Environment and Forests has framed environment-related policies, and the Central Pollution Control Board (CPCB) is acting as a monitoring agency. The Central Pollution Control Board as a statutory organisation was constituted in September 1974. As an implementing agency, the State Pollution Control Board has to inspect industries, monitor the pollution level and ensure the minimum standards. It is responsible for implementation of legislations relating to prevention and control of pollution. The Tamil Nadu Pollution Control Board was formed in 1982 and it employed a person in Tirupur in 1989. The State Board has direct control over the PCB in Tirupur. In Tirupur the Board is responsible for the supervision of the industries. The Tamil Nadu Pollution Control Board is the enforcing authority of the environmental laws by monitoring pollution levels and taking appropriate legal action against the defaulters. Further, the TNPCB has to ensure the suitability of the site for the near industrial establishment and the pollution control measures of the industry that meets the prescribed standards.

Since 1989, the TNPCB has been enforcing the following Pollution Control Laws/Rules relating to environmental protection in the State: (i)The Water (Prevention and Control of Pollution) Act, 1974 as amended in 1978 and 1988; (ii)The Water (Prevention and Control of Pollution) Cess Act, 1977 as amended in 1991; (iii) The Environment (Protection) Act, 1986; (iv) The Environment (Protection) Rules, 1986. According to the Water (Prevention and Control of Pollution) Act, 1974, Section 20 (2), 'A State Board may give directions requiring any person who in its opinion is abstracting water from any such stream or well in the area in quantities which are substantial in relation to the flow or volume of that stream or well or is discharging sewage or trade effluent into any such stream or well, to give such information as to the abstraction or the discharge at such times and in such form as may be specified in the directions'. According to the Water (Prevention and Control of Pollution) Act, 1974, Section 24 (1a) 'no person shall knowingly cause or permit any poisonous, noxious or polluting matter determined in accordance with such standards as may be laid down by the State Board to enter (whether directly or indirectly) into any [stream or well or sewer or on land].' According to the Water (Prevention and Control of Pollution) Act, 1974, Section 41(1) 'Whoever fails to comply with any direction given under sub-section (2) or sub-section (3) of section 20 within such time as may be specified in the direction shall, or conviction, be punishable with imprisonment for a term which may extend to three months or with fine which may extend to ten thousand rupees or with both and in case the failure continues, with an additional fine which may extend to five thousand rupees for every day during which such failure continues after the conviction for the first such failure'. Section 41(2) 'Whoever fails to comply with any order issued under clause (c) of sub-section (1) of section 32 or any direction issued by a court under sub-section (2) of section 33 or any direction issued under section 33A shall, in respect of each such failure and on conviction, be punishable with imprisonment for a term which shall not be less than one year and six months but which may extend to six years and with fine, and in case the failure continues, with an additional fine which may extend to five thousand rupees for every day during which such failure continues after the conviction for the first such failure'. Section 41(3) 'If the failure referred to in sub-section (2) continues beyond a period of one year after the date of conviction, the offender shall, on conviction, be punishable with imprisonment for a term which shall not be less than two years but which may extend to seven years and with fine'.

Further, this Act gives an option to the State Government to implement or exclude the particular area. The Water (Prevention and Control of Pollution) Act, 1974 became effective at the State level when the concerned State Assemblies adopted it. According to the Water (Prevention and Control of Pollution) Act, 1974, Section 19 (1), 'Notwithstanding contained in this Act, if the State Government, after consultation with, or on the recommendation of, the State Board, is of opinion that the provisions of this Act need not apply to the entire State, it may, by notification in the Official Gazette, restrict the application of this Act to such area or areas as may be declared therein as water pollution, prevention and control area or areas and thereupon the provisions of this Act shall apply only to such area or areas.

Environment (Protection) Act, 1986 (7) states that no person carrying on any industry, operation or process shall discharge or emit or permit to be discharged or emitted any environmental pollutants in excess of such standards as may be prescribed.

Based on the above Acts and rules, the State government also issued several orders. For instance, G.O. Ms. No: 213, Environment and Forests (EC-I) Department, dated the 30th March 1989: '…no industry causing serious water pollution should be permitted within one kilometer from the embankments of rivers, streams, dams etc. and that the Tamil Nadu Pollution Control Board should furnish a list of such industries to all local bodies. It has been suggested that it is necessary to have a sharper definition for water sources so that ephemeral water collections like rain water ponds, drains, sewerages (bio-degradable) etc. may be excluded from the purview of the above order. The Chairman, Tamil Nadu Pollution Control Board has stated that the scope of the Government Order may be restricted to reservoirs, rivers and public drinking water sources. He has also stated that there should be a complete ban on location of highly polluting industries within 1 kilometre of certain water sources'. In 1989, Tamil Nadu Government ordered that no industry causing serious water pollution should be permitted within one kilometer from the embankments of Noyyal River. Further, in 1998, government order further extended ban on setting of highly polluting industries within five km of the Noyyal River. It is evident that both the Central and State Governments has enacted several acts to control the pollution during the last three decades.

VI. Ecology and Environmental Damages

According to the Water (Prevention and Control of Pollution) Act, 1974, '"pollution" means such contamination of water or such alteration of the physical, chemical or biological properties of water or such discharge of any sewage or trade effluent or of any other liquid, gaseous or solid substance into water (whether directly or indirectly) as may, or is likely to, create a nuisance or render such water harmful or injurious to public health or safety, or to domestic, commercial, industrial, agricultural or other legitimate uses, or to the life and health of animals or plants or of aquatic organizers.'

Due to the demand both from within the country and rest of the world, the number of polluting industries has increased, resulting in environmental and ecological damages in and around the Tirupur and the entire downstream of Noyyal basin over the last three decades. These industries destroyed natural resources in two ways. One, they exploited natural resources like water and forest resources on a large scale, which resulted in decline of the water table; two the used water and other chemicals were discharged without any treatment affecting the ground water, productivity, health and the entire eco-system of the region, particularly after the 1970s. In 1980, there were only 26 units, which together consumed about 4.40 Million Liters Per day (MLD), whereas the number of units increased to 325 and consumed about 40.89 MLD in 1990. In 1997, the number of units further increased to 866, consuming about 106.91 MLD (Appasamy, 2000:38).

Not only has the number of water-consuming industries and the quantity of water increased, but the pollution load has also increased during this period. In 1980, the pollution load was 10,252 tonnes of Total Dissolved Solid (TDS) comprising 6,053 tonnes of Chloride, 420 tonnes of Sulphate,

482 tonnes of Total Suspended Solids, 413 tonnes of Chemical Oxygen Demand (COD), 169 tonnes of Biological Oxygen Demand (BOD) and 8 tonnes of oil and grease; this has increased to 1,74,201 tonnes of TDS, 91,404 tonnes of Chloride, 11,984 tonnes of Sulphate, 54,592 tonnes of TSS, 4,928 tonnes of COD, 714 tonnes of BOD and 40 tonnes of oil and grease in 2000. Between 1980 and 2000, the cumulative pollution load was 23,54,464 tonnes of TDS, 13,11,722 tonnes of Chloride, 1,25,775 tonnes of Sulphate, 97,152 tonnes of TSS, 90,160 tonnes of COD, 29,848 tonnes of BOD and 1,513 tonnes of oil and grease (Appasamy, 2000: 52)}. After using the water for processing purposes the industries discharge effluents into Noyyal River, and other small streams—Nallar and Jamunnai, other water bodies and agricultural lands, without treating it properly.

A recent estimate by NTADCL indicates that about 40 MLD of effluents discharged by the dyeing and bleaching units (City Corporate Plan—Tirupur, Tamil Nadu Urban Development Project-II, 1999:43). The total volume of effluent discharges from the 851 bleaching and dyeing units was about 60 million litre per day (Madras School of Economics, 1998:66). The effluents were discharged into Noyyal River, which has become a drain and the Orathupalayam dam downstream, a storage tank for these effluents. The groundwater has also been affected; and productivity and the land value have declined over the period in the Noyyal river basin. The Madras School of Economics (1998) conducted a study at Veerapandi and Orathupalayam villages near Tirupur revealing that about two-third of the landholders do not cultivate their land either in part or fully due to the non-availability or poor quality of irrigation water; 90 per cent of the landholders attributed that the productivity has declined due to the water pollution and 43 per cent reported that the land value had declined during the previous decade (Madras School of Economics, 1998:65, 111). A study by Jacob (1996) concluded that the indiscriminate discharge of the industrial waste effluents has led to ground- water contamination. Even the industries could not make use of that water. Further, it shows that the open wells had a higher level of contamination than the borewell (Jacob: 1996). At present, about 102 MLD waste water, containing bleaching powder, sulphuric dyes, inorganic catalysts and other chemicals are discharged daily (Robins and Roberts, 2000: 88). A recent report says that there are 800 bleaching and dyeing units in Tirupur, using 60,000 kg of chemicals and over 115 millions litres of fresh water per day.[2]

Although the knitwear industry provides employment for a large number of the people and earns huge foreign exchange, its consequence on environment and health not only affects those who are engaged in the activities, but the entire region immensely. The hospital reports substantiated: ' ... widespread incidence of skin diseases and pulmonological disorders' (Krishnakumar, 1998). Till recently, pollution control measures were inactive. Even now only, small quantities of industrial effluents are treated with the Common Effluent Treatment Plants (CETPs).[3] According to the Pollution Control Board's estimation, about 80.70 million litres of effluent water was discharged into Noyyal River daily from dyeing and bleaching units in Tirupur and its vicinity and another 30 lakh litres of untreated municipal waste water also found its way in the river (Gurumurthy, 2002). Even after the treatment, these industries' (TDS) levels remained more than double the permissible level of 2100 ppm (Gurumurthy, 2003). In addition to this, a large quantity of firewood is used to process the fabric. Studies reveal that around 3,600 truckloads of firewood per month come from different parts of the state from a minimum distance of 300 km.[4] Precisely, the dyeing and bleaching industries not only aggravate the water scarcity situation but also pose a great threat to the ecology, environment, human beings and other natural resources of the entire river basin.

VII. The Role of Civil Society and Judiciary Intervention

Since the early 1990s, pollution control initiatives have been conceived but been not implemented till recently. For example, in 1991, the Tirupur Dyers' Association formed the Tirupur Effluent Treatment Company Pvt. Ltd (TETCO) to establish four effluent treatment plants with the support of both State and Central Government (Krishnakumar,1998). However, there was no development and at the same time the number of the units have increased several-fold. In 1994, the Tirupur Dyers' Association had accepted to establish eight Common Effluent Treatment Plants (CETPs) for the 300 dyeing units and the others to have independent plants before January 1996. But they have not established the treatment plants.

The Honourable High Court of Madras has constituted a 'Green Bench' during the year 1996 as per the directions of Honourable Supreme Court of India dated 28.8.96 in W.P(C) No.914/1991. The Green Bench deals with cases pertaining to tanneries and other environment-related cases (TNSPCB).

In the meanwhile, the Karur Taluk Noyyal Irrigation Farmers' Association had filed a writ petition at the Chennai High Court in 1996, seeking to close all these industries. In response to the petition, the High Court ordered the Tamil Nadu Pollution Control Board (TNPCB) to issue notices to these polluting industries and to prepare a status report on the ETPs. Further, the High Court also ordered these polluting units liable for the past damages. In November 1997, the TNPCB submitted the status report saying that more than half of the units had not started the ETPs. According to the report, 866 dyeing and bleaching units in Tirupur, of which 288 were covered by 8 CETPs, 404 had Independent Effluent Treatment Plants (IETPs) and 114 units closed down (Sankar 2001:253). In a memorandum, they said 288 units are connected with the eight CETPs work that were under progress and 78 units, which had completed 100 per cent work on ETPs and around 243 units had completed 75 per cent of the works on ETPs. All other units will be closed. In the meanwhile, in March 1997, the High Court ordered the closure of 160 units, which have not started any Effluent Treatment Plant work. In January 1998, 108 units were closed. On February 11, 1998 in the presence of counsel for the petitioner, the joint Chief Environmental Engineer and Assistant Environmental Engineer of Tirupur agreed to complete 8 CEPTS and 386 IETPs on or before May 11, 1998. Accordingly, on March 10, 1998 the Chennai High Court granted time till May 11, 1998 to 609 dyeing and bleaching units to establish the CETPs and individual plants (Sankar, 2001:253).

The defaulting units have been subjected to legal action by the CPCB under Section 5 of the Environment Protection Act, (a) the closure, prohibition or regulation of any industry, operation or process; or (b) stoppage or regulation of the supply of electricity or water or any other service. Further extension of six months was denied by the High Court in April 1998. Consequently, the Tirupur Dyeing Factory Owners' Association (TDFOA) and Tirupur Exporters' Association (TEA) filed special leave petitions in Supreme Court to pass an interim stay for the High Court Order that was denied in May 14. The Supreme Court upheld the decision of the High Court Order to close the 860 add dyeing and bleaching units by May 11, 1998 (Palaniappan, 1998).

Even after establishing the CETPs, the dyeing and bleaching units did not treat the effluents at the prescribed level. At present, there are about 300 Individual Effluent Treatment Plants (IETPs) and eight[5] CETPs in Tirupur. According to TNPCB's latest record (during August 2000) 424 units have their own treatment plants, while 278 units have joined CETPs. One hundred and Sixty-four units, which were not connected either to an IETP or CETP, have been closed down by the order of the Madras High Court (Appasamy, 2000:4). The total effluents generated by 8 CETPs (consist of 278 units) and 424 individual units is about 83.14 MLD. Of this, about 37.98 MLD is treated by the CETPs and 45.16 MLD by the individual plants (Appasamy, 2000:49). But these industries are not treating them effectively (Appasamy, 2000: 4, 54)[6].

Even though CETPs and IETPs are built in Tirupur, they have not treated the effluents effectively because the industrialies discharging the effluents without treatment and the current technologies have not been designed to meet the Pollution Control Board norms. Hence, the discharged effluents contain salinity. The Pollution Control Board found that the dyeing units covered under the seven CETPs discharging the treated effluents without removing the salts used in dye-fixation that led to groundwater pollution and ecological damages in the Noyyal River basin (Appasamy, 2000: 4). The dyeing and bleaching units in Tirupur are using around 350 tonnes of salt.[7] Nearly 6,500 tonnes of salt and 570 tonnes of bleaching powder are used every month for processing (Appasamy, 2000: 37). In April 2002, the TNPCB has served notice to seven CETPs to bring down the total dissolved solids (through segregation of salts used in dyeing) to the permitted level of 2100 ppm in treated effluents by May 31. Further, it has warned that if they do not follow the deadline, they have to face action including closure of the units.[8] Studies reveal that the effluents of these industries caused serious health problems—skin diseases, stomach aches in general. The cattle in these areas also suffered from skin diseases and have became impotent or infertile in recent years.[9] The farmers of Perundurai and Kangeyam taluks of Erode district in the downstream of Noyyal River basin complain of discharge of effluents by dyeing and bleaching units processed by the knitting garment segments in Tirupur and seek compensation for the crop loss suffered on account of the contaminated ground water (Gurumurthy, 2002). Though most dyeing and bleaching industries employ some method of primary treatment that controls the effluent colour the salt content remains in the effluent. Despite civil society and judicial interventions over a decade, the pollution problem is continuing remain unsolved, which has led to social costs by reducing agricultural productivity, fish stock, drinking water, health problems and entire natural ecology and environment of this region.

VIII. The Political Economy of Knitwear Industry

The political economy of the knitwear industries plays a prominent role in the process of environmental damages in the Noyyal river basin of Tamil Nadu. The profits drawn from these industries are individual and the environmental consequences are either distributed on the public or expected the State should bear it. The knitwear industrialist lobby that also influences the state administration through the political representatives either not to follow the pollution norms prescribed by the pollution act or to seek support of the government in their favour. This is a clear indication that the State favours the industrialist's needs and ignores the consequences of environment, ecology and the people of the entire geographical region. In other words, the voice of these industrialists is taken into account while the voice of the major chunk of population is not even heard.

Let me explain how the political economy of the knitwear industrialists operates to extract resources and evaded the rules and regulations, which has led to environmental damages in these regions, in a historical perspective. The knitwear industrialists influenced the diversion of the water resources from the other river basins. Though Noyyal River runs through the town, the water problem in Tirupur city has been acute since the 930s (Saravanan and Appasamy, 1999a). In 1920s, the government proposed a scheme from the Koilveli infiltration gallery, located about five miles away from Tirupur city. This scheme was designed to cover water supply to only 20,000 people at 22.73 litres of water per head a day (G.O. No.3089 Mis, Health, dated 16th November 1954.). But, this was inadequate and served not even one third of the water demand of Tirupur city. Consequently the government has proposed several water supply schemes (Saravanan and Appasamy, 1999a: 172; Saravanan 1999b). In 1949, the government sanctioned investigation for a water supply scheme with the Bhavani River as the source to serve the Tirupur city and seven wayside villages, viz. Pogalur, Kurukkalaiyam Palayam, Annur, Karavalur, Nombiyam-palayam, Avanasi Karamadai and Tirumuganpundi. In 1962, water was diverted from Bhavani River at Mettupalayam. The capacity of this scheme was 7 MLD, of which 4.5 MLD was supplied to the Tirupur city. In 1993, a second scheme was developed with the capacity of 45 MLD, of which 24 MLD was supplied to Tirupur. As a whole, 28.50 MLD water from Bhavani River was diverted to the Tirupur town (Tirupur city Corporate Plan, 1999:34). The per capita water supply was 127 litres in 1996. At present, the knitwear industries are tapping about 600 tankers of water every day from a 35km radius of Tirupur town. Due to the ever-increasing water demand for the industries, a proposal was made for a new project to divert the water at the confluence of Bhavani with the Cauvery jointly financed by industries in Tirupur.

Due to the increase of demand for water, mainly for the industries, both government and private, a jointly initiated water project on Build, Own Operated and Transfer (BOOT) basis was set. The three partners, TACID, TEA and IL&FS, together designed the Tirupur Area Development Project (TADP) as a public – private partnership in 1995. The main objective of the project is to provide piped water supply to Tirupur Municipality and 21 wayside town and village panchayats and water supply to the dyeing and bleaching industries in the Tirupur Municipality Area. The project was launched in 2002. Under this project, water supply is to be provided for the dyeing and bleaching industries in Tirupur and the domestic consumers in Tirupur Local Planning Area (TLPA) comprising the Tirupur Municipality (TM), 15 village panchayats and 3 town panchayats. In addition, water supply is proposed to provide five wayside Panchayat Unions (Municipal Administration and Water Supply Department Policy Note 2001-2002 and 2003-2004). According to the project, 185 MLD of water will be drawn from River Cauvery at Anaiyinasuvam Palayam, near Bhavani town (Tenth Five-Year Plan 2002–2007; 222–223). Of this, 115 MLD will reach industries and the rest 62 MLD will reach Tirupur Municipality and other neighbourhood villages (Ninan 2003). This project is expected to be completed by April 2005 (Policy Note—Municipal Administration and Water Supply, 2003-04).

Under this project, a differential pricing system will be adopted. At present, it is proposed to be Rs.2.25 per/kl for wayside Panchayats; Rs.3.50 for town panchayats and Rs.5

for domestic consumers. For industrial consumption, the cost would be Rs.45 (Ramakrishnan, 2002). At present, the water charges for domestic consumption up to 24 kl/month are Rs.2.00 and above 24 kl/month Rs.4.00; and for non-domestic consumption, commercial use it is Rs.6.00 kl and industrial use is Rs.4.00 kl.[10] In short, the increasing number of industries in general and water-consuming industries in particular, have influenced the government to extract water resources without considering the consequences in other river basins. Besides, several other issues, like child labour, low wage rate, casualisation of labour, absence of unionisation, etc. are determined.

While diverting the water from the other basins, violation of the environmental pollution norms prescribed by the pollution control board led to environmental and ecological damages in the region. Since 1989, a branch office of Tamil Nadu Pollution Control Board has been located at Tirupur but it is ineffective in its role due to the nexus of knitwear industrialists and ruling political parties. The knitwear industrialists are buying times to oblige even the court orders Even after establishing the plants, they are not treating the effluents because the cost involved is high. Further, they are seeking the government support to clean the polluted areas. The above instances are sufficient to make my claim to argue that the political economy of the knitwear industrialists is very powerful to avail of all the necessary inputs from the state and evade its social responsibilities, which has led to disaster in the entire geographical region. In short, the political economy of the knitwear industrialists is operating in such a way that the profit is the individual's and the cost to the public, while the State also meet the interest of the pollutants at the cost of the public and damages natural resources.

IX. Concluding Observations

At the global level, decline in natural resources and its consequences on the environment, has led to a search for the sustainable natural resources. In recent decades, different countries have attempted to ensure sustainable natural resources management by framing appropriate policy measures and enacting different Acts. At the same time, the liberalisation and globalisation process has accentuated the threat to sustainable environment, particularly in the developing countries.

After a shift from the State-regulated market to open market regime in the globalisation process, the State became a agent for the corporate companies and industrialists. In other words, the State is playing a facilitator's role, particularly for the industrialists. In the context of Tirupur knitwear industries, the State is providing all kinds of infrastructural facilities required for the production and export to other countries by withdrawing the erstwhile restrictions on the quantity of export. At the same time, the State is unable to control the environmental damage which was created by these industries. The powerful industrialist—political nexus is successful in availing the facilities from the government while evading its social responsibility. The existing legal mechanism is also ineffective.

The government is extending all kinds of privileges to promote export-oriented industries in order to have a comfortable foreign exchange reserve while knowing the damages to the environment, ecology and the entire natural resources. Precisely, the State's approach towards the export-oriented knitwear industries, particularly dyeing and bleaching industries, will not ensure sustainable natural resources, but it will certainly sustain the polluted environment, not only for the present generation but also for several generations in the future.

References

Appasamy, P (2000) *"Economic Assessment of Environmental Damage: A Case Study of Industrial Water Pollution in Tirupur"*, Project Report ,Madras School of Economics , Chennai.

Cawthorne, P (1993), 'The Labour Process under Amoebic Capitalism: a Case Study of the Garment Industry in a South Indian Town', Development Policy and Practice *Working Paper No.20*, Milton Keynes, UK: The Open University.

Census of India, 1891, *"Imperial* and *Provincial Tables"*, Vol.XIV, British Territory, Madras, Government Press.

Census of India, 1931, *"Imperial and Provincial Tables"*, Vol.XIV, Part II, Madras, Government Press.

Census of India, 1961, Vol.IX, Madras, Part IIC(II) ii, *"Migration Tables"*, Madras: Government Press.

Census of India, 1971, Series 19, Tamil Nadu, Part IIB, "Migration *Tables"*, DVDVI, Madras: Government Press.

Census of India 1971, Tamil Nadu, *"District Census Handbook, Coimbatore"*, Series 19, Part II, Madras: Government Press.

Census of India 1981, Series 1, India, *"General Population Table"s*,Part IIA(i), Madras: Government Press.

Census of India 1981, Series 1, *"Primary Census Abstract, General Population"*, Part IIB(i), Madras: Directorate of Census Operations.

Census of India 1991, *"Primary Census Abstract, General Population"*, Part IIB(i), Chennai: Directorate of Census Operations.

Census of India 1991 *'Migration Tables'*, Tamil Nadu, Chennai: Directorate of Census Operations.

Census of India 2001, *"Slum Population, 2001"*. in *http:/gisd.tn.nic/incensus-paper2/statements/stat_8_1.htm*

Chari, Sharad (2000), 'The Agrarian Origins of the Knitwear Industrial Cluster in Tirupur, India', *World Development* 28 (3), pp. 579–599.

City Corporate Plan—Tirupur, Tamil Nadu Urban Development Project-II, 1999.

Cropper, Maureen L and Oates, Wallace E (1992),' Environmental Economics: A Survey', *Journal of Economic Literature*, Vol. 30 No. 2, pp. 675–740.

Government of Tamil Nadu, *"Policy Note on Environment - 2000-2001"*.

Government of Tamil Nadu, *"Environment and Forest Department Policy Note – 2003 – 2004"*.

Government of Tamil Nadu Tenth Five Year Plant 2002–2007.

Government of Tamil Nadu *"Tamil Nadu Urban development Project – II", City Corporate Plan- Tiruppu,* in *http://www.tn.nic.in/tnudp/Tirupur.htm.*

Gurumurthy, G. (2002), 'Dyers' Failure to Meet Norms—Pollution Notices to Tirupur Units' in *Business Line* in *http://www.blonnet.com/2002/04/19/stories/2002041900711700.htm*

Gurumurthy, G. (2002) ' Effluent Discharge into Noyal River—Erode Farmers Complain Against Tirupur Units' in *Business Line,* Saturday, March 23, in *http://www.thehindubusinessline.com/bline/2002/03/23/stories/2002032302471700.htm*

Gurumurthy, G. (2003) "Pollution in Noyyal River System being Assessed", *Businessline,* Wednesday, Jan 29, 2003.

Jacob, C.Thomson (1996), *'Impact of Industries on the Ground Water Quality of Tirupur and its Ethical Implications',* Unpublished Ph.D thesis, Department of Zoology, Chennai: University of Madras.

Kalaimani, M and Sathiah, R (1994), "Water resources: Constraints for Urban Growth of the Coimbatore Region', in *Sharing Common Water Resources,* Madras institute of Development Studies,Chennai.

Kalaimani, M (1995) 'Urban Environment of Coimbatore' in *Proceedings of the Seminar on Urban Environment of Coimbatore,* July 20, Madras Institute of Development Studies, Chennai.

Krishna Kumar, Asha (1995) "Troubled Tirupur", *Frontline,* April 7.

Krishnakumar, Asha (1998) 'A Pollution Challenge' *Frontline,* Vol. 15, No.13, June 20–July 03 in *http://www.frontlineonnet.com/fl1513/15130660.htm*

Krishnaswamy, C. (1989), 'Dynamics of the Capitalist Labour Process—Knitting Industry in Tamil Nadu', *Economic and Political Weekly,* Vol. 24 (24), pp. 1353–59.

Madras School of Economics (1998), *"Economic Analysis of Environment Problems in Bleaching and Dyeing units and Suggestions for Policy Action",* Madras School of Economics.,Chennai.

Municipal Administration and Water Supply Department Policy Notes (various years), Chennai: Government of Tamil Nadu.

Neetha, N. (2002) 'Flexible Production, Feminisation and Disorganisation: Evidence from Tirupur Knitwear Industry', *Economic and Political Weekly.*

Ninan, Ann (2003), "Private Water, Public Misery' in *http://www.waterobservatory.org /news/press.cfm?news_id=611*

Palaniappan, V.S. (1998), 'Colours of Pollution', *Indian Express*, Sunday, May 31 (Chennai edition).

Ramakrishnan, T. (2002), 'A Much-awaited Project for Knitwear Town' *The Hindu*, June 19 (Chennai edition).

Robins, Nick and Roberts, Sarah (2000), *The Reality of Sustainable Trade*, London: International Institute for Environment and Development.

Sankar, U: (2001), *Economic Analysis of Environmental Problems in Tanneries and Textile Bleaching and Dyeing Units and Suggestions for Policy Actions*, Allied Publishers Ltd, Delhi.

Saravanan, Velayutham (2007), "Competing Demand for Water in Tamil Nadu: Urbanisation, Industrialisation and Environmental Damages in the Bhavani and Noyyal Basins, 1880s-2000s" , Journal of Social and Economic Development, Vol.9, No.2, pp.199-238.

Saravanan, Velayutham, "Inter-Basin Water Transfer: Conflicts in Bhavani-Noyyal River Basins of Tamil Nadu, 1890-1970" (unpublished).

Saravanan, Velayutham (2006), "Competing demand and Management Practices' *The Book Review*, Vol.30 (4), pp.25-26 (*The Politics and Poetics of Water: Naturalising Scarcity in Western India* by Lyla Mehta Orient Longman Private Limited, 2005), Delhi.

Saravanan, Velayutham (2004), "Linking the Rivers: Nightmare or Lasting Solutions?' *Man and Development*, 26(3), pp.79–88.

Saravanan, Velayutham (2001), "Technological Transformation and Water Conflict in the Bhavani River Basin of Tamil Nadu: 1930–1970", *Environment and History*, 7(3).

Saravanan, Velayutham and Appasamy, P (1999a), "Historical Perspectives on Conflicts over Domestic and Industrial Supply in the Bhavani and Noyyal River Basins, Tamil Nadu", in Marcus Moench, Elisabeth Casperi and Ajaya Dixit (eds): *Rethinking the Mosaic: Investigations into Local Water Management*. Nepal Water Conservation Foundation, Kathmandu and Institute for Social and Environmental Transition, Boulder, pp.161-190.

Saravanan, Velayutham (1999b), "Urban Drinking Water Options in the Noyyal Basin: Population, Industrial Growth and Water Demand in *Coimbatore and Tirupur: 1881-1991*", submitted to the International Development Research Centre (IDRC), Ottawa, Canada, in December.

Saravanan, Velayutham (1998a), "Local Strategies for Water Supply and Conservation Management in the Bhavani and Noyyal River Basins,

Part *I- Domestic Water Supply and Industrial Sector"* submitted to the International Development Research Centre (IDRC), Ottawa, Canada, in August.

Saravanan, Velayutham (1998b), "Local Strategies for Water Supply and Conservation Management in the Bhavani and Noyyal River Basins, Part *II- Agriculture Sector"*, submitted to the International Development Research Centre, Ottawa, Canada, in August.

Sathiah, R. (1994), "Water Resources Management of Tirupur: Some Issues" in *Sharing Common Water resources,* Madras Institute of Development Studies , Chennai.

Sengodu, K.P. (1997) "Lower Bhavani Project Irrigation Water Resources Development Environmental Protection A View" in *Natural resources Accounting of Water Resources of the Bhavani Basin,* Madras Institute of Development Studies, Chennai.

Shanmugam initials (1994), "Industrial growth in Tiruppur" in *Sharing of Water Resources,* SIDA Report, Madras Institute of Development Studies, Chennai.

South Indian Hosiery Manufactures Association (SIHMA), Bulletin (1998).

Steele, David (2002), "Child Labour in Tamil Nadu: An Initial Survey", in *http://www.eti.org.uk/pub/publications/2002/05-chlab-lit/index.shtml# note3*

Study Report of CIRT, Pune, 1988.

Venkatachalam, L (1997), "National Resources Accounting for Water Resource: A Case Study of the Bhavani Basin" in *Natural Resources Accounting of Water Resources of the Bhavani Basin,* Madras Institute of Development studies, Chennai.

Verma, Samar (2000), "Export Competitiveness of Indian Textile and Garment Industry" *Working paper no. 94,* Indian council for research on international economic relations, New Delhi.

World Commission on Environment and Development (1987), *"From One Earth to One World: An Overview.* Oxford: University Press, Oxford.

Endnotes

1. *http://www.indigodev.com/ADBHBApxCases.doc*
2. 'New Initiative to Protect Garment Workers in South India' Labour behind the Labels' Bulletin19,July 2003 - India Special, *http:// www.labourbehindthelabel.org/newsletters/19.htm*
3. Different studies estimated that large quantities of industrial effluents were discharged by the industries. But only a low quantity (33.48

MLD) of effluents is being planned to treat by the CEPTs. They are: Angeripalayam (0.85 MLD), Andipalayam (4.50 MLD), Chinnakarai (5.00 MLD), Kasipalayam (3.65 MLD), Kunnangapalayam (3.68 MLD), Manickpurampudur (1.60 MLD), Mannarai (4.20 MLD) and Veerapandi (10.00 MLD) (City Corporate Plan – Tirupur, Tamil Nadu Urban Development Project-II, 1999, p.66).

4. *http://www.tirupurghgemissions.com/html/piassesment.htm*
5. Common Effluent Treatement Plants are located in Veerapandi, Chinnakkarai, Kasipalayam, Kunnangalpalayam, Andipalayam, Mannarai, Angeripalayam and Manickampurampudur.
6. *http://www.bologi.com/environment/15.htm*
7. *http://www.blonnet.com/2002/04/19/stories/2002041900711700.htm*
8. *http://www.blonnet.com/2002/04/19/stories/2002041900711700.htm*
9. *http://www.bologi.com/environment/15.htm*
10. City Corporate Plan – Tirupur, Tamil Nadu Urban Development Project-II, 1999, p.99.

Chapter 17

The Causality Between Environmental Degradation and Poverty: A Case Study of Delhi

Deepti Tandon

Introduction

Environmental degradation and poverty are major threats to the world. The two are entwined in a complex way by which each reinforces the other and makes it even more difficult to control both, particularly for poor countries that face resource constraints. Poor countries are, for the most part, biomass-based subsistence economies. Although many of us live in a high-tech urbanised society, we are as dependent on the earth's natural systems as our hunter-gatherer forebears were. This heavy dependence of the people on the natural resources has brought considerable benefits but at a great environmental cost.

Many studies and empirical results also show that there is a strong interplay between poverty and environmental resource degradation. Some studies conclude that poverty is the principal cause of environmental degradation while others dwell on the premise that it is the degradation of environmental resources which is the principle cause of poverty.

Coming to the basics first: what is poverty, what is environment and what finally is environmental poverty?

There has been and continues to be much debate on how poverty should be defined. Almost all studies conclude that poverty should be defined according to what is prioritised as a 'need', the catch there remaining what is and what is not 'need'. For measurement purposes therefore, food consumption has remained the most widely used indicator of poverty.

The environment is widely recognised as a broad term with many interpretations and definitions. The term 'environment' may be used narrowly, with reference to 'green' issues concerned with nature, such as pollution control, *biodiversity* and climate change; or, more broadly, including issues such as drinking water and sanitations provisions (often known as the brown agenda). For definition purposes, environment may be referred to as a natural resource base that provides sources and performs sink functions.

Environmental Poverty is thus referred to as, on the one hand, degraded environments being the major determinant of poverty; and on the other, poverty being an important factor in degrading the environment, where the health effects of degraded environment are taken as a measure for Environment Resource Degradation (ERD).

According to Jodha 1998, the lead line of reasoning behind this two-way Poverty—Environmental Resource Degradation, i.e. the P-ERD view, is that poverty and scarcity cause desperation, which in turn promotes over-extraction of resources, leading to resource degradation and still greater extent of scarcity and poverty, which further accentuate the above cycle.

i.e.

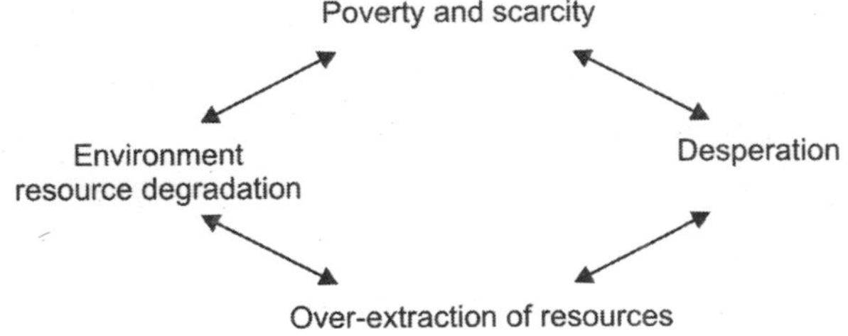

According to the World Bank, the following environmental factors have an impact on poverty:

- Degradation increasing the health burden of the poor;
- Quality of natural resource base;
- Access to natural resources;
- Air quality;
- Ecological fragility.

The Department for International Development (DFID), London, suggests that environmental factors are responsible for almost a quarter of all diseases in the developing countries. The people are most at risk, due to water-borne vectors, inadequate sanitation facilities and air pollution. The poor suffer more losses, illness, injuries and death as a result of resource degradation, natural disasters and pollution than the rest of the population because they are more likely to be dependent upon natural systems for their livelihoods.

There is a need to study this causation under different situations and contexts in order to reassess the inferences drawn from the earlier studies and capture the specificities of a local condition. This paper thus aims to analyse these linkages in Delhi through a primary study, in tune with what Ms. Wangari Mathai, a Nobel Laureate had said,

'It's very very important for us to take action at the local level. Because sometimes when we think of global problems, we get disempowered. But when we take action at the local level, we are empowered.'

An Asian Development Bank (ADB) paper categorises environmental poor into the following categories.

- Dry land poor—those living on arid and desert land areas;
- Flood-prone and disaster-affected poor—those frequently affected by flooding and natural disasters;
- Upland poor—those living in remote upland or mountainous areas;
- Coastal poor—those living adjacent to coasts and dependent on natural and/or marine resources;
- Slum poor—those living in substandard settlements, with high exposure to urban pollutants.

The paper suggests that environmental poverty in Asia and the Pacific is likely to increase, becoming the main form of poverty in near future. It is estimated that by 2020, more than three-fourth of the poor or the vulnerable population in the region will suffer from environmental poverty—up from one-half today.

Methodology

Coming on to the methodology of the paper, to establish the direction of the causality, the study tested the validity of the following two equations:

In the first equation, ERD has been taken as the dependent variable and the incidence of poverty as one of the explanatory variables.

Equation Ist:

$$E = \alpha_0 + \alpha_1 D_1 + \alpha_2 D_2 + \alpha_3 P + u_i$$

where,

E = Environment Resource Degradation, measured through health-environment indicators, i.e. total number of cases of diarrhoea and any acquired respiratory disease in the family in the last 30 days

D_1 = The type of fuel used for cooking, heating, etc., where the use of conventional fuels, an attribute of poor families in Delhi, accentuates the problem of degradation of environment

So,

$$D_1 = \begin{cases} 1 \text{ if using coal/ fuelwood/ dung/ kerosene} \\ 0 \text{ otherwise} \end{cases}$$

D_2 = The source of water for the household, where the common stream, used by poor families, whose water does not go to a water treatment plant, degrades the environment further

So,

$$D_2 = \begin{cases} 1 \text{ if using a water stream flowing nearby} \\ 0 \text{ otherwise} \end{cases}$$

P = Measure of Poverty based on the definition of urban poverty given by the Planning Commission, where an individual with per capita income of less than Rs. 455 per month, at current prices, Economic Survey (2001–2002) falls below poverty line

In this, the second equation, ERD explains the dependent variable, the incidence of poverty.

Equation IInd:

$$P = \beta_0 + \beta_1 S + \beta_2 D_3 + \beta_3 D_4 + \beta_4 R + \beta_5 G + \beta_6 C + \beta_7 M + \beta_8 E + \varepsilon_i$$

where,

P = Measure of Poverty based on the definition of urban poverty given by the Planning Commission (Planning Cell, 2001-02)

So,

$$P = \begin{cases} 1 \text{ if } \dfrac{\text{total household income per month from all sources}}{\text{Number of family members}} < \text{Rs. } 455 \\ 0 \text{ otherwise} \end{cases}$$

S = The level of education, measured as the total number of years of schooling of all adult members in the household

D_3 = Measure of homelessness

where,

$$D_3 = \begin{cases} 1 \text{ if living on streets/ pavements/ night shelters, etc.} \\ 0 \text{ if living in a house (owned/rented)} \end{cases}$$

D_4 = Type of house ownership [if living in a house]

where,

$$D_4 = \begin{cases} 1 \text{ if living in a rented house} \\ 0 \text{ if its owned} \end{cases}$$

R = The amount of rent paid [if living in a rented house]

G = Total gold and/or silver (in gms.) with the household

C = Total present value of consumer durables with the household

M = Number of non-earning (dependent) members in the household

E = Environment Resource Degradation, measured through health-environment indicators, i.e. total number of cases of diarrhoea and any acquired respiratory disease in the family in the last 30 days

The data for the study was collected through primary survey from two slum areas in Delhi, namely:

- Shahpur Jat Village near Siri Fort
- Wazirpur J.J. Colony near Ashok Vihar,

these being chosen because of their vicinity to Ashok Vihar and Siri Fort Pollution Monitoring Stations.

To analyse these **simultaneous equations**, Ordinary Least Square (OLS) could not have been applied because of the problem of 'simultaneity bias' Hence, after analysing the equations to be over-identified through the identification test, 2 **SLS** i.e. **two stage least square method** was applied to find the direction of causality.

Before going on to the results, one important characteristic that was observed in the sample surveyed needs to be emphasised—the aspect of migration. At least 73 per cent of the respondents were observed to be migrants. This aspect is of importance because these people were found to be sending some income back to their homes, which left lesser income per month for consumption of the members staying in Delhi. So, for the purpose of the present study, income per month of the household has been calculated as the income left after deducting from the total per month income that portion that has been sent back home as remittances. Then this leftover household income has been divided by the number of members in the family to get per capita income of the individuals.

Empirical Results

Equation 1st:

$$\hat{E} = 1.647 + 2.233E\text{-}2\ D_1 + 0.301\ D_2 + 1.012\ \hat{P}$$

S.E. = (0.092) (0.186) (0.217) (0.385)

t = (17.89)* (0.12) (2.276)** (2.686)*

*, ** : Values significant at 1 per cent and 5 per cent levels, respectively

Equation 2nd:

$P =$	-0.790	$-6.56E\text{-}2\ S$	$+0.481\ D_3$	$+0.347\ D_4$	$-2.85E\text{-}2\ R$
S.E. =	(0.468)	(0.40)	(0.176)	(0.208)	(0.066)
t =	(-1.688)	(-1.656)***	(2.737)*	(1.663)***	(-0.435)

$+7.46E\text{-}2\ G$	$-2.6\ E\text{-}2\ C$	$+0.119\ M$	$+\ 0.456\ E$
(0.091)	(0.058)	(0.038)	(0.284)
(0.818)	(-0.452)	(3.101)*	(1.605)

As explained by the first equation, the level of poverty bears a positive relationship the level of Environment Resource Degradation. This value is highly significant at the 1 per cent level. This is in conformation with the theory that poverty is one of the important factors explaining ERD. The rationale is that poor people cannot adopt the environment-cleansing techniques that the rich can as the opportunity cost for them is too high. These people attempt to alleviate their poverty by means of an intensification of resource extraction. In this process, therefore, not only do the rates of resource depletion exceed their rates of regeneration, but also the generation of wastes increases in quantity and toxicity, thereby leading to environmental degeneration.

Contrary to expectation, the second equation reveals that ERD is not a significant variable in explaining the level of poverty. Going back to the literature, the effect of ERD on poverty is found to be through the following two links:

(a) With an increase in the ERD, the health of the people deteriorates. These diseases have to be cured, which increases the medical bill and so the out-of-pocket expenses of the households rise. As a consequence, the income left for non-medical expenditures, especially food, is inadequate to meet the desired calorific intake. This link most seriously affects the households that are on the margin of the poverty line and, as a result of these medical expenses, have the tendency to slip below poverty line.

(b) Also, if the wages of the people are fixed on a daily basis, the deteriorated health has a consequence on the income-earning capacity of the individuals. With bad health, the productivity of the individuals decline and the so-called 'sick days' increase. As a result, the per month income of an individual becomes inversely and proportionately related to the 'sick days'.

Coming back to this paper, both the links were found to be missing in the present study. This could be due to the following reasons:

(i) Given the habits of the poor that were revealed at the time of data collection, most of the environment related health problems, like diarrhoea, breathing problems go untreated. The sampled people preferred to ignore their health problems than bear the expenses to get them cured as long as the disease was not very serious, in their perception. Hence, the first link gets broken and in spite of being sick, their income for food consumption does not go down.

(ii) Majority of the people surveyed were not of the category of daily labourers. They had fixed monthly sources of income, which were not productivity based. Thus, the link was again broken because being sick had no negative impact on their monthly incomes.

Thus, because of the above-cited reasons, the coefficient of ERD was insignificant in explaining the level of poverty.

Conclusion and Policy Recommendations

This two-way analysis thus showed that, in the context of the present study, poverty was a highly significant variable in explaining the environment resource degradation. However, as the second equation revealed, the environment resource degradation affecting poverty was still an open-ended statement in the context of the present study. This may necessitate the need to increase the sample size or to redefine certain variables or devise better techniques of data collection. Like any other primary study, this one too is not free from the problem of under-reporting or non-reporting of information.

Based on these conclusions, certain policy recommendations can be suggested and appropriate policies can be designed like:

- ERD not having a significant impact on the poverty levels was an indicator of medical problems of the poor going untreated. Hence, free medical facilities and dispensaries should be opened exclusively for the poor people and awareness should then be spread among them to make use of the same.
- Since the washing and bathing habits of the poor have been found to be a significant cause of environmental degradation, steps should be taken to make water available to these people through the Delhi Jal Board (DJB) tanks so that they do not rely on natural streams flowing by.
- It was established that it is not the habits of the poor that result in high SPM levels in Delhi's air, which we all know are above the critical limits. Keeping this in mind, intensive steps must be taken by the Delhi government to reduce the vehicular and the industrial air pollutants.
- The high dependency ratio was also found to be a significant factor in reducing the per capita availability of income, thereby leading to poverty. The newly passed Employment Guarantee Act can

help somewhat in tackling the problem of high dependency ratio. However, the government should ensure that its benefits are being reaped only by the targeted group and not the non-deserving people.

It is now high time to address the problems of environment degradation right from the community up to the global level, else the flora and fauna which is our geography would soon become history.

References

C.S.E. (1990), *"Human-nature Interactions in a Central Himalayan Village: A Case Study of village Bemru"*, Center for Science & Environment, New Delhi.

Dasgupta, P. (1995), "Population, Poverty and the Local Environment", Scientific American, February, pp.41-45

Dasgupta, P. & Mäler, K.G. (1991), *"Poverty, Institutions, & the Environmental Resource-base"*, Oxford University Press

Dasgupta, P. & Mäler, K.G. (1994),*"The Environment and Emerging Development Issues"*, Oxford University Press

Dietz, T. and Rosa, E.A. (1997), "Poverty and the Environment", Effects of Population and Affluence, National Academy of Science

Duraiappah, Anantha K. (1996), "Poverty and Environmental Degradation: A Literature Review and Analysis", CREED Working Paper, Series No. 8, London: IIED

Economic Survey of Delhi (2003-04), Planning Department, Govt. of NCT of Delhi.

Economic Survey (2001-2002), Planning Department, Govt. of NCT of Delhi, pp 166–167.

Hodgson, G. & Dixon, J. (1992), *"Sedimentation Damage to Marine Resources: Environmental & Economic Analysis"*, Taylor & Francis, London.

Jodha, N.S. (1998) *"Poverty and Environmental Resource Degradation—An Alternative Explanation and Possible Solutions"*, EPW, September, 1998, Pp. 2384-2390

Khan, Nisar Ahmad (2000), "Exposure to Air Pollution and Socio-Economic Characteristics: A Case Study of Delhi", M. Phil Thesis, South Campus, University of Delhi

Killeen, Damian & Rehman, A. Atiq (May 2001), "Poverty and Environment", in World Summit on Sustainable Development

Lash, Jonathan, *"Environment, Poverty and Development"*, Pacific and Asian Journal of Energy, Vol. 3

Lester, Brown R., (1991), *"Eco-Economy: Building an Economy for the Earth"*, Earthscan Publications Ltd., London.

Nadkarni, M.V. (2000), "Poverty, Environment and Development: A many Patterened Nexus", EPW, April, 2000, pp. 1184-1190

Prakash, S. (1997), *"Poverty & Environment linkages in Mountains & Uplands: Reflections on the poverty thesis"*, CREED Working Paper Series No.12, International Institute of Environment & Development, London

Sen, A.K. (1985), "Concept and Measurement of Poverty" in "Poverty and Famines: An Essay on Entitlement on Deprivation " Oxford University Press, London.

Shyam Sunder, Priya (2002), "Poverty—Environment Indicators", Environmental Economic Series, Paper No. 84, Jan.

Srinivasan, T.N. (2000), "The Environment, Economic Development & International Trade: Some Issues", Pacific and Asian Journal of Energy, New Series

The Asian Development Bank working paper, "Environmental Poverty: New Perspectives and Implications for Sustainable Development in Asia and the Pacific".

"Why the Environment matters to People Living in Poverty", Commission on Urban Poverty and Environment, Part 2.

Chapter 18

Economic Loss Due to Displaced Monkeys: Results from a Pilot Survey*

A.K.M. Nazrul Islam, Salma Sultan and Bazlur Rahman Khan

Introduction

India is a global hub of biodiversity.[1] It is one of the mega-diversity countries[2] in terms of biodiversity. Kothari (1997) pointed out that there are around 81,000 animal species and 45,000 plant species found in India and many more yet to be discovered. Although India 2 per cent of the world's landmass, it represents around 6.5 to 7 per cent of the recorded species worldwide; 14 per cent of the world's birds are discovered in India; 33 per cent of the flowering species and 18 per cent of the total species of the world are found in India. The North-eastern states, Western Ghats, Himalayan ranges and the Andaman & Nicobar Islands are home to the maximum of these species. Despite many policies, strategies, agreements, seminars, conferences, and symposia, the pace of biodiversity loss is alarming in India. This is indeed a matter for concern as human survival is largely dependent on biological resources.

---

* This is a revised version of the paper entitled: "When Tragedy of Commons Turns into a Tragedy of Common People: Results from a Pilot Survey", ICFAI Journal of Environmental Economics, 5(4), November 2007.

Large-scale degradation of forest resources in India started from the British period due to logging, making railways and for other commercial purposes. Even after Independence, the country could not stop this trend because of many reasons—poverty, population growth, food scarcity along with corruption, ignorance, lack of appropriate policy measures and other related institutional failures. The importance of forest resources in economic, as well as other life-support facilities is immense. Although the community forestry in India was by and large a success and in some States it made the total forestlands increased in figures, if we look at the actual state of Indian forests, the panorama is not so rosy. Due to deforestation the Indian biodiversity is at a stake. We are not only losing the plants, trees, animals, birds, reptiles and micro-organisms, but also our rich sources of medicinal needs, diversified forest-based foods and fruits, recreational facilities, means of research, development and innovation are also getting lost. The long-term consequences of deforestation are even more significant. Directly or indirectly, it will affect our economic, social, recreational interests, besides disrupting other life-support facilities. Global warming, weather change, soil degradation, shrinking and water-level, floods are some of the larger consequences of deforestation.

Due to over-pressure from the growing population, scarcity of agricultural land and poverty, people naturally depend upon nature for their survival, and this is one of the major causes of deforestation. The situation can be worse if it is accompanied by major institutional failures. This deforestation and human interruption in the forestlands can lead to direct economic loss and larger social problems. In Assam, although the percentage of forestland to total area is better as compared to many other States in India, the situation is rapidly changing. One of the affected areas is the Patharia Range of Karimganj district. The present study is based on a pilot survey on this range tries to estimate the direct economic loss due to displaced monkeys from their safe-habitat, which is caused mainly due to deforestation and human-interruption in the range. It also focuses people's opinions about the causes, consequences and probable preventive measures of this manmade disaster. The article has been arranged into four sections: Section I presents a brief picture of endangered species in India, Section II is devoted to the nature of the current problem, whereas Section III presents the survey results other than economic loss. Section IV highlights the nature and extent of economic loss due to monkeys and the summary and policy suggestions are presented in the previous section.

I

Degradation of Commons in India

The general condition of the common property resources in India is not very encouraging. Loss of forestlands, degradation of water-bodies and wetlands, etc. has further aggravated the situation. The loss of precious biodiversity is a big concern. Once lost, these species cannot be reproduced any more, which will have negative consequences on our life and interests. The latest 'Red List' of the International Union the Conservation of Nature (IUCN) (2006) identifies 569 species in India which are categorised as 'threatened'. Many of the endemic mammals, birds, reptiles, amphibians, fishes, mollusks, plats, and trees come under this category. Although the 'Red List' category is based on a fewer percentage of the total described species, it may really shake us to see the pace of loss of Indian biodiversity. The following Table (18.1) presents the IUCN data on 'threatened' biodiversity of India:

Table 18.1: Threatened Indian Species

Sl. Number	Species	Number
1.	Mammals	89
2.	Birds	82
3.	Reptiles	26
4.	Amphibians	68
5.	Fishes	35
6.	Molluscs	02
7.	Other Inverts	20
8.	Plants	247
	Total	**569**

Source: Red List 2006, IUCN.

The list also categorises these species of plants and animals into different states on the basis of their present situation. Of these 569 species, many are considered as 'extinct', 'critically endangered' or 'vulnerable'. A tabular presentation of the IUCN data can give us a more appropriate idea about the state of these species:

Table 18.2: Threatened Species by Category

Species	EX	EW	Sub-Total	CR	EN	VU	Sub-Total	LR/cd	NT	DD	LC	Total
Plants	07	02	09	45	113	89	247	01	22	18	68	365
Animals	01	00	01	42	92	188	322	09	157	142	1413	2044
Total	08	02	10	87	205	277	569	10	179	160	1481	2409

Source: *Red List 2006, IUCN.*

(EX = Extinct, EW = Extinct in the Wild, CR = Critically Endangered, EN = Endangered, VU = Vulnerable, NT = Near Threatened, DD = Data Deficient, LC = Least Concerned, LR/cd = Lower Risk/ conservation dependent)

Table 18.2 shows that as many as 10 species are already extinct, whereas 569 species are either 'critically endangered', 'endangered' or 'vulnerable'. This shows how Indian biodiversity has been losing its rich animal and plant species. More systematic and comprehensive policy measures along with all stakeholders' participation are very critical in the preservation of these species.

II

Nature of the Problem

Assam, compared to many other Indian states, has quite a high percentage of forestlands. It represents the transition zone between the Indian, Indo-Malayan and Indo-Chinese bio-geographical regions as well as the meeting zone of the Himalayan Mountains and Peninsular India (NBSAP, 2005). But the forestlands are gradually decreasing, although the rate of decline may be slow or sometime show reversing trends. The Table 18.3 can give us an idea of the declining trends of forestlands in the state.

Table 18.3: Forest Area in Assam (in sq. km)

Year	Area	% of total area
1974–75	28,924.00	36.83
1977–78	28,608.00	36.43
1982–83	27,337.00	34.81
1987–88	21,513.20	27.39
1989–90	21,522.60	27.40
1991–92	21,684.52	27.61
1993–94	22,175.85	28.24
1999	23,688.00	30.20*

Source: National Biodiversity Strategic Action Plan, 2005.

In the front of biodiversity, the State is one of the richest not only in India, but considered as one of the most biologically diverse areas in the whole of South Asia. A glance at Table 18.4 will reveal this picture.

Table 18.4: The approximate number of family, genera and species in Assam

Name of group	Family	Genera	Species
Angiosperms	207	1111	3010
Dicot	170	824	2251
Monocot	37	287	759
Gymnosperms	4	4	7

Source: National Biodiversity Strategic Action Plan, 2005.

The forests are extremely rich and diverse, containing significant primate, carnivore, herbivore and bird communities. About 193 species of mammals and more than 958 species and subspecies of birds are so far reported from Assam. The state possesses 16 important wildlife areas, which houses nearly 44 types of endangered and rare species of mammals and 14 types of reptiles and amphibia. Among these, the Hoolock gibbon, Assamese macaque, Pigtailed macaque, Golden langur, Hanuman langur, Spectacled monkey, Capped langur are the main types. The Golden langur, the entire population of which is found only in Assam, is confined to the western part of the State mainly along the Bhutan border. The Pharye's leaf monkey, is on the other hand confined to the forest of Cachar. As many as 19 cat families are reported to be found in the state. Moreover, Assam holds the entire known world population of the Pigmy hog, 75 per cent of the world population of the Indian rhinoceros and Wild water-buffalo and a sizable population of Asian elephants and tigers. Assam harbours 17 endemic species belonging to three families. Of the total 50 RDB species belonging to 20 families of the north-eastern states, the state houses 45 species belonging to 19 families. Assam shows the presence of one species of Crocodylia, 19 species of Chelonia and 77 species of squamatas (NBSAP, 2005). The state is indeed a rich one in terms of its biodiversity. But destruction of natural habitat for commercial felling, encroachment for settlements and cultivation, jhum (shifting) cultivation in the hill-slopes and various development activities has put serious threat to the biodiversity of Assam.

The Karimganj district of Assam is a Borak-valley district. The area of the district is 183,900 hectares, comprising 46,328 hectares of forestlands representing 25.19 per cent of the total land. According to the 2001 Census, the density of population in the district is 555 persons in per square kilometre. The last home of dwindling species, e.g. churaibari, and duhalia of south Karimganj for threatened ferns like *Angiopteris evecta, Asplenium vidus, Cyathea gigantea, Cyathea brunoniana, huperziasquarossa, helminthostachys zelianica,* etc. and some areas of Duarbandh for an endemic species of fern *dipteris wallichis,* forest areas of Jirighat for bamboo, orchid- *Arundhina graminifolia.* The forestlands of this district are gradually falling into human grasp. Patharia Range is one the most important forestlands in the Borak-valley area. It is located in Karimganj District in the southern tilt of Assam. It is also called the Adamail Range. The range marks the western border of the district forming the international border with Bangladesh. Running from the south to the north, its length is about 28 miles and breadth about 7 to 8 miles. The highest point of the range is about 800 feet above sea-level. This range is full of 'hillocks' and thin green cover. There is more than a dozen villages around the range, which are directly or indirectly dependent on the range for their fuelwood, timber and forest-based products. But the overgrowing population, high incidence of poverty, ignorance and institutional failures make the range a degraded one and the situation is getting worse day by day. In the last few decades the situation has been changing mainly due to deforestation.The once green range is now rapidly losing its greenery along with rich biodiversity. If this process goes on, Patharia Range will only become a part of history for the new generation to come. Many of the animals, birds, reptiles of the area are either extinct or under threat from human interference. Many fauna, like the Golden langur, Rhinoceros, *sishoo,* treesnake, stork (*hargila*) are listed as endangered species in the Karimganj district. The National Biodiversity Strategy Action Plan (2005) has identified indiscriminate killing, illegal trade, feather market by unscrupulous people as some of the reasons for such threats. Many of the surviving species are out of their safe havens and displaced in the nearby areas. These displaced species losing their habitats are now easy prey for humans. But the most interesting part is that now these displaced species have become a source of serious animal revenge.

The present study tries to highlight a problem which arises due to these displaced animals. Thousands of monkeys, a common species in the Patharia area, now have no shelter, food and playground for their survival because of the human interruption in the jungles. Losing their natural survival sources, the monkeys have no option but to share human interests for their sustenance. Crops, fruits, vegetables, kitchen foods and other household foods become their easy prey, which causes much economic loss to the people. It not only causes economic loss, insecurity, frustration and other social problems but also makes the life of the people pathetic in Patharia Range. Often these monkeys attack cattle, frightened school-going children or even bite them. The frustrated people have formed local organisations and NGOs to attract governmental attention for a proper solution. The 'animal revenge' makes the life of the people of Patharia Range a misery. It can also give us an idea about the side-effects of losing commons. This is indeed a very unique and serious problem, which requires appropriate and well-designed policy plans and proper implementation integrating all possible stakeholders.

III

Results of the Survey

There are more than a dozen villages which are around the range and people are directly or indirectly dependent on the range for various needs. The pilot survey was conducted on 60 randomly selected respondents from these villages adjacent to the Patharia range. Our survey represents different section of the people living in these villages ranging from educated or uneducated, male or female, Hindus or Muslims, farmers, businessmen, private service-holders, government employees and from different income groups. A brief profile of the respondents is given here:

Figure 18.1: Respondents by Sex

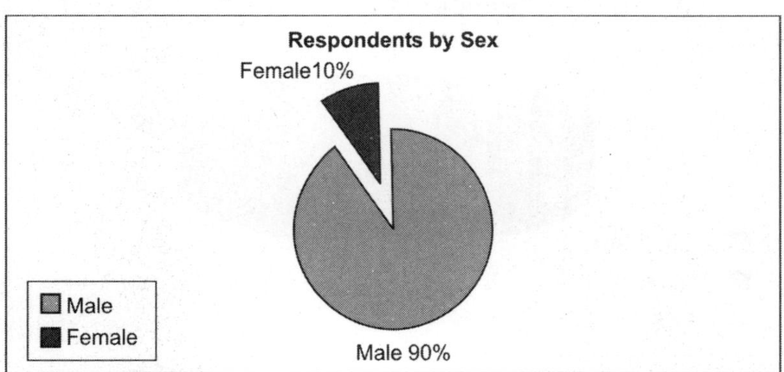

Figure 18.2: Respondents by Level of Education

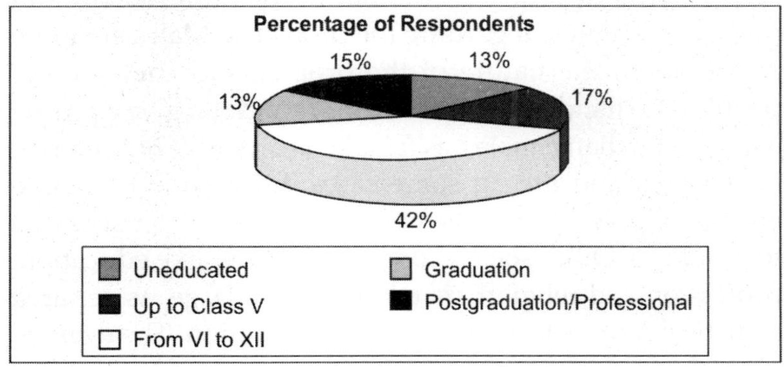

Figure 18.3: Respondents by Profession

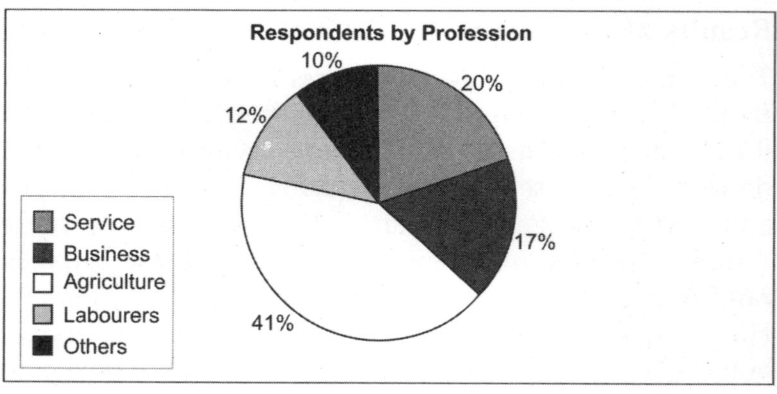

Figure 18.4: Respondents by Religion

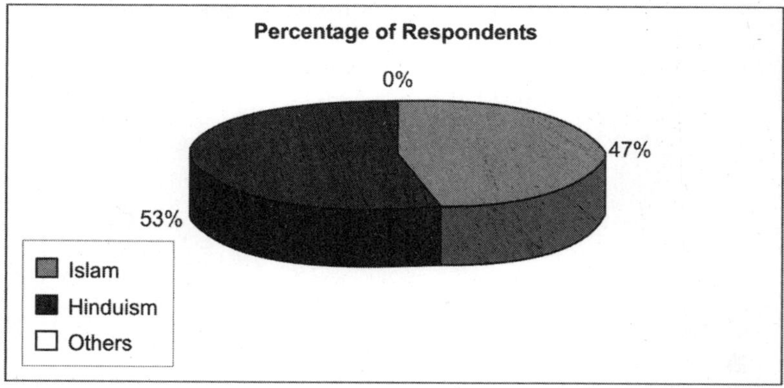

The rationality of selecting more male against female respondents depends on the nature of the problem and proper knowledge regarding the problems. Males are more aware about the nature of the problems as they are the people who handle the situation more and can give a proper justification than females, as females are more of housewives in the area and are, in some cases, fully knowledgeable enough to give a justifiable value of economic damages due to monkeys. Other socio-economic features, like education, profession and religion are taken, by and large, as per area statistical figures to give a proper justification. The average family members of the respondents was calculated as 7.53 persons and average family income per month was estimated Rs. 6,192.29. The Joint family system and lack of proper family planning are the main reasons of large family size, whereas per house monthly income represents that they are mainly middle-class or lower-middle class people living in the traditional Indian villages.

Figure 18.5: Types of problems people face

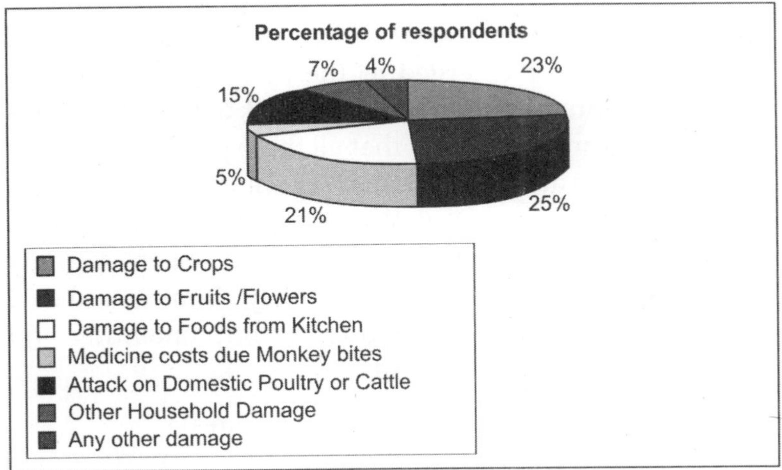

Figure 18.6: Causes of problems

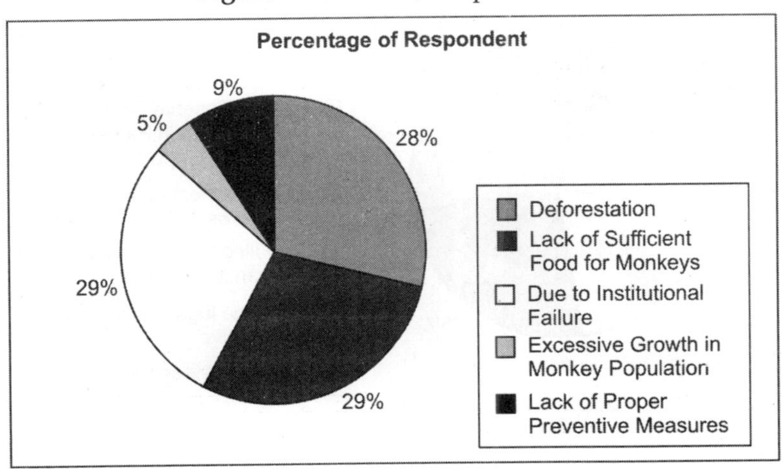

If we look at the survey results of the types of problems people face due to monkeys, then three most important problems arise from the survey-damage to fruits/flowers from fields/gardens, damage to crops and kitchen foods. These are followed by some other minor problems, like medicinal costs due to monkey-bites, attacks on poultry, cattle and panic among school going children. Types and causes of the problems identified by the respondents are shown in the pie charts. According to our survey, institutional failure, scarcity of food for monkeys and deforestation have emerged as the three most important causes of the present problems. If we look at these problems carefully we can understand that all these problems are due to deforestation and deforestation is because of institutional failure, poverty, overpopulation and many more reasons. So a policy addressing all these associated problems can only give us a sustainable solution. Our policy planners and civil society have equal responsibility for implementing and making such policy successful.

Figure 18.7: Preventive Measures

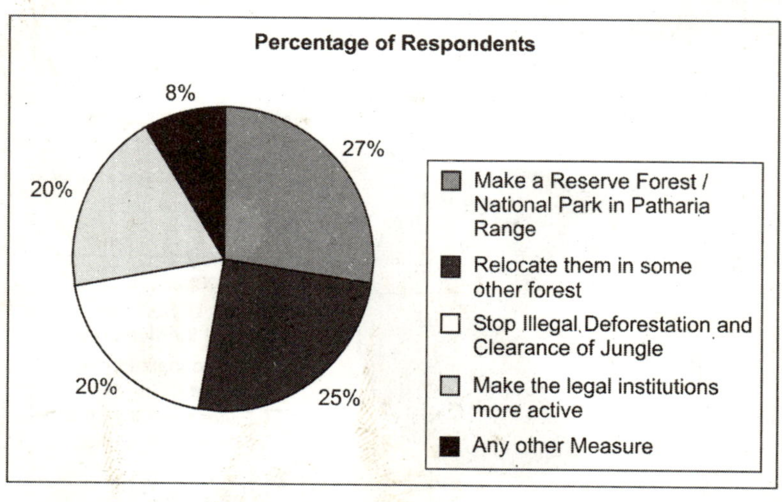

Figure 18.8: Who Should take Preventive Measures?

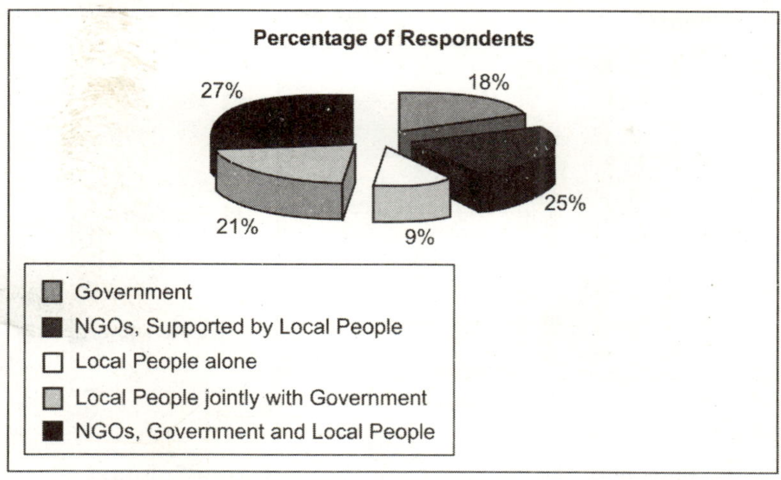

The questions regarding protective measures and by who should be responsible for such measures got mixed responses from the respondents. Reserve forests, relocation of monkeys and institutional arrangement are considered favourable preventive measures. But, above all these possibilities, people want a proper and permanent solution to the problem. According to the respondents, the role of NGOs, people's participation along with governmental interference can solve the problem efficiently. This once again proves how important it is to integrate all stakeholders in solving the problem of degradation of commons. Many important researches around the world and within India too have highlighted the importance of integrated policy measures involving all stakeholders for preventing the degradation of commons.

Figure 18.9: A Permanent Solution

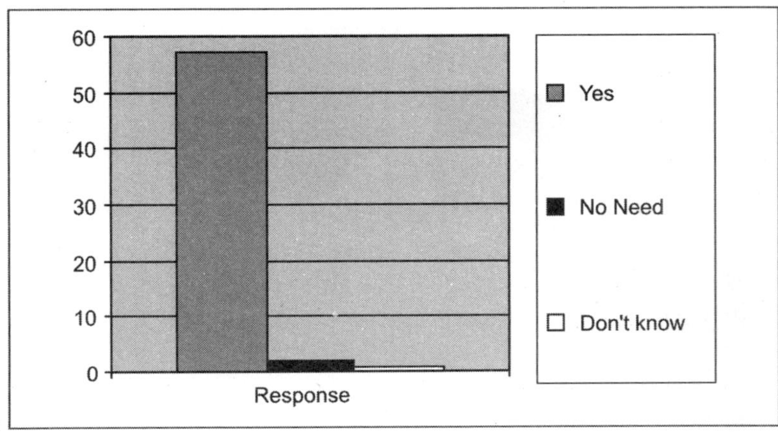

Figure 18.10: People's Contributions for a Solution

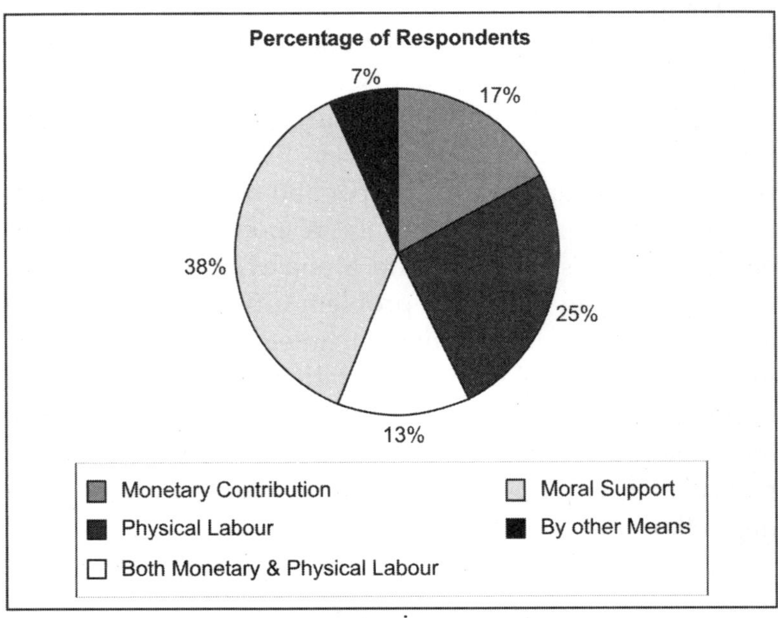

On a permanent solution of the problem respondents overwhelmingly support such a step. This step will not only preserve these endangered species from extinction but will generate long-term benefits for all stakeholders. The survey findings also show that more people are ready to support morally, i.e. support from outside without contributing monetarily to solve the problem, although some people are ready to provide physical labour hours, if needed. Only 17 per cent of the respondents are ready to contribute monetarily, although many of these respondents want to contribute a meagre amount which may not generate sufficient money for making any step towards a proper solution without governmental or other third-party participation. This reveals the importance of generating an alternative resource base for any steps to solve such problem.

IV

Economic Loss

The main objective of the present survey was to estimate the economic loss people have to bear due to different kinds of damages by monkeys. Besides social, mental and physical problems people face, most important threat they are facing from these homeless monkeys is loss of valuable crops, fruits, flowers, foods, other household chores, medicinal costs due to monkey-bites, etc. The Table 18.5 shows the survey results.

Table 18.5: Annual Average Economic Loss per Family due to Monkeys

Types of Loss	Crops	Fruits/ Flowers	Kitchen Foods	House-hold Cores	Medicine	Others	Total
Average Economic Loss	1288.33	1084.58	637.50	421.67	121.67	153.33	**3707.08**

Source: Data Compiled from the Pilot Survey.

The average economic loss estimated from survey (a whopping Rs. 3,707.08 for the poor and middle-class people living in this area) seems to be quite high because of the extent and nature of damages by the monkeys. This leads to a disastrous situation in the villages surrounded by the Patharia Range. Losing their safe-habitats in the hands of human-being, monkeys now become so aggressive that it not only causes economic loss but also creates social problems among the villagers.

V

Conclusions

The findings of the present study give us an interesting and unique case of a problem which can directly cause much economic loss and social problems. The 'tragedy of commons' can also turn into a 'tragedy of common people' if we ignore the loss of forestlands, particularly biodiversity, besides many other far-sighted negative consequences. Conservation of biodiversity has attracted worldwide attention from the 1980's onward and since the 1992 'Earth Summit' (Rio de Janeiro, Brazil) it is fast emerging as a critical area of theory, research, policy across the academic disciplines. Maintaining biodiversity is more than just a question of preserving living beings for their own sake; it is one of ensuring our present and future well being, perhaps even our very survival. Government of India has made many important policy documents and signed many international and regional treaties for the preservation of nature. Our national policies—forest policy, biodiversity policy, environment policy, policy on wildlife along with other related policies should address these micro-level problems also as they can emerge as a bigger problem . Proper institutional mechanism—like regulations, legal provisions, property rights, along with people's participation and more awareness of environmental degradation and their negative consequences must be addressed properly. So, appropriate and comprehensive policy measures, integrating all stakeholders-governments, local people and organisations, NGOs must be taken into consideration to protect endangered environmental resources and make their use more sustainable so that a greater benefit can emerge from these once-abundant but now scarce resources.

Reference

Hardin, G. (1968). *"The Tragedy of the Commons"*, Science, 162, 1243-8

http://www.iucn.org/redlist.htm

http://www.karimganj.gov.in/geog.htm

http://www.karimganj.nic.in

Kothari, Asish (1997). *"Understanding Biodiversity: Life, Sustainability and Biodiversity"*, Orient Longman, New Delhi

National Biodiversity Strategy and Action Plan (NBSAP, 2005). Ministry of Forests and Environment, Government of India, New Delhi

Survey of the Environment" (2002). The Hindu (Supplement), New Delhi

Endnotes

1. According to convention on Biological Diversity (CBD). "Biodiversity means the variability among living organism from all services, including interalia, terestrial, Marine and other acquatic eco-systems and the ecological complexes if which these are a part, this includes diversity within species, between species and of eco-systems".
2. Countries likely to contain the highest percentage of the globel species richness.

Chapter 19

Socio-Economic and Environmental Aspects of Migration: A Case Study of Cuttack Slum Dwellers

T.A. Baig and M.A. Baig

1. Introduction

The cities are growing rapidly in India, unfortunately slums growing many times faster. Even though the phenomenon is not new to the urban areas, there is concern over the acceleration of the problem in recent years. Owing to a wide range of economic and political factors, the rural poor have been leaving for destinations within the State and across States. Whatever may be the reason, whether you call it manifestations of our iniquitous society or faulty planning, population explosion added by increasing migration leads to an imbalance in the ecosystem and many socio-economic problems.

There is growing concern from many dimensions about the migration of the rural poor to the urban areas. In India, migration is more visible from poor states to developed states and within a state from rural to urban regions. There has been a noticeable change in the nature and pattern of rural migration. It is interesting to note that, in the last century, people have also been migrating from the developed states to poor states. It is not only economic factors that matter but other factors like socio-cultural, political, and psychological are responsible for migration. India is still in the initial stages of urbanisation, given that about 70 per cent still lives in rural areas. Major cities are already showing clear signs of 'over-exhaustion'. In the coming decades, rural to urban migration is likely to become much more serious even in small cities. While migration is seen as a livelihood diversification strategy, it is also important to be seen as an economically and environmentally destabilising process.

In Orissa, the Cuttack city[1] is arguably the most seriously affected by the migration of rural poor. They have come from different parts of the country, but mostly from within the states and from neighbouring states. The migrants are of different race, caste, religion and language. Even the socio-economic and political status of the migrants is of a varied nature. The economic motive has probably been dominant at all the times. Non-economic factors do play a significant role in all population movements. However, the underlying social-economic determinants of migration, which establish different patterns of migration, remain mostly unexplored. The city is generally overcrowded and has better job opportunities for the poor people. The migrants are mostly slum dwellers. Most of the unidentified slums are in the canal sides or in the bank of the rivers surrounding the city. They live in unhealthy conditions and pollute the local environment.This is mostly found in an identified colony. The old slum dwellers construct a house and sub-let it to fellow migrants.

Migration to the city has become a controversial issue, as it is a trade-off with the political wills and economic concerns. The magnitude and effects of migration of rural poor to the city constitute an important place, but research and information on issues relating to their migration has been scanty. It is therefore deemed necessary to undertake research to provide evidence of the magnitude of the problem to enable policy makers to plan, using accurate information. The present study seeks to provide detailed information about migration patterns, the reasons for migration and in broader terms its effects on the quality of life. The study also examines the role of the government in the betterment of the migrant slum dwellers of Cuttack. The findings of this study can be used to develop human resource policies and strategies to strengthen the capacity of employment to deliver efficient and effective services. It is a timely response to the expressed needs for better information in this area from within. There is need for urgent attention and immediate redress of the different problems associated with the slum dwellers of the city.

In this backdrop, many research issues arise, like the driving force behind migration; the effects of migration of slum-dwellers on economic, environment, and social delivery system; the economic consequences of migration and steps to mitigate the effects of migration; application of policies and strategies to reduce inward/outward migration of slum-dwellers; the roles of governments, donors, partners and the communities in addressing the situation; and others. Nevertheless, the present study has a very limited objective: to explore the socio-economic and environmental aspects of migration with special reference to the Cuttack slumdwellers, who are migrants. The specific objectives of the study are (i) to understand the socio-economic aspects of migrants, (ii) to understand the environmental aspects of migration and to study the linkages between migration and environment, and (iii) finally to provide some policy suggestions to improve socio-economic condition of the migrants and breaking the vicious circle of migration and environmental degradation.

In order to meet the above-stated objectives, the study tries to find the motivational drives behind migration. The emotional, political, educational aspects of migrants are also studied as corollary to the socio-economic aspects of migration. The rest of the paper is organised as follows. Section 2 deals with some selected literature survey. Section 3 deals with the methodology of the study. The analysis of the data and discussion of the results are presented in Section 4. Finally, Section 5 provides some policy implications.

2. Select Literature Review

Migration has become a global issue and being studied from various perspectives. In recent years, internal migration has emerged as a major issue of investigation in the economic and sociological literature. The massive waves of migrants moving from rural to urban area is being explained by various types of theories. However, the explanations to internal migration need close attention.[2] The 'global movement' most commonly stems from the desire to achieve more both personally and professionally. This aspiration of the migrants is combined with factors that emanate from the political, economic and general living conditions within each country.

The available literature consistently reports that many rural poor are dissatisfied with their current living condition in their original place. The common reasons for dissatisfaction found in Baines (1998) are: non-availability of the job, over burden on agriculture, burden of personal debt and an inability to afford the basic necessities of life. Consequently, these rural poor often migrate to urban places with the hope of more satisfaction both financial and intrinsic. Migration comprises predominantly short distance, and sometimes, it is a temporary move (Baines, 1998).

Haberfeld et al. (1999) in their study on seasonal migration in Dungarpur, one of the less developed districts of India, closely examine both the determinants and impact of seasonal migration. The study finds that rural households in India use migrant labour offered by their members to improve their well-being both by reducing the impacts of inferior conditions and by raising household's income levels. Migrant households are characterised by lower education levels, lower levels of income from agriculture, and by an inferior geographical location. However, those households sending migrant labour are found to have higher income levels than those not sending migrant labour. Income from migrant labour accounts for almost 60 per cent of total annual income of households sending at least one migrant laborer.

Samaddar (1999) in his book tries to explore the twilight zone of legality and illegality, exceptionality and normalcy, infiltration and migration. He applied the participant observation technique to study the lives of migrants. The book questions the 'nation' above other forms of community and attempts to discover other existences that might cut across national boundaries, and thus presents an anthology of various existences. To him, the people here are displaced due to ethnic unrest, political repression and armed conflicts besides failures of government and development policies, health and environmental disasters, changes in cropping patterns and economic deprivation.

Migration is usually viewed as having a negative impact on the economy. Arjan de Haan (2000) identifies possible negative aspects of migration, including increasing inequality and other effects on those who stay behind, but the emphasis is on the positive role migration plays for poor households. Building on new literature on sustainable livelihoods, he argues that we need a better understanding of the capabilities and strategies of poor people, in their own perspective, and that this will help to improve development policies. This helps us to derive positive benefits of migration.

Bijwaard (2001) finds that migration is a dynamic phenomenon, many migrations have a temporary and repetitive character. Ignoring the dynamic aspects of migration will lead to underestimating the economic assimilation of immigrants and their economic impact on the host country. Important factors of economic assimilation, like the earnings profile and the accumulation of human capital also fluctuate over time and the choices on these issues are intercorrelated with the (re-)migration decisions.

Mukherji (2001) regards the syndrome of poverty, distressed migration and urban involution in India, with a new and novel research perspective, as well as new and alternative planning prescriptions. Concrete plans and their effective implementation for the benefits of the poor migrants are necessary. Then only will we be able to reduce poverty, human misery, agony and pain from the lives of the poor masses, and hopes and aspirations in their lives for their sojourn towards a life with human dignity and upward transformation.

While migration enhances socio-cultural integration, the unintended consequences of these exchanges, such as increased migration, require greater regulation, which could stifle regional economic growth and development. Migration issues present important impediments to regional economic growth, especially by establishing barriers to the development of a labour market. Yet it is exactly the underdevelopment of the regional economy, especially as experienced by certain groups, that creates hostility towards economic migrants. Rogaly et al. (2001) in their study on scale and pattern of seasonal migration for rice work in West

Bengal find that people who migrate temporarily for manual work are not usually unionised and are often unprotected by effective legislation against travel and workplace risks. West Bengal's gangs of mobile rice workers are recruited directly by individual employers at busy labour market places or in migrants' home villages. Seasonal migration contributes to the socio-economic and political transformation and it is influenced by changing social identities (Rogaly et al., 2002).

Sinn (2004) analyses the possibilities to defend the wages of the poor against the low-wage competition of migrants. He shows that fixing social standards harms the workers and that fixing social replacement income implies migration into unemployment. Defending wages with replacement incomes brings about first-order efficiency losses that outweigh the budget cost to the government. By contrast, wage subsidies involve much smaller welfare losses. Defending social standards or replacement incomes are shown to be very inefficient ways for the government to proceed. This means that the replacement strategies should be abandoned and that governments should make no particular attempt to defend social standards. Only a limited programme of income subsidies to the poor would be compatible with a roughly efficient migration pattern in Europe.

Eapen (2004) attempted to raise an issue, which has always concerned feminist scholars—the sex segregation of jobs and its perpetuation over time to the disadvantage of women workers, in the context of Kerala. More important, he emphasises the need to include women's domestic work as a category of work in such an economic analysis, arguing that a growing proportion of women (or 'working' days of women) moving into the activity 'not in the labour force' whether voluntary or involuntary, reduces their mobility.

The state boasts of the high(est) female literacy rates among all states of India; yet, as recent studies have shown, it scores poorly in terms of what are termed as non-conventional indicators attempting to capture power and subordination (Eapen, 2004).

Jha (2005) finds that in the poverty-stricken tribal areas of Orissa, recent shifts in migration trends have revealed the increasing movement of young women towards urban centres in search of work. The 'push' factor is responsible for such migration, but as a recent workshop revealed, the prospects of such work offers leave much to be desired. Living conditions are unhygienic, the salary poor and tribal women are vulnerable to exploitation by unscrupulous agents. The study revealed the exploitative and brutal character of tribal migration in Orissa. Sexual exploitation, trafficking of women and sometimes poor health and disease appear as the consequences of such migration. Yet tribal families of Orissa can do little to stop migration, as migration is not a matter of choice but often a compulsion to avoid starvation.

Migration is a very old problem for any country and has been recorded quite early in literature. Several attempts have been made to find an appropriate theoretical framework to understand the pattern of migration. Attempts are on to find out the reasons for migration, despite much research being done. The factors determining migration are many and different to diverse type of migration. It is different for internal as well as external migration and different at different point of time. There are many studies available on international migration, but very few available on internal migration.[3] No serious research has been conducted especially on slumdwellers in the Indian context, particularly in Orissa. From our limited search of the literature, we do not find any research study in Orissa on the environmental aspects of poor migrants who are basically slumdwellers.

3. Data and Methodology of the Study

The study uses both primary and secondary data. The secondary data were collected on the variables pertaining to understand the profiles of the migrants of Cuttack city and correlate the empirical results with the facts. The secondary data are collected from other sources: Municipality Office, Cuttack; Project Office, Cuttack; Statistics Office, Cuttack. The whole analysis is based on the primary data.

The present research study is a descriptive and diagnostic design. The study uses purposive sampling method while selecting the slums. The sample of slums is chosen on the convenience of the researcher, keeping the objective of the study in mind and the number of respondents is selected at random. The questionnaires and guidelines used in the study were pre-tested. Both qualitative and quantitative methods were used in the study. In this research work, direct personal interview with the schedule is conducted.[4] The study makes theoretical and empirical reviews to understand the issues involved in the literature on migration. The results are discussed in the light of authors' own experience gathered in the field, along with the data collected from the above-mentioned sources.

4. Empirical Analysis

The secondary data were collected on relevant information about the slumdwellers and presented in Table 19.1.[5] The total population of slumdwellers is approximately 1,34,994 which is about 25 per cent of total population of the city. According to Cuttack Municipality data, there are 364 slums, of which 106 are identified and 258 are unidentified. The land accrued by the identified slumdwellers is 415.65 acres.

It is important to note that more than 50 per cent of the slum dwellers live on the riverbank and canal side. We have chosen eight identified and unidentified slums with high representation from identified slums. The sample of 50 households is taken for the study.[6] The sample units are chosen on the assumption that the chance of migration of the people in the early age group is very high as compared to the aged people. The 66 per cent respondents are in the age group of 30–40 (see Table 19.3).[7]

In this paper, we attempt to understand the different aspects of migration and broadly discuss then under three important headings, namely, sociological, economic and environmental aspects.

Socio-Economic Aspects of Migration

The overwhelming majority of the respondents are from within the State and from close distance. The number of the respondents within the State is 43 and outside the State are 7. The migrants outside State are from West Bengal and UP. The migrants within the State are from within Cuttack and its neighbouring districts (see Table 19.4 for detail). The migrants are mostly Hindus. Out of this 70 per cent Hindu migrants are lower-caste Hindus (Sudras). If we see the comparative statistics in relation to the population, migration of Muslims are very alarming, which is around 30 per cent. An interesting feature is that those who have migrated from UP are all Muslims. Only 4 per cent Brahmins are migrants, out of 92.7 per cent Hindu slumdwellers.

Looking at the educational background of the migrants, the situation is not bad. The distribution of the sample in terms of the level of education is given in Table 19.5. It is quite interesting that 38 respondents are found to be literate. The number of those who have studied up to primary level, middle english and matriculation are respectively 19, 10 and

8. Only one respondent has education above matriculation. It is interesting to observe that all the slumdwellers are have awareness of the importance of education. This is evident from our finding that all of them send their children to schools.

Socially, the slumdwellers are not excluded from politics, medical, electricity and drinking-water facilities. They are either directly or indirectly involved in politics. Around 54 per cent slumdwellers owned voting cards whereas 46 per cent did not. The response to the political participation in favour is 40 per cent. (See Figure 19.1 in appendix for a clear view).[8]

Economic Aspects of Migration

The economic profile of the respondents is judged from their nature of occupation, income, house type and ownership, holding of BPL card, use of medical, electricity and water facilities, and others.

The slumdwellers are mainly daily wage labourer, Rickshaw Pullers, and Sweepers (see Table 19.6). Around 28 per cent migrants are daily wage labourer, 20 per cent Rickshaw Pullers, 12 per cent sweepers and 40 per cent involved in any other occupation.[9] The distribution of monthly income of the migrants is presented in the Table 19.7. Around 24 per cent of migrants belong to the Rs. 500–1000 income group, 40 per cent belong to the Rs. 1000-2000 income group, 34 per cent belong to Rs. 2000-3000 income group, and only one family has above Rs. 3000 income. The mean monthly income of the household is Rs 1,640. Around 76 per cent slumdwellers are satisfied with their present job.

The migrants have either *kaccha* or *pakka* type houses. Around 52 per cent respondents stay in kaccha houses whereas 46 per cent stay in pakka houses (See Table 19.8). This helps us to understand the lifestyle and economic status of the migrants in a well-balanced manner. It is interesting to find that the migrants are equally distributed based on their house/home status, that is, 50 per cent people have their own house (government land) whereas, 50 per cent stay in rented house.[10]

In our sample, 28 migrants hold BPL cards whereas 22 migrants do not. However, holding of a BPL card may not reflect the economic status of the slum dwellers. It is rather dominant by other non-economic factor. Most of the migrants use government medical facilities. Again, 88 per cent respondents enjoy electricity and all enjoy pure water. Around 70 per cent use the Municipality tube well and only 30 per cent use a personal well. During health problems, most of the migrants use government medical facilities (86 per cent), whereas, only 10 per cent contact private health professionals. Only 4 per cent respondents say that health workers of both Govternment and NGOs visit them (see Table 19.9).

Most of the slumdwellers are poor. Given the income estimation downwardly biased, with low cost of living and mostly nuclear family structure, no water and electricity charges, they could meet their basic needs, hence they are satisfied with their current economic status.

We have tried to determine the main reasons for migration. Economic factors, like inadequate job facilities in the native place emerged as the main reason whereas social factors like family quarrel and joining family members are also important. In our sample, 40 per cent people have migrated because of inadequate job facilities, 16 percent for family quarrel, and 14 per cent for business purposes. To join their family members, 14 per cent people have migrated. It is very significant to note that no respondents are found to migrate for the reason of better education for their children (see Table 19.10). It is interesting to note that the slumdwellers are satisfied with their present job, and they do not want to return to their native place.[11] Due to the availability of required facilities, 90 per cent of the respondents do not want to return to their native place and they also do not want to migrate any other place. However, they are still in touch with their place of origin. Around 92 per cent of the migrants miss their native place. Around 68 per cent of the migrants visit their native place while 32 per cent do not visit their native place at all. It is found that 20 per cent visit their native place monthly, 20 per cent visit yearly, and 28 per cent twice a year. As per the data, 80 per cent do not make any remittance to their natives and 90 per cent do not bring any goods from their village. See Figure 19.2 (a) through 19.2 (d) for a detailed analysis.

Environmental Aspects of Migration

The slumdwellers live in unhealthy conditions and they pollute their local environment. With the growing increase in the migrants, the pressure in the city is getting aggravated. With rapid growth of city population and illegal establishment of slums, the imbalance is created between the ecosystem and human population. The highest density of the slum is 3,814 in Oriya Bazar. The slums, particularly unidentified slums, hardly get any attention of the municipal authorities. There is no proper drainage facility and no proper sanitary facilities. Unidentified slums near river or canalside use river water for drinking purposes and use river basins for their daily duties. They fling paper, plastic bags, and other domestic waste near their surroundings or into the river. Sometimes, they throw dead bodies of the domestic animals and newborn babies in the river water. Besides the rest of the city, all the slums make their separate deities on every festival and make *'bisarjan'* in the river. These are some of the major causes of water pollution in the bank of two major rivers of the city, i.e. Mahanadi and Kathojodi.

The city is now planning to open industrial units. There is an alarming increase in the use of motor vehicles in the city. Even our study finds that all the migrants use electronic equipments and some of them have motor bikes. Barring a few, most of the slumdwellers use open space for cooking purposes and mostly fuelwoods. Some of them use kerosene and inefficient biomass stoves. In the process, on the one hand there is deforestation, and on the other there is increased air pollution.

It is not surprising but a matter of concern that slumdwellers live in very unhygienic conditions but are not aware of this. When we asked them about their opinion on living conditions, most of them (84 per cent) say that they live in healthy conditions. When we tried to find out from them whether specific attempts are being made to protect environment, 68 per cent did not understand, whereas others are equally divided on 'yes' and 'no' (see Figure 19.3 (a)). They hardly get sanitary and or psycho sanitary attention, either from the government or NGOs. In addition to their active participation in election, it is ideally expected that the government play an important role in the developmental activities of the slumdwellers. Nevertheless, when we asked the people whether they were happy with the government participation in the developmental activities, about 24 percent respondents were not aware of the role of the government or NGOs in the developmental activities of the slums, 58 per cent were not satisfied with the governmental effort. This is presented in Figure 19.3(b). From close observation, we found that these slums are having better road and transport facilities, and they enjoy electricity, water supply. Government does play an important role in the slum to undertake preventive measure before the outbreak of an epidemic.

These slum dwellers add more to the environmental pollution of the city in the form of soil pollution, water contamination, and air pollution. This has repercussions on health, economy and environment at both micro and macro level. The slumdwellers frequently get health problem. Chronic cold and skin diseases are very commonly found among the slum dwellers. Because of soil and water contaminations, the productivity of agricultural land decreases, as river water is the major source of irrigation. As a result, people migrate from nearby places to the city for alternative source of their livelihood. In our study, we found that most of the migrants are from the areas of close vicinity to the city in the range of 100 km. Therefore, a cyclical pattern of migration and environment is clearly visualised from our study.[12]

5. Conclusions and Policy Implications of the Study

This is a study on the slum dwellers (migrants) of the Cuttack city. The study uses general information on the slums and slumdwellers in the city from the secondary sources, to have a broad understanding about their socio-economic profiles. It is found that push and pull factors are important for migration but push factors are the source of drive for migration. The economic conditions of the migrants have improved after migration, but there is a rapid increase in the slum population. The social discrimination is not serious and politically they are inclusive.

Migration is not a matter of choice but often a compulsion to avoid starvation, to repay debts, among others. Under such circumstances, rural people can do little to stop migration. It is the responsibility of the government to control migration. The political parties should rise above petty political interest of creating slum vote-banks. They should rather cooperate the with the administration to put a check on migration. Government should appeal of all the political parties to rise above this cheap political interest. The local administration must pay attention to the unidentified slums as well. Local administration and government should not be the threat to poor slumdwellers, rather they should find some permanent resettlement. Government with the help of local administration must discourage migration and provide proper rehabilitation to the existing slum dwellers.

In our study, lack of opportunities to earn bread and butter in the native places is the main reason for migration. Government should provide alternative job opportunities at rural places. Though government is doing well in launching and funding different programmes in this direction, it should closely monitor the implementation of such programmes. In addition, the government should try to provide vocational training to enhance their productivity. The basic facilities like water, electricity, hospitals, etc. should be made available in different parts of the villages so that people are not forced to shift to other places.

Most of the slumdwellers send their children to school. This indicates the awareness towards the importance of education, hence government has a solid foundation but needs to focus on the high dropout rate at secondary and higher level of education. The government should help students by providing fellowships, stipends, books and other study materials.

However, unawareness of the people is the main cause of their living in unhygienic conditions and contributing more to environmental pollution. Government must campaign for the importance of a clean environment and make them aware about how they are unintentionally polluting the environment, which is the cause of their own health problems, and those of others. Here, NGOs must come forward and join hands with government and local administration to ensure better socio-economic life to the slum dwellers and save the environment.

The New Environment Policy (NEP), 2006 over-emphasises the trade-off between growth and industrialisation. Micro aspects of macro problems of environmental degradation are overlooked in the policy documents. The NEP lays down very broad and vague reforms and actions plans. Hence, there is a need for specific plans to deal with specific problems. There is a need for serious attempts to control migration. Finally, active participation of the governments and involvements of NGOs are necessary to control migration, improve the standard of livings of the people and prevent environmental degradation.

Bibliography

Anderson, M. (1990), "The social implications of demographic change", in E.M.L. Thompson (ed.), *The Cambridge Social History of Britain, 1750-1950,* Vol. 11,: Cambridge University Press, Cambridge, pp 1-70.

Arjan de Haan (2000), 'Migrants, livelihoods, and rights: the relevance of migration in development policies', *Social Development Working Paper,* No. 4 Department for International Development, UK.

Awases, M., A. Gbary, J. Nyoni and R. Chatora (2004), "Migration of Health Professionals in Six Countries: a Synthesis Report", *Division of Health Systems and Services Development,* WHO Regional Office for Africa, World Health Organization.

Baines, Dudley (1998), 'The economics of migration in nineteenth-century Britain', *Refresh,* Vol. 27, Autumn 1998.

Bijwaard Govert E. (2001), 'Dynamic economic aspects of migration', PhD Thesis entitled 'Rank Estimation of Duration Models', at the Free university and Tinbergen Institute, Amsterdam, in June 2001

Castle, S. (2003). "Towards a sociology of forced migration and social transformation", *Sociology,* 37:13-34.

Cutler, David M., Edward L. Glaeser and Jacob L. Vigdor (2005), "Is the melting pot still hot? Explaining the resurgence of immigrant segregation", *National Bureau of Economic Research Working Paper,* No. 11295, *http://www.nber.org/papers/w11295,* April 2005.

Eapen, Mridul (2004), 'Women and work mobility: some disquieting evidences from the Indian data', *CDS Working Paper*, No. 358, May 2004.

Fergany, Nader (2001), "Aspects of labor migration and unemployment in the Arab region", *Almishkat Center for Research, Egypt (www.almishkat.org)*.

Frey, William H. (1995), "Immigration and internal migration `flight' from US metropolitan areas: toward a new demographic balkanisation" *Urban Studies*, Vol. 32, Nos. 4- 5, pp. 733-57.

Gabriel, P. E. & Schmitz, S. (1995), "Favorable self-selection and the internal migration of young white males in the United States", *The Journal of Human Resources* 30: 460–471.

Gallanter, M. (1984), "*Competing inequalities : law and the backward classes in India*", Berkeley, CA: University of California Press.

Haberfeld, Y., R.K. Menaria, B.B. Sahoo and N. Vyas (1999), 'Seasonal migration of rural labor in India', *Population Research and Policy Review*, Vol 18, pp. 473–89.

Jha, Vikas (2005), 'Migration of Orissa's tribal women: a new story of exploitation', EPW Commentary, *Economic and Political Weekly*, April 9, 2005.

Ministry of Environment and Forests (2006), "National Environment Policy: 2006", Government of India.

Mukherji, Shekhar (2001), 'Low quality migration in India: the phenomena of distressed migration and acute urban decay', *24th IUSSP Conference*, Salvador, Brazil, August 2001.

Prabhakara, N. R. (1986), "*Internal migration and population redistribution in India*", Concept Publishing Company, New Delhi.

Productivity Commission Circular, "Economic impacts of migration and population growth", *http://www.pc.gov.au/study/migrationand population/index.html*.

Rao, M. S. A. (1986), "Some aspects of sociology of migration in India", in: M.S.A. Rao (ed.), *Studies in Migration*. Manohar Publication, New Delhi:

Rogaly Ben, Jhumá Biswas, Daniel Coppard, Abdur Rafique, Kumar Rana, Amrita Sengupta (2001), 'Seasonal migration, social change and migrants' rights: lessons from West Bengal', *Economic and Political Weekly*, December 8, 2001, pp. 4547–59.

Rogaly, Ben, Daniel Coppard, Abdur Rafique, Kumar Rana, Amrita Sengupta, and Jhuma Biswas (2002), 'Seasonal migration and welfare/illfare in eastern India: a social analysis', *Labour Mobility And Rural Society,* pp.89–114.

Samaddar, Ranabir (1999), '*The marginal nation: trans-border migration from Bangladesh to West Bengal*' Sage Publications, New Delhi, 1999; p. 228.

Sinn, Hans-Werner (2004), 'Migration, social standards and replacement incomes: how to protect low-income workers in the industrialised countries against the forces of globalization and market integration', *National Bureau of Economic Research Working Paper,* No. 10798, *http://www.nber.org/papers/w10798,* September.

Appendix

Table 19.1: General Information on Cuttack Slum

Indicators	Details
Slum Area	415.65 Acres (1.66 Sq. Kms.)
Number of Slums	364 (106 – Identified, 258 - Unidentified)
Locations	Tank Bund/River Bank/Canal side (51.03%), Main Road (34.69%), Sewerage Canal Bank (8.16%), Railway Line (6.12%)
Density	Density of Population: 212.17 per Acre Highest density - 3814 persons, Oriya Bazar; Lowest density - 50 persons, Bidanasi
Population	Total Slum Population: 1,34,994 (Male – 74010 and Female – 60984)
Average Per-capita income of a household	Rs.129/- per month
Sex Ratio	824 (Female/1000 Male)
Literacy Rate	48 percent
Occupational Pattern (in percent)	Daily wage (30.66), Rickshaw Puller (10.34), Maid Servants (04), Weaving/Basket making (03), Piggery (02), Govt. Services including local body (18), Trading & Business (10), Household Industries & other skilled jobs (02), Others categories (20)
Language, Caste and Religion (in percent)	Language: Oriya (74), Telugu (16.7), Hindi Speaking (6.7), Bengali (1.8), and Others (0.8)Caste: S.C. (41), S.T. (2) and Others (57)Religion: Hindu (92.7), Muslim (6.6), Christian (0.4) and Others (0.3)

Source: Authors' compilation from Municipality Office and Project Office, Cuttack.

Table 19.2: Distribution of Respondents over the Sample

Sl. No.	Name of the Slum	No. of Respondents	No. of Respondents (%)
1.	Bidanasi	1	2
2.	Hairanpur	5	10
3.	Sidheswar Sahi	4	8
4.	Gora Kabar Muslim Sahi	6	12
5.	Mahanadi Sahi	8	16
6.	Kathojodi Vihar	7	14
7.	Kadam Rasol	9	18
8.	Kafla Pana Sahi	10	20
	Total	**50**	**100**

Table 19.3: Age Group of Respondents			
Sl. No.	**Age group**	**No. of Respondents**	**No. of Respondents (%)**
1.	18–25	5	10
2.	25–30	5	10
3.	30–35	19	38
4.	35–40	14	28
5.	40 and Above	7	14
	Total	**50**	**100**

Table 19.4: Distance from Origin, Religion and Caste			
Sl. No.	**Variable**	**No. of Respondents**	**No. of Respondents (%)**
By Origin			
1.	**Out State**	**07**	**14**
2.	**Within State**	**43**	**86**
2(a).	Below 100 km	27	54
2(b).	100–200 km	11	22
2(c).	200–300 km	12	24
	Total	**50**	**100**
By Religion			
1.	Hindu	35	70
2.	Muslim	15	30
	Total	**50**	**100**
By Caste			
1.	Brahmin	2	4
2.	Kshatriya	4	8
3.	Sudra	25	50
4.	Any Other (Incl. Muslim)	19	38
	Total	**50**	**100**

Table 19.5: Educational Background			
Sl. No.	**Educational Background**	**No. of Respondents**	**No. of Respondents (%)**
1.	**Illiterate**	**12**	**24**
2.	Primary	19	38
3.	Middle English	10	20
4	Matriculate	8	16
5.	Intermediate	1	2
	Total	**50**	**100**

Table 19.6: Distribution of Respondents by Occupation

Sl. No.	Occupation	No. of Respondents	No. of Respondents (%)
1.	Daily Wage	14	28
2.	Rickshaw Puller	10	20
3.	Sweeper	6	12
4.	Any Other*	20	40
	Total	**50**	**100**

*: Police Constables, Maid Servants, and Shopkeepers

Table 19.7: Distribution of Income of Respondents

Sl. No.	Monthly Income (in Rs)	No. of Respondents	No. of Respondents (%)
1.	500–1000	12	24
2.	1000–2000	20	40
3.	2000–3000	17	34
4.	3000–4000	1	2
	Total	**50**	**100**

Table 19.8: House Type and House and Ownership Status

Sl. No.	Ownership and type of house		No. of Respondents	No. of Respondents (%)
1.	Type of Houses	Kachha(Jhugi)	26	52
2.		Tile	1	2
3.		Pakka	23	46
	Total	**50**	**100**	
1.	Home Status	Owned	25	50
2.		Rented	25	50
	Total	50	100	

Table 19.9: Access to Medical Facilities

Sl. No.	Medical Access	No. of Respondents	No. of Respondents (%)
1.	Private Doctor	5	10
2.	Govt. Doctor	43	86
3.	Health Worker	2	4
	Total	**50**	**100**

Table 19.10: Distribution of Respondents by Reasons for Leaving Native Place			
Sl. No.	**Reasons**	**No. of Respondents**	**No. of Respondents (%)**
1.	Inadequate job facility	20	40
2.	Family Quarrel	8	16
3.	Better Education	0	0
4.	Business	7	14
5.	To join family member	7	14
6.	Any Other	8	16
	Total	**50**	**100**

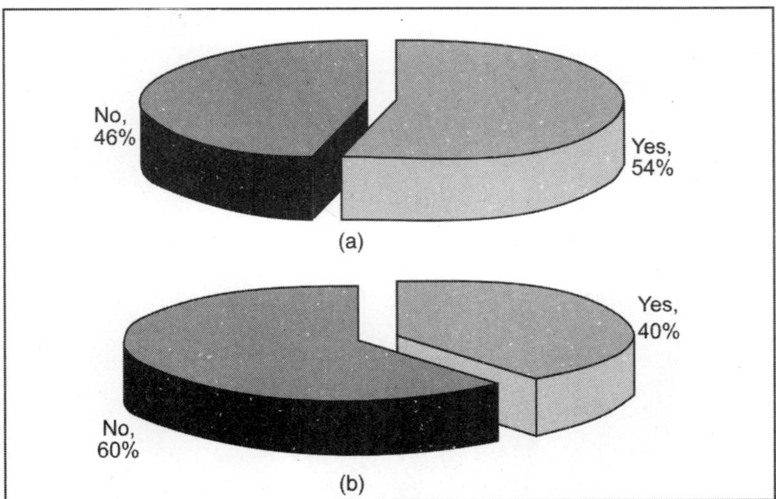

Figure 19.1: (a) Response to Holding Voting Cards and (b) Participation in Politics

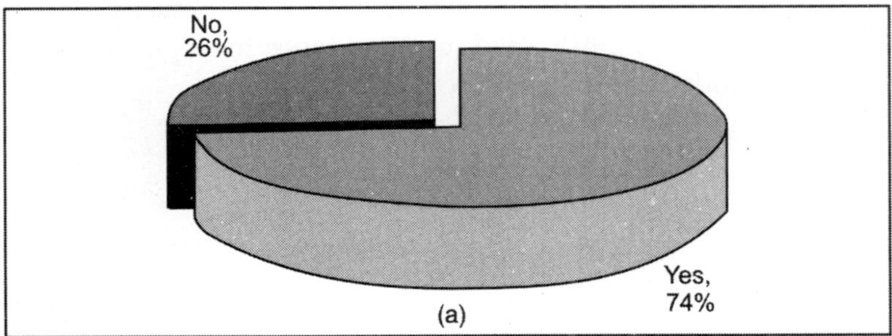

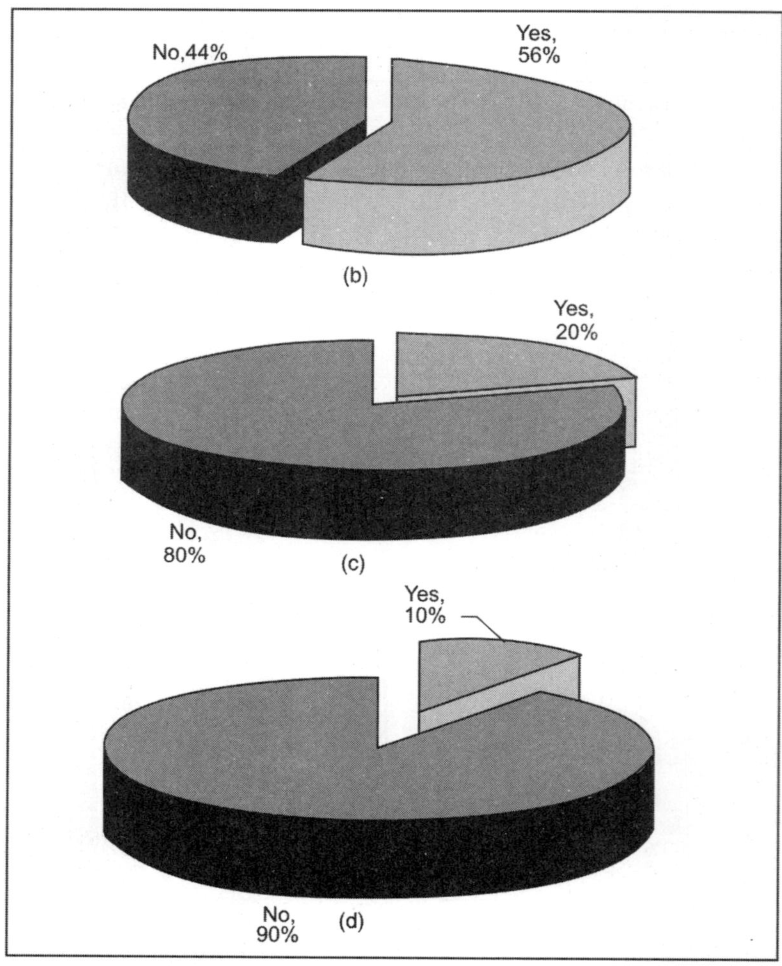

Figure 19.2: (a) Response to satisfaction with present job (b) Response to holding BPL card (c) Ramittance to origin place (d) Bringing goods from origin place

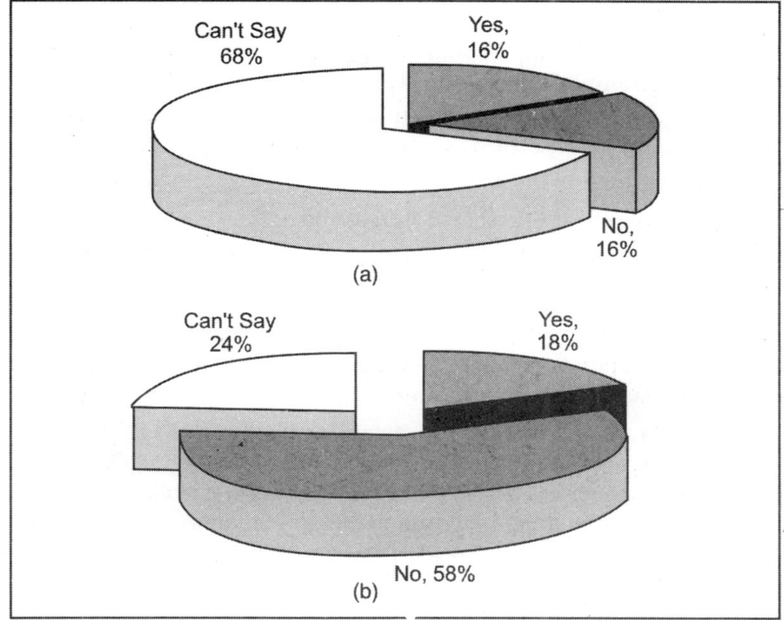

Figure 19.3: (a) Response to Preventive Measures to Protect Environment (b) Response to Developmental Efforts of Govt and NGOs

Endnotes

1. Cuttack city is generally known as slum city and its culture is called 'Basti Culture'. The migrants are influencing the economic, political, cultural and social aspects of the city.
2. Those studies designed to examine processes of internal migration made use of explanations developed originally in order to explain international migration. Gabriel & Schmitz (1995), for example, offered the positive self-selection explanation to understand migration of white men from one American city to another.
3. We have made attempts to survey the literature both on internal as well as external migration. Attempts have been made to understand the determinants of migration, but it is difficult to come out with specific set of determinants for internal migration of different location
4. The interview schedule was developed in order to collect desired information in accordance of the objective of the study

5. All the tables used in the study are presented in Appendix.
6. The name of the sample slums and the number of respondents chosen from the sample are given in Appendix, Table 19.2.
7. In our sample, interestingly, all the respondents have migrated in the age group of 18–40. Around 85 per cent people have migrated in the age group of 18–30.
8. All the figures are presented in Appendix.
9. Others include police constables, maid servants, and shopkeepers.
10. This is mostly found in identified colonies. The old slumdwellers construct house and sub-let it to fellow migrants.
11. Only 11 respondents out of 50 (22 percent) wanted to return to their native place whereas 78 per cent prefered to stay back.
12. This is a very weak way of establishing the linkages between migration and environmental degradation. A strong analytical and empirical exercise is required to understand and establish the vicious circle of migration and environmental degradation. This is beyond the scope of individual research without funding. However, the established relationship between migration and environment does not undermine policy implication of the study.

Chapter 20

Pricing of Potable Drinking Water in Delhi: A Contingent Valuation Approach

Farhan Ahmed

Introduction

Water wars are a reality now, right from the micro level to the global level. These conflicts are the result of cumulative effect of gross negligence and mismanagement of water resources over the years. The problem is not due to absolute shortage of water but due to the absence of a proper mechanism for conservation, distribution, and efficient use. Water will continue to be a leading policy issue in the 21st century and beyond.

Institutional factors are the key to improving the efficiency of the water distribution system. For shifting towards demand management (through pricing and introducing new technologies), an institutional agreement ought to be in place. A mere increase in the price may not result in financial sustainability of the system unless the institutional factors are in order to recover the charges. It has so often been suggested that deficiencies in water management are attributed to the alienation of water users from the process of planning and implementation. Maintenance and management of water systems through participation of user societies is expected to contribute to an efficient and equitable distribution of water resources.

Water-pricing policies have become a hackneyed and ritualistic term in the Indian context. Everybody supports but nobody implements it. Artificially kept low water price fail to provide any incentive to improve the systems, technically or institutionally, as the economic transaction costs go against it. To promote the efficient use of water, user charges on the basis of financial cost of providing access to water should be levied. These costs include Operation and Maintenance (O&M) of publicly provided schemes, capital costs, overheads such as the administration and the support required to operate, allowances to provide for depreciation, replacement or refurbishment, catchments management costs and social and environmental costs.

In the case of drinking water, basic human needs should be provided free of cost or at a very nominal cost. The price of water should be equivalent to the O&M cost, up the next slab of consumption. The National Water Policy document not only mentions covering the O&M cost but also the scarcity of value of water. Water should be treated as a social good as long as it fulfills only basic need, beyond which it should be treated as purely as an economic good. Pricing should be accompanied by efficiency improvement in the system. The consumer should not be penalised for the inefficiencies of the department in terms of distribution losses and unaccounted water. The task, however, is to rationalise the pricing structure so that at least O&M costs are recovered fully in a phased manner.

Of late, 'Willingness to Charge' has appeared as the main obstacle. Pricing of water on cost basis is essential because it now helps in resources generation but also results in efficient usage of water and discourages wastage. A multifaceted policy approach is necessary to move on the accelerated growth plane. This is more so in the case of the water sector as it is closely associated with other sectors.

It is widely understood that inefficient use of water resource can pose a threat to human beings and sustainable development through deteriorating quality and depleting quantity. Consideration of water as a free good and consequent under-valuation and improper incentive in the decision making leada to its inefficient use. But, due to the special non-substitutable nature of water, its allocation has remained outside the conventional market structure and standard theoretical prescriptions of achieving efficiency in allocation for quite a long time. In the wake of the rising demand for water quantity and quality on the one hand, and need for efficiency to achieve sustainability on the other, rethinking in terms of applicability of economic principles has arisen. Quite often Willingness to Pay tries to get quantification through a non-rigorous consumer survey on the stated preference and leads to a misleading conclusion towards water pricing and possible imposition of market structure for a good that merits special attention. In theory, it is easily shown that full-cost principles, if implemented, can facilitate transition to sustainable use of water.

According to the full-cost principle, all users of environmental resources should pay their full cost. Though this sounds rather obvious, implementing the principle would represent quite a change from the traditional practice of low lump sum or tax. Implementing the full-cost principle would not only end the implicit subsidy but would change the whole regime, since environmental costs need to be embedded too. This will not only call for correct estimation of environmental cost or scarcity value but also address the issue of huge hike in price of water in the face of debate over water as basic necessity and right. Simultaneously, it is true that when the level of economic activity was small, the corresponding subsidy was also small. Since the economic activity has grown, however, the subsidy has become very large indeed, ignoring its leading to significant resource distortion. Again wrong incentive and inefficient use has led to usable water privately, which may not be the case for all the consumers. All these conflicting issues do complicate the efforts towards achievement of efficiency in water use in reality through implementation of theoretically sound policies. It would be useful to know the loss of consumer welfare in the existing regime and how consumers are coping with the situation, to help policy makers to arrive at some alternative regime where quality of water may be ensured for all the users of water.

The urban water supply situation in Delhi is deteriorating and the urban population continually adjusts to the worsening to the situation. Many people have water in their pipes only for a few hours a day. Households who that can afford it drill their own wells and install both under ground and overhead storage tanks. They increasingly buy bottled water or treat their drinking water. They spend scarce resources on medical care for water-borne diseases that result from groundwater infiltration into the piped distribution system. State government can no longer afford to subsidise municipal providers. The status quo is bad, but the situation is deteriorating and the future for municipal water supply services looks even worse. The population continues to increase. Lack of maintenance results in broken pipelines and higher levels of contamination and unaccounted-for water. The groundwater table falls and people have to drill deeper to reach water. Cities go farther and farther away to find new, more costly sources of water.

One is not sure whether to pose this question as an economic good or as an ethical one. The ethical issues are related to equality, morality, cooperation, which are uniquely difficult to measure and quantify in their own way.

Economic logic, for a change, is simple here. Water is like any other natural resource, it does not have a market but it definitely has a utility and it is this utility value that can be the guiding rule for its development within the framework of the Fundamental Rights.

The water tariffs currently in use in Delhi are not accomplish their principal objectives. They are not generating sufficient revenues to ensure that utilities can recover their financial costs. They do not send correct economic signals to the consumer, i.e. that water is scarce and must be treated as a valuable commodity.

The basic issue that we address in this paper is that good quality water has emerged as a scarce resource. Valuation of water per se has importance in the context of sustainable development paradigm. It would be useful to know how the domestic end users value safe drinking water, from their observed behaviour. The objective is to address the issue from the consumers' point of view, for they are beneficiaries. The scope of this paper is limited to drinking water quality but the same can be extended to other types of water demand.

Issues/Problems

Water in India is not treated as an economic good where the price reflects the demand and supply for its competing uses, in fact, it is considered as a free good. Without an efficient pricing and regulatory mechanism, the cost of water pollution can not be factored out. Many issues/problems have come up regarding water pollution and it pricing which has made it a bone of contention among scholars. Some of the fundamental questions that needs to be answered are related to the problem of market failure. Promotion of water market, where the scarce resource is priced, thereby adding a cost to all those who have been using it as a free natural resources without giving a thought to its depletion. The Constitution of India, legally under Article 243G provides the right to The People of India to govern the use of number of basic resources, including water. Water is a common property resource characterised by rivalrous consumption and non-excludability principle. Here comes the problem of market failure where, if one person reduces total consumption, everyone else has the incentive to capture the benefit.

The current legislation assigns the property right of natural resource, with all those people who own the land above the aquifer however there is no limit specified on the amount of water to be extracted. Everyone else has the same right to extract water, leading to inefficient results. The famous 90–10 rule of diminishing returns comes—to play leading to depletion of groundwater. The result is poor water quality.

Water, such an important basic resource, fails to enter into the National Account except when it is used as an input for industrial sector. This fact alone hints at the fact that water, its use, its consumption benefits, its depletion and degradation are not dealt with at all in the National Income Accounting.

Externalities exist when a person does not receive all the benefit or bear all the cost of the actions undertaken. An example will make this point clear; the acute shortage has forced many individual to go in for private arrangements ranging from dug wells, tube well, pump set. They value water in terms of their capital cost and their maintenance. However, they are not concerned about the depletion rate and the exhaustibility. Additionally, a high powered pump may prevent others in the adjoining area to extract water. Here the cost of water extraction excludes the social impact as the individual has little incentive to switch to eco-friendly practices. Absence of any legal/institutional set-up allows one to go scot free, for the simple reason that any attempt by one to reduce the exploitation of water is not respected as others have the same right to exploit the groundwater.

Any environmental problem is in example of the tragedy of the collective. Garret Hardin (1968) popularised this idea in an article called 'The Tragedy of the Commons'. Hardin logically pointed out that it was the absence of any definite owner that led to environmental degradation, what we label as the problem of 'open access'. Since resources belong to everyone, and thus de facto to no one, everyone uses them, but no one has an incentive to conserve them, in spite of knowing that such non-discretionary use would leave nothing for the future. This is because wise use by one person does not guarantee similar behaviour by others, nor does it guarantee returns for good behaviour tomorrow.

In contrast to treating water as an economic good, some experts argue that the issue of water supply is between providing water as a basic need whilst also protecting it as a scarce resource. He argues that it is the State's responsibility to ensure all the citizens have access to sufficient water supply and sanitation facilities. Therefore, water has to be treated as an environmental resource rather than as a public good or a private good and is to be protected for long-term collective consumption. Others are of the opinion that free access to a resource leads to excessive use and that charging leads to sustainable water management. It is known that the environmental goods and resources do not generally have a market and are referred as 'missing market', where resources are treated free and are vulnerable to misuse.

The pricing of water services is, however, a controversial, perhaps an important reason why there is a lack of consensus among the academic group.

Who should pay?

The political economy of setting water tariffs requires that one should carefully consider the claim that water is a basic *"human right"* and should be provided free. When someone claims a *right,* they attempt to impose a corresponding obligation on other. The presumption may be that the individual for whom the right is claimed is too poor to pay for water himself, and that some one else should pay these costs. The 'someone else' is typically one of the three groups: (i) the state, (ii) the international community, or (iii) industry.

The real question is not whether this can be done, but whether such a 'free water' policy is a good idea. It is of course impossible for everyone to avoid the costs of providing water services to customers; someone must pay the financial and environmental costs. If the cost recovery objective is ignored, financial resources have to be diverted from others sectors of the economy. Given the economic system prevalent, such entitlement policies are sure to create massive macro-economic distortions as the economic system invariably struggles to meet such unfunded liabilities. Again, providing water free means the correct economic signals are not being sent out to the potential customers, means inefficient use of the resource.

So, a policy of 'free water' not only meant that the cost recovery and economic efficiency objectives were not met, but also that service quality itself declined. The 'free' service provided thus ironically became less valuable to the households.

The role of the State in public finance has often been looked at in the context of public and private goods in order to determine who should pay for the kind of services. It is argued that, because the benefits from water supply, such as time saving, amenity and health benefits accruing to the households, the public finance principle dictates that the bulk of the costs should be borne by the household themselves. By the same principle, because there are strong positive externalities in sanitation whereby the benefits accrue to the community as a whole from safe excreta disposal, the cost outside the household should be borne by the State.

There are reasons why water markets could be preferred to administered efficiency pricing. First, the reduction in the information costs since buyers and sellers of water generate the necessary information on the value of marginal product and opportunity costs of water. As in the case of other markets, different values are measured and compared by prices, and the way price signals coordinate dispersed information and preferences is one of the great advantages of water markets.

Finally, the administrative solution presumes far-seeing, incorruptible, influence-free administrative bodies that are able to design and implement the 'correct' prices. In practice, this may often not be the case: these bodies could be captured by interest groups or they may be shortsighted and unable to estimate future demand, or they may be unable to set and collect appropriate water charges.

For water markets to work, property rights to water must be private, exclusive and transferable. In this context, secure ownership provides an incentive to invest in greater productivity of the resource, while freedom to exchange provides the flexibility to reallocate the rights according to changing demand and other conditions. The role of the state should be minimal in this setting and should be restricted to protecting property rights, enforcing contracts, and reducing transaction costs and barriers to exchange. In fact, it can be argued that much of the current inefficiency in the water sector in India is due to excessive state regulation and subsidies which have distorted patterns of water use.

Another important rationale for water markets is the relationship between markets and liberty. In contrast to non-market allocation, which gives the state leverage in non-economic spheres as well, private property creates a space for individuals where the state cannot trespass.

In this context, by creating entitlement where none existed earlier, these markets can potentially be a tool for empowerment. Holders of water rights would be sought after (irrespective of their socio-economic status), by those who would like to buy these rights. Similarly, environmentalists can also purchase water rights in order to reserve a value wetland or to increase a waterway's flow. Without a market mechanism, environmental groups would have to depend on the state to achieve the same end.

Delhi—A Case Study

Delhi Jal Board (DJB) is the sole government agency responsible for meeting water demands of the city. The Recent literatures is full of evidence as to how drinking water has in fact become a luxury in Delhi.

The DJB tariff design has not accomplished its principal objective of generating sufficient revenues to cover its financial costs. This is evident from low recovery of water bills, number of defaulters and free riders which have been on rising curve. With the population rising, there is increased pressure on the existing infrastructure facilities; cases of broken pipelines, higher level of contamination, falling water level are some of its outcomes.

Pure drinking water is a luxury in India. However, in case of Delhi, it is just drinking water. Thus, people have begun to realise that in order to enjoy this luxury, they have to pay a price. *Delhi, the capital city of India* cannot boast of supplying sufficient water, a basic necessity to all its citizens.

Every year people have to spend thousands of rupees to acquire drinking water. Purity takes a back seat in this severe crisis.

The Delhi Jal Board (DJB) supplies 630 MGD, including 80 MGD of ground water, out of which about 40 per cent is lost to pilferage and transmission. The requirement of 13.8 million population residing within the area of supply of the Board, has been estimated at 830 MGD. The gap thus comes out to be 450 MGD. As per the Central Ground Water Board (CGWB), individuals and DJB together take about 300 MGD of groundwater. In a recent study, the French Institute has come out with higher figures; 120 MGD for DJB and for individual households. The Municipal Corporation runs 1,620 borewells that feed the system. Unfortunately, no record or reliable information is available on private extraction. Different studies, nonetheless, tend to suggest that while the gap between demand and supply has been increasing over the years the water table has been going down systematically.

It is estimated that Delhi has 40 per cent of the population that lives 'illegally' in the slums and other unauthorised areas. Officials simply pretend that these people do not exist. On the other hand, successive governments have accused slumdwellers of polluting the Yamuna. It is true that their hutments look dirty, but it is common sense that that the poor are not the main source of pollution. Water is polluted by those who use it and Delhi's poor get only a very small share of it.

The city officials supply 3,600 million litres of water to its people every day. But roughly half of it reaches households. The rest is officially accounted as the 'distribution losses'. The water that is supplied creates inequality and waste. 70 per cent of Delhi gets less than 5 per cent of water, while quarters where government officials and rich and affluent people reside get a staggering 400–500 litres per capita daily.

From the supply side, Delhi meets its water requirement through three sources, namely, the river Yamuna, which almost meets 90 per cent of the city requirement, the share of Delhi as per inter state agreement is mere 4.6 per cent, for a major share is diverted to Haryana and Punjab, the other two being groundwater and rainwater. According to a study *Ground water in Delhi—2003*, groundwater contributes around 100 MGD to the water supply system of DJB. It is being extracted through a system of network of tubewells and other ways. Another 256 MGD is being exploited by the individual private tubewell to meet the demand—supply gap. The growing population has certainly added undue pressure on the groundwater resources; given the fact that the water table has been going down. It is important to note that it is not only quantity which is being adversely hit a but also the quality, as its chemical composition gets adultered.The third source is Rainwater; Delhi annually receives a rainfall of 611.8 mm in 27 days.

According to the 2001 Census, the population of Delhi is 137.83 lakh. Decennial population from 1991 to 2001 in percentage terms has been 46.31 per cent. With the projected population of Delhi by 2011 being 195.1 lakh which is expected to cross two crore by 2021, concrete steps are to be taken to reverse the present condition. When DJB fails to quench the residents of the city, people resort to alternative means which are both legal as well as illegal. After all, DJB does not leave them with much option. These may range from boring their own tube-well, purchasing water from private tankers or purchasing bottled drinking water.

Invariably, every colony in Delhi has a borewell/tubewell to augment water supply. Legally, the bore well has to be registered with the Central Ground Water Authority

(CGWA); this law, however, is seldom practised. Presently there are around 1,00,000 registered borewells in Delhi though the authorities at the DJB believes that this number could be as high as 300000.

The estimates of extraction of groundwater given by various agencies differ significantly but these underline the problem of the declining watertable in and around the metropolis .The watertable has gone down to 33 metres in south Delhi, in some area the figure being as high as 40 metres. South-west Delhi is also seriously affected by large scale extraction of groundwater, the watertable here being 14 metres. Further, as per the projection of the CGWB, water would not be enough in this area even for a couple of year at the present rate of exploitation. Importantly, there are over 500 water bodies located mostly in south and south-west Delhi where the watertable has gone down alarmingly in recent year. Recharging of these holds the key to the water problem in the city. Banning the use of ground water for watering lawns and permitting its extraction only for drinking purposes has been proposed.

Not long ago, the concept of buying water appeared quite confusing. People were not used to drinking water that had to be bought. They, however, are now getting used to paying more than money for water than they do for milk every day.

Review of Literature

Ahmed et al. (2002), found that out of 2,430 sample respondents, 76 per cent were willing to pay a positive amount towards securing a better water supply system, the estimated mean WTP for stand post is Tk 51 per month and Tk 960 towards initial capital cost; the estimated WTP measure for poor households is Tk 44 per month and for non-poor were significantly as expected. Of those not willing to pay, 10.5 per cent cited income as their constraint in doing so. Also, an interesting result that emerged was that the respondents believe that piped water supply would not only relieve them of arranging water from far-off sources but would also mean better quality and convenient.

Bhandari and Bajpayee (2001),followed a very interesting format for deriving at WTP figures. Instead of GNP figures of the households, it extracts the required information through asset ownership like TV ownership, mode of commuting, newspaper subscription giving due recognition to their disparities assigning dummy variables to them. Further, it demarcated six potential groups namely: low needs–low economic status, low needs–medium economic status, low needs–high economic status, high needs–low economic status, high needs–medium economic status, high needs–high economic status. Of the 31,323 sampled households, about 40 per cent of the respondents were willing to pay some positive amount. Another 30 per cent were willing to contribute through their labour. Thus in total close to 70 per cent of the respondents were willing to contribute in one way or the other to improve the present condition.

In Hasler et al. (2005), significant results were obtained which shows that Danish people have a strong preference for clean, safe drinking water. The estimated figure for potable drinking water using WTP methodology worked out to be around 253 EUR/year. The estimated figure were significantly higher than the amount which the residents were already paying as a part of their water bill which in the present case was 188 EUR/year.

Jalan et al. (2004), used the data from the National Family Health Survey (NHFS). They included only those households with 15 or fewer members. Accordingly, they used 20,681 samples to come to the conclusion that the controlled Willingness to Pay more than doubles from Rs. 158 in the bottom wealth quartile to Rs. 329 in the top quartile. Median Willingness to Pay increases ten-fold between the bottom and top wealth quartiles from Rs. 44 to Rs. 451. Median Willingness to Pay in the lowest category for the maximum female education, no education, is Rs. 81, while for the highest category of 14+ years of education, it is Rs. 497. It is important to note that as the number of education of female increases from 3.5 to 10.5 years, the WTP figures inflate by almost 50 per cent, where as the corresponding figure for the male comes out to be around 21 per cent. These numbers probably support the view that men tend to judge health risks to be much smaller and less problematic than females do.

Kadekodi et al. (2002) in their paper found out Willingness to Pay of water for drinking purposes in Kumaun region. Using the Contingent Valuation Method, out of total sample survey of 248 respondents, 23.72 per cent were not willing to pay at all for one reason or the other, whereas 64.62 per cent were willing to pay for drinking water. Of those cases who were not willing to pay, 30 per cent of them cited income as their major constraint. On the basis of this study, it was concluded that average consumption of water for drinking and other household purpose for entire region works out to be around 106.58 litres per day. The annual willingness to pay worked out to around Rs. 42.58 per family per year.

Kramer, et al. (2001),–Hecht of Nischolas School of Environment and Earth Sciences, Durham, while estimating the economic value of water quality improvement in the Catabwa river basin, used, 1,085 samples. They used a split-sample survey design, an interesting format of phone – mail – phone approach was followed. Respondents expressed a mean willingness to pay of $39 for improvement in water quality and quantity at its current level.

Roy and Sahu, (2004), followed the Contingent Valuation Method to assess the Willingness to Pay for water at number of places.

Dehradun, in 1996, consumers were willing to pay more than twice the prevailing rate. The study revealed that, on average, households were already paying up to Rs. 10 per cubic meter in, coping costs arising from irregularity and unreliability of the supply.

Baroda, in 1995, households with income below Rs. 1,500 per month were willing to pay up to Rs. 275 per annum for a reliable service while the wealthier families with income between 4,500 to 6,000 were willing to pay up to Rs. 440.

Delhi, in 1998, households could pay anything up to Rs. 2,000 per year in indirect and direct costs to cope with the irregularity and unreliability of existing supplies. This potential source of revenue is not captured by the formal providers, but paid directly to the unregulated private sector interests.

Roy, et al. (2004), estimates the expenditure, as a measure of willingness to pay of the households, the information required on household expenditure in water purification and other socio-economic details is generated by primary surveys carried out for a sample of 240 households chosen randomly to be representative enough. It is found from their observed behavior that 77 per cent of the households surveyed in residential areas, have a positive willingness to pay for good quality drinking water since they have resorted to one or more water purification techniques. The estimated average averting expenditure is Rs. 168.72 per month per household, estimated from a range where the minimum and maximum expenditure are Rs. 1.97 per month and Rs. 887.50 per month respectively.

Tapvong and Kruavan (2003), was a Contingent Valuation study of the Chao Pharaya river. With a sample size of 1,100 households, a two stage stratified random sampling was used. The outcomes of the study have made it quite clear that the deteriorating water quality in Bangkok is one of the major problems and treatment is inevitable. They used both logistic and OLS regression to understand and study the WTP for the waste water treatment. Both the methods indicated factors like education, knowledge and importance of project as some of the variables affecting respondents' behaviour. The mean value of the fee for the treatment of water quality 1 and water quality 2 were found to be 100.81 and 115.03 baht per month respectively.

Viscusi et al. (2004), study used the stated preference method to estimate the economic value of protecting water quality. With a sample size of over 1,000 respondents, the relevant statistics of the amount people are willing to pay had a mean value of $23.17 for one percentage point improvement in water quality. The independent variables that were significant enough to affect WTP included income, age, education, and perceived water quality.

Whittington (2003), in a survey in Katmandu, estimated that out of 80 per cent of the 991 sampled respondents were willing to pay a positive amount to secure a better quality and increased quantity of drinking water than they avail at present. In fact, they were willing to pay four times than their current bill warranted. The observed variables that were found significant were substantial amounts of money spent on treatment of water-borne diseases, income-earning capacity, opportunity cost in arranging alternative option to secure water supply.

Whitehead and Hoban (1998), 'Willingness to Pay and Drinking Water Quality' used the logistic regression to estimate the WTP. Consumers who were concerned with their health due to water-related problems were found to pay 1.67 times more than others. Similarly, those who rate the current state of water as fair or poor were found to bid 2.4 times more. On the same account, the bid level given by the educated class was significantly higher than uneducated class. The cost in exploring alternative options for drinking water was also a significant variable.

The objective of the present study is to estimate the Willingness of Delhi residents to pay for improved drinking water supply through the Contingent Valuation Method. A price which is close to the real cost of supplying drinking water may actually become a key component if the present condition are to be reversed.

Methodology

The present Paper describes the details of methodology employed for achieving the set objectives and tests the hypothesis.

The primary objective of a Contingent Valuation Method (CVM) is used to discover the value placed on the improved provision of non- marketed good and services by household.

The conventional definition of WTP is the maximum sum of money that an individual is willing to pay for a certain good rather than do without an increase in that good.

The economic concept of value that one uses to evaluate WTP has its foundation in the neo classical welfare economics. At the heart of welfare economics is the premise that the purpose of economic activity is to increase the well being of individuals who constitute society and that each individual is best judge of how well off he or she is in a given situation (Freeman, 1993).The level of welfare that an individual attains depends on the consumption of goods and services available in an economy. These goods and services are either available in markets (as private goods and services) or provided by the government (as public goods) or flow from the resource-environment system that is neither private good nor pure public good. If we focus only on natural resources and environmental good then it can be said that the basis for deriving measures of economic value of changes in resource-environment systems is their effect on human welfare.

Specifically, we can write:

$$U = u(X, Q)$$

Where, X is a vector of the quantities of market goods and q is the vector of environmental goods. It is assumed that this utility function is increasing and all its argument, is continuous, convex and twice differentiable. The consumer faces a set of given prices for all goods that the consumer wants to include in the basket and a given quality of environmental goods. Now, the consumer's task is to choose optimal quantities of all goods given his monetary income or budget such that this utility is maximised.

Mathematically,

Maximise $U = u(X, Q)$

Subject to $\Sigma_i P_{i.} X_i = M$

Where, X is a vector of quantities ($X = x_1, x_2, x_3 \ldots\ldots x_n$) and P is the corresponding price vector ($P = p_1, p_2, p_3 \ldots\ldots p_n$). The solution to this problem leads to a demand function of a certain quantity as a function of its own price, the price of other commodities and the available disposable income.

Substituting the expression for x_i as function of P and M into the direct utility function, we get indirect utility function which implies that consumers' utility is a function of commodity prices and money income assuming optimal choices of goods i.e. $U = v(P, Q, M)$.

Let us assume that the resulting level of utility that he attains is U_o. Now, let there be an improvement in environmental quality to q_1, where $q_1 > q_o$ such that the individual reaches a higher level of utility at U_1. What is the value of this change in environmental quality? The most popular method of valuing the change due to quality improvement is to determine the amount of money, in lieu of the action, that would have the same effect on utility. Compensating surplus and equivalent surplus capture the welfare effect of quality change.

The amount of money that would keep the individual at the original or initial level of utility with the change is called the compensating surplus (CS). This compensating surplus is equivalent to compensating variation for market goods. Alternatively, CS measures the maximum amount of money that can be taken from an individual for quality improvement such that he is not worse off than what he had started with . This also gives the maximum willingness to pay of an individual for change in quality. Technically, we can define the term CS using expenditure function. Expenditure function (E) is the minimum expenditure that is incurred to achieve a certain level of utility with a given vector of prices of commodities and income.

Hence, $CS\ (q_o, q_l)= E\ (p, q_o, U_o)- E\ (p, q_l, U_o)$ where p is the price vector for all other goods in the basket. In other words, CS is equal to the area to the left of the Hicks-compensated demand curve between the two prices.

On the other hand, equivalent surplus (ES) is the amount of money that would move the individual to the new level of utility without the change. Hence, it is the change in income that would have the same (equivalent) effect as the quality change, without the quality change actually taking place.

Technically this can be estimated as

$ES\ (q_o, q_l) = E\ (p, q_o, U_o) - E\ (p, q_l, U_1\ )$ or

$ES\ (q_o, q_l) = E\ (p, q_o, U_o) - E\ (p, q_o, U_1)$.

Willingness to pay for an improvement to ql from qo is measure by compensating surplus

$[(WTP=E\ (p,\ q_{o},\ U_{o)}) - E\ (p,\ q_{l},\ U_{1})]$

It is equally to note two important things:

Maximum WTP is bounded by the discretionary income: WTP<Y.

Negative WTP is assumed to hold no meaning; 0<WTP<Y.

In the theoretical model of the CVM, a household's WTP was given as a function of the household's characteristics.

An attempt is made to establish the factors that influence the utility change experienced by the household as a result of provision of non-marketed good, these factors being inducted into the model.

For the present purpose, it is done through a simple linear form:

$$WTP = a + bX$$

Where, X is the set of independent variable, for this project it includes expenditure on arranging alternative water sources, water-borne diseases, affect on work, out-of-pocket expenses on water-borne diseases, importance of water, perceived water quality, credibility of the scheme.

Data Requirement and Sources

The official figures report that almost all households in the capital, Delhi have a metered water supply. Although there is acknowledgement of problem of water supply in certain areas, there is no clear information about the magnitude of the problem.

Primary Survey through Questionnaire has been done to know the WTP for improved drinking supply, taking into account their socio-economic characteristics. Delhi was divided into four zones namely north, south, west and east. Each of these zones is characterised by heterogeneous population. A stratified random sampling method was followed, where three homogenous strata were formed; namely high income bracket, middle income and low income bracket. Random sampling thereafter was followed in these selected strata.

A total of 80 samples have been collected from different areas of the Delhi.

Empirical Results

Model	Variables Entered	Variables Removed	Method
	Affect on work, Imp of safe water, sum of medical expenses, expenditure on arranging water, perceived water quality, credibility of the scheme, water borne disease(a)		Enter

(a) All requested variables entered.
(b) Dependent Variable: willingness to pay

ANOVA[b]

Model	Sum of Squares	df	Mean Square	F	Sig.
Regression	1884725.8	7	269246.546	24.609	.000a
ResidualTotal	787742.93	72	10940.874		
	2672468.8	79			

(a) Predictors: (Constant), affect on work, Imp of safe wtr, sum of medical expenses, expenditure on arranging water, perceived water quality, credibility of the scheme, wtr borne disease
(b) Dependent Variable: willingness to pay

Model Summary

Model	R	R Square	Adjusted R Square	Std. Error of the Estimate
1	.840(a)	.735	.700	104.599

(a) Predictors: (Constant), affect on work, Imp of safe water, sum of medical expenses, expenditure on arranging water, perceived water quality, credibility of the scheme, water borne disease

Coefficients[a]

Model	Unstandardized Coefficients		Standardized Coefficients	t	Sig.
	B	Std. Error	Beta		
1 (Constant)	207.242	81.356		2.547	.013
Imp of safe water	35.009	32.585	.087	1.074	.286
Water borne disease	87.505	48.031	.182	1.822	.073
Sum of medicalexpenses	.051	.014	.277	3.538	.001
Credibility of thescheme	118.411	38.645	.276	3.064	.003
Expenditure onarranging water	.535	.077	.512	6.938	.000
Perceived waterquality	59.465	22.984	.214	2.587	.010
Affect of work	68.615	34.013	.172	2.017	.047

(a) Predictors: (Constant), affect on work, Imp of safe water, sum of medical expenses, expenditure on arranging water, perceived water quality, credibility of the scheme, water borne diseases.

Summary Results

Average Willingness to Pay per month : Rs 300.625
(Highest Bid : Rs 800 ; Lowest Bid : 0)

Average Family income : Rs 20,318

Expenses Incurred on arranging alternative water sources : Rs 254

Out of Pocket expense borne on water borne disease (past six months record) : Rs 1138.56

Adjusted R-square : 0.700

Significant Variables: credibility of the scheme, sum of medical expenses, expenditure on arranging alternative water sources, adverse effect on work.

Interpretations

The model explains 70 per cent of the variation in WTP. This means that WTP is explained by the variables namely importance of safe drinking water, occurrence of water-borne diseases, out-of-pocket expenses incurred on water-borne diseases, credibility of the scheme, expenditure incurred on arranging alternative water sources, perceived water quality, adverse affect on work . The remaining 30 per cent is explained by variables which are left unaccounted in the model.

F- value of 24.609 is statistically significant at 7 and 72 df at a 0 per cent level of significance which means to say that the probability of committing Type [2] error is nil.

Equation:

$$WTP = \beta_0 + 0.87X_1 + 0.182X_2 + \mathbf{0.277X_3} + \mathbf{0.276X_4} + \mathbf{0.512X_5} + \mathbf{0.214X_6} + 0.172X_7$$

Where, the letters in the bold format are the ones which were found to be positively and significantly related to the WTP.

Policy Prescriptions

The water tariff currently in use in DJB has not accomplished its principal objective. It is not generating enough revenue to ensure that utilities can recover their financial costs. They are not sending correct economic signals to households, i.e. water is scarce and ought to be treated thus.

The findings have important policy implications: Majority of the Surveyed respondents are already incurring expenditure on arranging drinking water, a proper price mechanism characterised by revenue sufficiency, economic efficiency, equity must be placed.

Revenue Sufficiency

From the supplier's point of view, the main purpose of tariff is often cost recovery. The revenue from water users should be sufficient to pay the operation and maintenance costs of the water utility's operations, repay loans, undertaken to replace and expand the capital stock, provide a return on capital at risk and maintain a cash reserve for unforeseen events. The revenue stream must be adequate and should be relatively stable and not cause cash flow or financing difficulties for the utility.

Economic Efficiency

Economic efficiency requires that price be set to ensure that consumers face the avoidable cost of their decisions. In other words, water prices should signal to the consumers, the financial and other costs that their decision to use water impose to the rest of the society. From an economic efficiency perspective, a tariff should create incentive that ensure, for a given water supply cost, that user obtain the largest possible benefits.

Equity

This means that water tariff treats similar customers equally, and that customers' indifferent situations are not treated same. This would usually be interpreted as requiring users to pay monthly water bills that are proportionate to the costs they impose on the utility by their water use.

The basic solution is to put these resources back in the hands of the people—convert the informal arrangements that had worked before into formal and legally enforceable rules and contracts. Transfer management and ownership of the commons to local communities and user groups. Simultaneously, the state must withdraw from the day-to-day management and focus on the overall policy and legal framework within which people will own and manage their common resources.

The principles recommended are age-old wisdom: user rights, user ownership and responsibility for managing collective services and common resources. The single change of legal ownership and genuine perception of people addresses many of the anomalies of the current state controlled management. Ownership brings responsibility, state subsidies, which largely went to the richer groups, will continually decline, as people will take charge of the provision, full-cost recovery will reduce overuse and misuse, rights-based allocation decisions will become more immune from political calculus.

In fact, the National Water Policy 2002, acknowledges the changed realities and recommended policies are: water should be treated as an economic good instead of a free service and the government should act as a facilitator and not a service provider. However, non-WTP among low income respondents must be taken care of through alternative price mechanism like cross-subsidisation. However, the detail needs to be worked out, which beyond is the purview of current research.

References

Ahmed K. Junaid, B.N. goldar, M. jakariya and Smita Misra, (2002), Willingness to Pay for Arsenic-free, safe Drinking Water in Rural Bangladesh-Methodology and Results.

Bhandari, L, Bajpayee, P (2001) " *Ensuring Access to Water in Urban Households"*, Economic and Political Weekly, September 29.

Central Pollution Control Board [CPCB] (1990), "Status of water Supply & Wastewater Collection, Treatment and Disposal".CUPS/30/1989-90,New Delhi.

CPCB (1998), "Status of water Supply & Wastewater Collection,Treatment and Disposal in Class I cities", New Delhi.

Daga, S (2003) " *Private Supply of Water in Delhi"*, Discussion Paper, Centre for Civil Society,

Dasgupta, P : "*Valuaing Health Damages from Water Pollution in Delhi; A Health Production Function Approach"*, IEG.

Dasgupta, P : "*Household Water Services in Incomplete Markets: Managing Water Stress for Urban Poor"*.

Freeman A.M. (1993), "Non-use values in antural resolve damage assessments" in R.J. Kopp and V.K. Smith (ed) Valuing natural assets, the economics of natural resource damage assessments Washington, D.C.

Garret Hardin (1968), "The Tragedy of the Commins", Science 162, 1243–8.

Gleick, P.H : "*Water in Crisis: A guide to fresh water resources"*.

Hasler,B, Martinsen, L, Thomas, S (2005). " *Economic Assessment of the Value of drinking Water in Denmark"*, Discussion Paper Geneva, Jalan, J, Somanathan, E, Choudhary, S : *Adoption of Safe Drinking Water Practices;Does Awareness of Health Matter*, SANDEE Policy Brief Paper, 4-04,2004.

Jalan, J, Somanathan, E, Choudhary, S(2004 *"Awareness and Demand for Environment Quality Drinking Water in Urban India"*, SANDEE working Paper. 4-04.

Kadekodi, Gopal, K.K.S.R. Murthy and Kireet Kumar (2000), "Water in Kumaon's Ecology, Value and rights," Institute of Economic Growth, Delhi, Centre for Multi-Disciplinary Development Research Dharwal and G.B. Pant Institute of Himalayan Environment and Development, Almora's Gyanodaya Prakashan, Nainital.

Kramer R.A. and J.I. Eisen-Hecht (2001), "Estimating the Economic Value of Water. Quality in the Catawba River Basin", Water Resource Research, Vol 38. pp 1–1.

Kundu, A, Thakur, S : *"Access to Drinking Water in Urban India : An Analysis of Emerging Spatial Pattern in the Context of New System of Governance"*.

Littlefair, K (1998), *Willingness to Pay for Water at the Household Level*, Occasional Paper No.26,.

Managing Water Resources : Communities and Markets, CCS Briefing Paper, March 2003.

Mohanty, M, Gupta, S: "Breaking *the Gridlock in Water Reforms through Water Markets: International Experiences and Implementation Issues for India"*.

Narain, S (2003) *"India's Water Problem needs Local Solutions"*, Centre for Science and Environment, Report.

Oca, G, Bateman, I.J, Tinch, R " *Assessing the Willingness to Pay for Maintained Drinking Water Supplies in Mexico City"*, CSERGE Working Paper ECM 03-11.

Pushpangadan, K *"Drinking Water and Well-being in India; A Data Development Analysis"*.

Roy, J, etal (2004),*"An Economic Analysis of Demand for Water Quality ; A case Study of Kolkata"*, Economic and Political Weekly, January 10,.

Roy, J and Sahu (see P. 456) (2004), *"Willingness to Pay ; A Policy Tool"*, Jadavpur University, Paper Presented at the Conference on Market Development of Water and Waste Technologies, May 28,.

Shah,C.H (2005), *The Economic Analysis of a Drinking Water Project in Andhra Pradesh,* Economic and Political Weekly, January 29.

Shah, P.J(2005), *"Is Privatization and Water Pricing a Sensible Policy",* Discussion Paper at Centre for Civil Society, Published in The Financial Express, March 22.

Tapvong, C. and Jittapatr Kruavan (2003), "Water Quality Improvement: A Contingent Valuation Study of the Chao Pharya River (Bangkok)", EEPSEA reserach report series.

Viscusi, W.K. and Joel Bell (2004), "The Value of Regional Water Quality", Harward Jhon. M. Oblin Centre for Law and Economics, Discussion Paper number 477,06,2004.

Whitehead, J.C, Hoban, T.J (1998) *Willingness to Pay and Drinking water Quality, A Economic Research Project,* California University.

Whittington, D . (2003),*"Municipal Water Pricing and Tariff Design; A Reform Agenda for South Asia"*, Paper Presented at University of North Carolina, January 31.

Chapter 21

Consumption of Public Water for Domestic Use: Micro Level Evidence from North Delhi

Sapna Kaul

Introduction

The management of the limited freshwater resources and their efficient usage has emerged as a vital field of teaching, research and policy making in recent years. This is precisely because of the fact that no other resource, except perhaps oxygen, is so inextricably linked to human survival. Water is an essential element of the basic human requirements and thus acts as an important input for the sustained socio-economic development of every country. Water scarcity affects food production adversely, thus leading to famines; increases the incidence of poverty and thus undermines economic development; while also harming the ecosystem. It can lead to graver consequences like violence for any society, particularly for overpopulated countries, like India. No wonder, therefore, access to safe drinking water is one of the major goals of the Millennium Development Goals (UNDP, 2000).

Most of the world's water, about 97.5 per cent exists as salt water in oceans and seas. Of the world's 2.5 per cent of freshwater, roughly 99 per cent is either trapped in glaciers and ice caps, or held as soil moisture or located in watertables too deep to be accessed. Thus only about one per cent of the world's freshwater supply is readily available for use by human beings and animals for various purposes (Smith and Gross, 1999). Therefore, the world's freshwater supply is finite. Though water is a renewable resource, its availability is variable and limited.

On 2.3 per cent of the world's land, India supports 16 per cent of the world's population, with 4 per cent of world's freshwater resources. The demand for water is increasing rapidly throughout the country due to population growth, urbanisation and economic development. At Independence (1947), India's population was less than 400 million and the per capita availability of water over 5,000 cubic meters per year (Raghuram, 2005). Now, with population growth rate of 2.1 per cent per annum and absolute population over a billion, per capita availability of water has fallen to about 1,900 cubic meters per year (Singh, 2002). Total gross available and utilisable water resources in India are assessed at 2,384 billion cubic meters (BCM) and 1,086 BCM respectively (Planning Commission). The major sources of water in the country are rivers, rainfall, groundwater, dams, etc. (MWR, 2006).

India, with a geographical area of about 329 M. ha, a land of many rivers and mountains, has a varied water supply across different regions and states. Sustainable water management becomes important in the country where population growth is at 2 per cent per annum and water resources are dwindling fast, both quantitatively and qualitatively. India has witnessed rampant population growth since its independence. From a population of 343 million in 1947 to 1,000 million in 2000, the growth rate has been as large as 2.04 per cent per annum. Increasing population directly leads to increase in the demand for water from all sources. Greater dependence on agriculture, where the net sown area has risen from 119 million hectares (1947) to around 150 million hectares now and high cropping intensity, has pushed the demand for water from this sector relatively in the greatest proportion. The water requirement has increased from 12 per cent to about 20 per cent in the above-mentioned time frame. The country's total water requirement by the year 2050 will become 1,422 BCM, which will be much in excess of the total utilisable average water resources of 1,086 BCM. At the national level, it would be a very difficult task to increase the availability of water for use from the 1990 level of approximately 520 BCM, to the desired level of 1,422 BCM by the year 2050 as most of the undeveloped utilisable water resources are concentrated in a few river basins such as the Brahmaputra, Ganga, Godavari, and Mahanadi (MWR, 2006).

Thus the water management scenario in the country clearly indicates three major complex issues, namely: population growth and increasing needs, depleting water resources and disparity in respect of quantity of available water, as well quality of water across different regions and groups of people.

The Situation in Delhi

Growing urbanisation, improvements in living standards and increasing population are some of the contributing factors behind escalation in the demand for water. The population of Delhi is around 15 million at present. The city, at the moment, requires 3,324 million litres of water a day (MLD) while what it gets is about 2,034 MLD. Average water consumption in Delhi is estimated to be 240 litres per capita per day (lpcd), the highest in the country. People spend thousands of rupees in order to acquire drinking water which seems like a luxury in the metropolitan city. Both qualitatively and quantitatively, water situation in Delhi is in the vortex of severe crisis. Delhi receives its water from three sources:

(i) Surface Water—86 per cent of Delhi's total water supply comes from surface water, namely the Yamuna River, which equals 4.6 per cent of this resource through interstate agreements.

(ii) Sub-surface water—Ranney wells and tubewells. This source, which is met through rainfall (approx. 611.8 mm in 27 rainy days), and unutilised rainwater runoff, is 193 MCM (million cubic meters).

(iii) Graduated Resources—It is estimated at 292 MCM. However current withdrawal equals 312 MCM. Salinity and over-exploitation has contributed to depletion and drastically affected the availability of water in different parts of the city. According to a report released by the Central Ground Water Board (CGWB), Delhi's groundwater level has gone down by about eight meters in the last 20 years, at the rate of about a foot a year.

Apart from groundwater, Delhi gets its water from the Ganga Canal, the western Yamuna canal, the Bhakra canal and the Yamuna. The main supplying organisation is the Delhi Jal Board (DJB), which is primarily a government agency. It supplies water in bulk to the New Delhi Municipal Corporation, Delhi Cantonment Board and Municipal Corporation of Delhi for further distribution in their respective areas. There is a number of water treatment plants set by the DJB, wherein clarifloculators and filtration techniques are employed to clean water from rivers and canals. The treated water is supplied to the city through a network of about 9,000 km of water mains or lines (Daga, 2004).The total supply capacity of DJB is around 650 million gallons of water per day (MGD). Of this, around 20–30 per cent is lost due to transit inefficiencies, thus ultimately only around 500 MGD water reaches the consumers (Maria, 2004).

The current demand for water for all users is 363 (800 MGD) liters per capita per day from the developed localities and 100 lpcd from the underdeveloped localities (Daga, 2004). The total use (in Ipcd) comes from: (a).Domestic—172 (b) Industries—47 (c) Floating population—52 (d) Fire Protection—31.

At the macro level, the total demand from all sources surmounts to about 800 MGD. The total public water supply is around 500 MGD, thus the shortfall is about 300 MGD. The story does not end here; there is a huge diversity in the supply of water across different colonies in the city. It has been seen that the supply is tilted towards the high income residing colonies and the unplanned areas suffer from irregular and at times no water supplies (Daga, 2004). The shortfalls are met by the residents either boring their own borewells, purchasing water from private tankers or relying on mineral water.

Water tariff in Delhi consists of fixed access charges and water use charges. Fixed access charges are payable by all registered consumers to meet the cost of access to the network and operation and maintenance costs. Volumetric water charges are based on actual consumption by the different consumer categories. Forty per cent of water supplied by DJB is not billed and 10 per cent is supplied free of cost to JJ colonies and fire department. Since tariff rates do not even cover the O&M costs and only fifty percent of total water supplied by DJB is billed, the total recovery from the consumer categories is a small proportion (24 per cent in 2003–04) of the cost of production (DJB, 2006). The price paid varies according to the residentship and the kind of locality (i.e. planned or unplanned).

Macro issues in water management

A number of studies has been done on the factors that affect the water security in any region. These have indeed enriched our understanding of the current water problem that our country is facing. These studies have also identified macro issues that need to be kept in view while analysing the water situation in the economy. Some of these issues are:

"Water as a Public Good" (Winpenny, 1994)

The essence of a public good is that it is available to all and nobody can be denied access to it. Efficient water allocation does not take place and usage efficiency is low due to the problem of commons Therefore, it is often difficult to recover costs as the problem of free riders exists. Underpricing of water acts as a major economic distortion. The pricing policy should be such that it gives the consumers some incentive to conserve water. Often it is seen that water rates and its consumption bear no relation. Whether water is a *social* , an *economic good* or both has to worked out before framing any national or regional water policy.

Groundwater and its over-exploitation

Due to the ever-growing demand for water, groundwater has been harvested beyond the natural revival process. Water extraction exceeds the annual replenishable rate in various parts of the country. Water rights automatically give the ownership of groundwater to the landowner, which acts as a major deterrent to sustainable water management (Raghuram, 2005). For example, in Delhi 11 per cent of DJB's production is through groundwater extraction. According to Maria (2004), groundwater represents around 45 per cent of the water consumed. Therefore, management of groundwater becomes an important policy issue in the National Capital Region.

Water pollution and health impacts

Water gets contaminated either at source or while passing through pipes which are poorly laid and maintained, or at homes due to poor storage facilities. Major water pollutants are:

(a) Industrial waste/effluents

(b) Domestic waste and municipal effluents(75 per cent)

(c) Agriculture runoff.

Water-borne diseases are a direct consequence of poor quality drinking water. The World Bank and WHO (1993) estimated that 21 per cent of all the communicable diseases in the country were water related. These diseases directly affect the health indicators, like expectancy at birth, infant mortality rate, maternal mortality rates, which show the extent of development in any region.

Utilisation Gaps

The kind of techniques being used in India for water management are not in general resource friendly. India has more than 59 million hectares of net irrigated area (the largest in the world) but the average yield is at a dismal low level of 2.5 tonnes/hectare (Singh, 2002). Canal irrigation efficiency, which is around 30–40 per cent is well below the international (Iyer, 2001). Underutilization occurs in almost all sectors, as already pointed out, under pricing does not allow the consumers to cut on wasteful water consumption.

Inefficiency also exists in the form of ill-maintained pipelines. Every year in Delhi, more than 20 per cent of drinking water is lost due to leakages from the pipes (Daga, 2004).

Financial Constraints

Though our country has reaped great benefits from investment in hydraulic infrastructure but much needs to be accomplished. The sector is suffering from insufficient amount of revenues for maintenance and operations. Staffing levels are ten times the international norms, and most of the scarce funds are wasted in feeding the administrative machinery (Briscoe, 2005).

The India Water Vision 2025 presented at Hague Forum postulated an investment of Rs. 5000 billion in 25 years or Rs. 200 billion per year. The National Commission's estimates of amount needed for spill over effects in this sector is around Rs 70,000 crore in the Tenth Plan. This surely is difficult to come by as the sector is incapable of generating huge revenues by itself (Iyer, 2001).

Demographic effects

India has been experiencing population explosion, urbanisation, industrialisation and large-scale inter-state migration. A direct consequence of the above is an increasing and diversifying demand for water. As mentioned earlier, the population is currently growing at the rate of 2.1 per cent per annum which implies that not only does the demand for domestic water use increase but also for agricultural and industrial use. Due to large-scale industrialisation, MNCs are coming up in large parts of the country. MNCs like PEPSI and COCA-COLA have been accused of depleting ground water at many places, like Kala Dera in Rajasthan and Thane in pune (Corpwatch, 2003), as the bottling plants require more than 30 lakh litres of water every day. Widespread migration adds to the problem of water scarcity in our country.

Public to User-Based market allocation (Privatisation)

Traditional State Management is mostly complied with in almost all countries, but now a few policy makers are supporting the move towards market allocation. Though state is often the best to deal with economies of scale and social welfare, the private parties have a general advantage in collecting information about localised variations in demand and supply. Thus, often market allocation is more efficient when the good being considered is an economic good (Meinzen and Mendoza, 1996).

Water Disputes and River linking

Water being a state subject, often the State Governments are reluctant to share water with others. The problem does not end here; India also faces complex political issues on river sharing with neighbouring countries like Pakistan (Indus River) and Bangladesh (Ganges) (Briscoe, 2005). Water disputes and conflicts lead to a general social menace (Iyer, 2001).

Review of Related Literature

Over the years, a number of studies has been done on water scarcity and the problems arising in different countries because of limited supplies of water. These studies have enriched our understanding and improved our capacity to seek answers to fresh questions or replicate the framework, which has guided such studies. A review of these studies will therefore act as a backdrop and benchmark to the present study. It is with this objective in mind that the present chapter reviews some of these studies. It is not possible to cover all the studies but this does not any in way mean that studies excluded are less significant.

Doria (2006) lists factors that contribute to the use of tap water alternatives in the paper titled as 'Bottled Water versus Tap Water: Understanding Consumer's Preferences'. In the present context, consumption of bottled water has been increasing consistently, even in the countries where tap water is considered to be of excellent quality and regular also. The question arises: why do people choose an alternative that is more expensive and at the same time the quality is no better or no worse than the tap water? Traditionally two factors have been stressed on:

1. dissatisfaction with tap water (organoleptics)
2. health concerns

However, the author argues that the reasons may not be as straightforward as the above two. Rather, there is a complexity of factors that leads to switching of consumption from tap to bottled water. Currently, bottled water market is about $ 22 billion. The identified factors were as follows:

(a) Dissatisfaction with the tap water in terms of its characteristics, taste, odour and sight of collection. Taste is particularly relevant.

(b) Water with a high mineral content (bottled and bore hold water) is preferred over low mineral concentration (distilled and local tap water).

(c) Healthier options and risk averseness play an important role.

(d) Availability of tap water is a crucial determinant. If the supply is untimely, people turn towards private water supply.

(e) Brand cutting and style of bottled water.

(f) Demographic variables—patterns may vary according to region. Country variables like ethnic groups, age, income and occupation they all hold responsible in water drinking and using habits, e.g. in US, bottled water sales are higher amongst African Americans, etc. which typically have low incomes as compare to whites.

(g) Locations—whether it's urban or rural where the supply situations vary at large.

Thus the conclusion remains that the quality of water may be perceived to be good in the case of bottled water but there are other factors working with equal force at the same time that determine such a preference function.

In the paper titled 'Emerging Issues in Water and Health Management", Ford (2006) provides a summary of priority areas for water quality in Developing Countries, Transition Economies and Developed Countries. In developing counties major concern is on water quality and quantity and at the same time there is absence of, or deteriorating infrastructure. It is argued that, in view of the growing population, lack of education in hygiene and sanitation, wastewater collection and treatment should be adopted and more and more low cost and maintenance technologies should be used. In transition economies, the focus shifts to the quality of water, infrastructure, failing technologies, surveillance and chemical content (including pesticides, nitrates, metals, etc.). The developed countries have to build an effort to check the deteriorating infrastructure, chemicals, disinfection by-products, and come up with new molecular techniques. An area of concern for the developing countries is the emergence of new water-borne diseases. While concluding, the author gives a priority list that needs attention in the current context—training modules for small water systems, surveillance in terms of opportunities, physicians and other health education, health linkage and data integration and determination of water borne diseases and development of new technologies like corrosion free piping materials, detection methods testers, etc.

Mazzanti and Montini (2005) carried out an empirical study on 'The Determinants of Residential Water Demand for a Panel of Italian Municipalities'. The area under supervision was Emilia-Romagna, representing around 7 per cent of the Italian population and for the provinces, municipal-level data was used to determine income and price elasticity. The authors argue that estimation of price and income elasticity are necessary steps, both for private and public management of water resources. A Panel dataset of 125 municipalities was considered for a period of over four years (1998–2001). The water demand function was estimated as w=f (P, I, Z), where W is the independent variable (water consumption per capita), P the water price, I the municipal income were the explanatory variables and Z (a vector of socio-economic variables is both a control and additional explanatory variable). Z included municipal specific data, like altitude, share of rural areas, household size, population age, density of commercial enterprises, etc. The econometric model used is based on the fixed effect model. The estimated water demand price elasticity is negative showing values between -0.99 to –1.33, never significantly different from 1. Income elasticity comes out to be positive, though lower than 1. The role of other socio-economic factors is less relevant, except for altitude as an additional explanatory variable. The study concludes that elasticity measures thus obtained give important insights into whether price acts as a demand driver and if price-induced changes are greater than income induced changes. For private utility managers, price elasticity is relevant to predict the marginal revenue impact of price increases aimed at funding water infrastructure. The high price elasticity means that there is a rationale for profit maximisation in a monopolistic market for water. The regions where tariff rates are lower, proper pricing may entail a demand restructuring in favour of true and accurate water demand.

Chaudhary et al. (2004) worked on "Demand for Water in Behrampur; Issues and Tools for Management" and analysed factors that influence the demand for water in Behrampur town of Orissa state in India. They have provided a brief profile of the socio-economic and administrative system of the town. The study processed primary data collected through a random sample of a cross-section of about 100 households canvassing a specifically designed questionnaire during March–June (2002). A demand function using both simple regression and multiple regression models is evaluated. The dependent variable, i.e. water consumption is regressed by incomes, property, family size, number of adults, literate members, educated women per family, etc. The outputs of simple regression models show that no single independent variable sufficiently explains changes in dimensions of water consumption, though in multiple regression analysis there is a positive relation between the explanatory and dependent variables. Thus the results suggest that several socio-economic variables explain demand for water in Behrampur and all the variables are mostly significant. Therefore the policy prescription is that public agencies should aim at doing more investments; property-linked ability assessment and literacy and family planning should also be emphasised.

Daga (2004) in her paper titled 'Private supply of water in Delhi', talks about the demand–supply gap in Delhi and the way this gap is fulfilled by authorities and the public in general. The total supply of water from DJB is around 520 MGD and the total demand for water from all users is close to 830 MGD. Thus there is a shortfall of about 300 MGD water. The author has analysed DJB's supply to different parts of the city which reveals highly unequal supply. She states that about 10 per cent of people have no access to piped water and 30 per cent of the population gets very low supply of water. The shortfalls are met by private sources of water like borewell, tankers, packaged water etc. Daga has also given the cost of such sources. The impact of high demand for water in the city has led to a greater reliance on private sources, which threatens the already stressed groundwater level.

Maria (2004) compiled a paper entitled 'Standing on blue gold? Alternative scenarios of integrated water resource management in Delhi', which deals with water-related problems in Delhi. The unreliable and unequal water services have been critically analysed and the demand – supply gap has been calculated. She has categorised different uses of water in the city and quantified them. The author has questioned as to why there is a shortfall in the city when the availability is the highest in the country. She gives a whole range of alternatives which can be reckoned, like dam building, privatisation, waste water reuse, rain water harvesting and river interlinking.

Renwick and Green et al. (1998) worked on 'Measuring the Price Responsiveness of Residential Water Demand in California's Urban Areas'. There have always been debates on whether demand for water is price responsive or not. Some advocate a price of water that reflects the marginal cost as a means of reducing demand during limited water supply availability. Others argue that since water is primarily a public good, residential demand is price inelastic and thus Demand Side Management policies fail to curb the demand. In this study, potential of price policy as a measure of residential water resource management has been assessed using an econometric model. The econometric model incorporates alternative demand side management policy instruments, endogenous block pricing schedules and a Fourier series to capture the effects of seasonality and climate on residential demand. Agency-level cross-sectional time series data for eight water agencies in California representing around 7.1 million people approximately 20 per cent of the total population had been used for the evaluation. To assess the potential of price policy to reduce demand the two stage least square method is used for the estimation. The magnitude of the income variable implies that a 10 percent increase in income will increase average household monthly water demand by 2.5 per cent. The own price elasticity of demand equalled -.16 which implies that a 10 per cent increase in price will reduce the aggregate quantity demanded by 1.6 per cent. It was seen that the aggregate single family household demand was responsive to price changes. In addition, the non-price demand side management policies were relatively effective residential water resource management tool.

Renwick et al. (1998) in the paper titled 'Demand Side Management Policies for Residential Water Use: Who Bears the Conservation Burden?' assessed the potential for urban demand side management (DSM) policies as a water resource management tool. They analysed the extent to which price and alternative policy instruments (such as use and quantity restrictions and subsidies for water efficient technologies) reduce residential demand and their distributional implications by the type of household. An econometric model of household water demand, which explicitly incorporates technological change, is specified and estimated to identify reduction in aggregate demand attributable to specific policies and their distributional implications. The household water demand model is composed of two basic components:

(a) technology adoption equations

(b) water demand equations.

Using the detailed level household panel level data for two California communities, the results suggest that the ultimate effect of DSM policies in terms of the reduction in aggregate demand and distribution of water savings amongst households classes depends both on policy instruments selected and the composition of aggregate demand. Results indicated that DSM policies were effective in reducing demand. The magnitude of reduction varied significantly with respect to the policy instruments used.

1. Overall, household demand was responsive to price changes. Price responsiveness though was different for various income groups. Price policy will achieve a larger reduction in a lower income community than the higher income community, thus price burden will be more on the shoulders of the lower income groups in the community.
2. In non-price policies, socio-economic and structural characteristics of households influence the outcome. Landscape irrigation and technological change reduce demand more among the households with larger landscaped areas. Again, the same policy instrument will bring about different results in different communities.

A key policy implication of the paper was that planners need to develop an efficient demand side water policy to reduce aggregate demand, given the equity considerations, regional water demand needs to be disaggregated based on selected characteristics of the community. Once the characteristics are understood, policy makers can use the results of household level residential water demand analysis to get to the optimum reduction with emphasis on the distributional aspects.

Batchelor (1975) in the paper titled "Household Technology and Domestic Demand for Water" has attempted to construct a model of household demand with primary focus on technology within which the water is used to satisfy particular needs. The model is used to show how accumulation of consumer durables and advancing household technologies affect the level of demand over time and simultaneously cause systematic changes in its sensitivity to income and price. He suggests that for accurate growth accounting, such models should be used to make estimates flexible over time otherwise the projected demand and growth rates may be lower. Demand behaviour has been analysed for United Kingdom and the income and price elasticity for demand for water has been calculated for USA. The model revolves around advancements in pure theory of consumer demand where three equations have been used. X = AY, showing how goods in certain vector x are combined with some activity y; z = by showing how a set of characteristics z are extracted from activities U = U (z), where U is the maximum for consumption function. The requirements like bathing, clothes, dish washing, etc. have been taken into account and for these; three determinants have been taken, namely the level of activity at which it is carried on, price of water input to each process and the technology in which it participates. The demand function has been estimated taking into account both dynamic and static factors and it was seen that the wealth effects are statistically significant and positive. The effect of family size is less consistent and the wealth effect says that not only will the consumption be higher in more capital intensive households but is also sensitive to a given change in wealth. The author argues that full effect of economic growth over time (dynamic estimation) cannot simply be described as consumption being a function of wealth since technology is also a function of wealth and as water consumption largely depends on technology the total effect has to be accounted for.

Wong (1972) wrote a paper on 'A Model on Municipal Water Demand: A case study of North-eastern Illinois'. The purposes of the study were to answer several questions related to demand for water.

(a) How do price, income and average summer temperature affect the water demand function of Chicago and her outside communities' overtime?

(b) What impact do price and income have on the municipal water demand function among the various community groups at a point in time?

(c) Is municipal water demand responsive to price changes—if so to what extent?

(d) How much is the income elasticity of municipal water?

(e) How do price and income elasticity compare with some previous studies on urban water demand?

Data was collected for Chicago and her outside communities from different state organisations. The basic statistical model of least squares multiple regressions was used, where average per capita municipal water demand is a function of price per unit, average household income, average summer temperature. The results were as follows:

(a) Over time, both income and average summer temperatures have a significant impact on Chicago's water demand. Price is found to be insignificantly different from zero; the negative sign of elasticity coefficient is consistent with the consumer behaviour theory.

(b) Price has no significant impact in Chicago; it was significant at 5 per cent for outside communities. For income variable no significant relation was found. In the city of Chicago, average summer temperature has the most significant impact on the per capita water demand of outside communities.

(c) Price variable was found to have a significant effect and income was insignificantly different from zero.

(d) Price elasticity ranged from -0.26 to -0.82, the income elasticity varied from .48 to 1.93.Price elasticity for Chicago in time series analysis was -0.02 and that for outside community was in the order of -0.28

(e) In general, the results lend support to the results of several other earlier studies conducted and thus it is evident that households outside Chicago on the whole were more responsive to price changes than those in Chicago.

Thus it seems desirable that if the urban water consumption of Chicago was to be reduced, the price policy of the city was to be reformed.

Key issues that emerge from the above review of literature are:

- Factors determining the demand for water - determinants of residential, industrial and agriculture demand for water;
- Appropriate water price policy;
- Quality of public water, private supplies of water and water markets;
- Consumers' preferences for water, elasticity of demand for water for different uses, water issues and health management;
- Integrated water resource management and the alternatives available, demand side water management policies;
- Water related issues in developing and developed countries;
- Water related stress in villages and urban cities, water conflicts;
- Household technology and domestic water demand.

Water management entails management of both the supply of and demand for water from all sources. It is argued that since supply of water cannot be increased beyond a limit, demand management is the need of the hour. To properly understand the demand for water, more studies should be undertaken at the micro level for different uses of water, since demand is mostly case sensitive, depending on a number of factors. Uniform demand management policies cannot be implemented simply because the socio–economic environment, policies and attitudinal factors that affect consumption of water, vary at large across different places. Against this background, the present paper attempts to study the factors that determine the consumption of public water for domestic use in some parts of North Delhi. What may be applicable to the sample considered may not necessarily be applicable to other parts of the city.

In this study, I have considered factors like household size, number of males, females, adults in the household, number of literate members, total monthly household income, expenditure on public water, plot area of house, number of rooms, number of vehicles owned, garden area, technology (number of washing machines, dish washers, etc. perceived quality of water, cost of installation of tubewells and borewells, quantity and price of bottled water, juices, cold drinks, etc. have been considered. The significance of all such factors was checked in explaining the variation in consumption of public water for domestic use.

Objectives

Against the preceeding backdrop, the specific objectives of the present study are:

1. Factors affecting the consumption of public water for Domestic use in North Delhi.
2. Significant factors across different income groups explaining the variation amongst quantity of public water consumed.
3. Policy Suggestions for Delhi Jal Baord (DJB).

Hypothesis

In consonance with the above objectives, the following hypothesis was tested:

Null Hypothesis: "Consumption of Public water for Domestic Use is not affected by Family size, Female literacy, Income and Cost Incurred on Water".

Alternative Hypothesis: "Consumption of Public water for Domestic Use is affected by Family size, Female literacy, Income and Cost Incurred on Water".

Methodology and Data Sources

Data source

Since we attempted to study the factors effecting public water consumption for domestic use at the household level, data to be used therefore had to be specific, relating to the needs of the set objectives and hypothesis. Since the specific data was not available through secondary sources, primary data was generated to cater to the problem. In North Delhi, there are a number of residential colonies. Of the available number of colonies, three were selected at random by lottery method. The colonies were—Dr Mukerjee Nagar, Model Town and Burari village.

Estimation of regression equations required data on various variables that was collected through a cross-sectional survey. In the three mentioned colonies, 75 households were surveyed *randomly* using a closed-ended schedule that yielded information on the chosen variables that were spanned from the review of literature.

From each household, information was collected on following variables:

- Family size, number of adults, number of children, number of women, number of literate women, number of working females;

- Income of the household, assets of the household—like plot area, vehicles, etc.
- Sources of water—DJB, tubewells, private water supply, natural spring water;
- Price paid for water per month to DJB and to private sources of water.;
- Quantity consumed per month;
- Fixed cost of installing tubewells at the time of installation, depth of tubewell installed;
- Quality perception of water—whether it has good taste, odor and colour;
- Number of private water bottles, cold drinks consumed per month;
- Presence of water using technology i.e. whether they owned washing machines, dish washers, water harvesting techniques;

Methodology

To assess the impact of different factors on the consumption of water for domestic use, the linear multiple regression analysis has been employed in the present context. In this analysis following were the dependent and explanatory variables:

- Dependent variables: Quantity of Public water consumed per month by each household (water from Delhi Jal Board).
- Explanatory variables (per household): Household income, family size, plot area, literate females, number of adults, quantity of soft drinks consumed, expenditure on water, cost of cold drinks, garden area and water using technology.
- The regression equation is

  $$Q = \beta + \beta_0 y + \beta_1 E + \beta_2 Lf + \beta_3 f + u$$

Here Q is the average quantity of water consumed in the past months per households, y is household income, E is the expenditure on water consumed, Lf is the number of literate females per household and f is the family size. u is the error term. ß is constant, $ß_0$ is the partial coefficient of regression of y on Q Similarly β_1, β_2 and β_3 are partial coefficients of regression of E, Lf and f on quantity consumed per month respectively.

The following steps were observed to check the significant factors:

- Correlation amongst explanatory variables. Variables that were highly correlated were considered for analysis.
- Only significant variables were taken into consideration at the later stage. Level of significance was taken to be 5 per cent.
- For testing the hypothesis, t tests and ANOVA were used.

Results, Analysis and Inferences

Sample profile (overall)

- Sample size—75 households, average family size—4.40
- Total population—325 (average family size * total households)
- Population under different income groups—As shown below, LIG accounted for around 24 per cent of total sample population and MIG and HIG were 50.66 per cent and 25.33 per cent respectively. Since the objective of comparative analysis was added at a later stage, random sampling was resorted to. Otherwise, purposive sampling would have been done, giving equal weights to the three income groups.

- Average Household income was Rs. 22,633.33 per month. Since two sample areas are relatively well off, the average household is high. Though, if we notice these figures at the various income groups (in the next topic), average income varies at large.
- Literate females as a percentage of total females were 67.7 per cent. The above argument holds here as well, this figure may not a true representative of total sampled population. Working females as a percentage of total females were 69 per cent. This is synonymous with the literacy rate above.
- Female literacy rate (as a per cent of total population) was 24.6 per cent.
- Average size of the plot area was 117.2 yards; 73.3 percent had no gardens in their house premises.
- Water-using technology—In the sample considered, 82 per cent owned washing machines. A large number of storage tanks signifies that water needs to be stored as public water is largely irregular. People also resort to water purifiers since quality of public water is not considered to be as good as private supply. No one used rainwater harvesting in the sample areas.
- Eighty-four per cent of the total households paid monthly bills to DJB and sixteen per cent who did not pay any bills were from the low income group.
- Average quantity of public water consumed per month 33.09 Kilo Liter
- Average cost incurred on public water per month Rs. 149.80
- In low-income groups, not all households had public water connections. Of the total households surveyed in this group, only around 33 households had access

to public water taps. The rest 67 had no metered water connections. These households either dug their own tubewells or collected water from the leaking pipelines of DJB. This reflects the severity of problem of water supply to unplanned areas of Delhi where all households do not have any access to direct public water supply

- On the regularity front, 42.7 per cent rated water as regular and 57.3 per cent as irregular. This shows that water is not available throughout the day.

Results

First correlation was checked between all the related factors to avoid the problem of multicollinearity that makes precise estimation difficult as the t ratios become insignificant whereas R square may show high values.

Correlation Results

Correlation between total family size (as shown in Table 2.11), number of adults and number of females was seen and it was found to be significant at 1 per cent level. Thus only family size was retained for further analysis.

Similarly, correlation among household income, vehicles and plot area was checked and it was seen that the three variables were significant at 1 per cent level. Thus only one of them i.e. household income was taken to show a complete picture.

Lastly, correlation between the number of literate females and working females was seen and again it was found to be significant. Only the number of literate females was taken into consideration for further analysis.

Regression Results

Overall results

For the main results, the SPSS package was used to estimate

regression equation with dependent variable as quantity of water used by household per month and explanatory variables as:

- Family size
- Number of literate females in a household
- Total household income
- Monthly expenditure on water
- Number of bottled water used monthly
- Number of cold drinks per household consumed daily
- Cost of installation of tubewells and borewells
- Average price paid for cold drinks
- Number of rooms in the house
- Water-using technology was assessed by taking
  - Number of washing machines
  - Number of dish washers
  - Number of water storage tanks

Model Summary

Model	R	R Square	Adjusted R Square	Std. Error of the Estimate
1	.988(a)	.975	.974	4.581

(a) Predictors: (Constant), Lit fem, fam size, main income, expenditure

ANOVA (b)

Model		Sum of Squares	df	Mean Square	F	Sig.
1	Regression	58087.275	4	14521.819	691.952	.000(a)
	Residual	1469.072	70	20.987		
	Total	59556.347	74			

(a) Predictors: (Constant), Lit fem, fam size, main income, expenditure

(b) Dependent Variable: quantity

Coefficients (a)

	Unstandardised Coefficients		Standardised Coefficients	t	Sig.
	B	Std. Error	Beta		
(Constant)	3.503	1.958		1.789	.078
expenditure	.166	.007	.977	25.187	.000
fam size	3.095	.503	.186	6.153	.000
income	.000	.000	–.116	–3.424	.001
Lit fem	–2.864	.949	–.081	–3.018	.004
a Dependent Variable: quantity					

Observations

- All the variables that were not significant were dropped, like water-using technology, number of adults, price of cold drinks, number of females, etc .and only the significant variables were analysed for precise estimation.
- It was seen that variables like: number of bottled water used, monthly number of cold drinks per household consumed daily, cost of installation of tube wells and borewells, average price paid for cold drinks, number of rooms in the house, water-using technology was assessed by taking the number of washing machines, number of dish washers, number of water storage tanks could not significantly define variability in the dependent variable. Thus they were not taken as the explanatory variables and only those significant were taken into consideration.
- The beta coefficients of the variables used in the regression variables were tested using the ANOVA and t tests.
- Expenditure and family size were significant at 0 per cent level, income at 1 per cent and literate females at 4 per cent level of significance.

- ANOVA showed significant f values; clearly suggesting that simultaneously also the coefficients can not be zero.
- Thus the null hypothesis is rejected in favour of the alternative hypothesis.

Inferences:

- Family size has a direct impact on water consumption as more is the number of people in a household, more would be the consumption of water for domestic use.
- Similarly, water quantity also depends on the expenditure done on water-supplying techniques, maintenance etc.
- The impact of household income on public water consumption is quite unique. In my study, it was seen that as there is a move from lower income to higher income group, the consumption of public water actually declines.
- There is a negative relation between household income and consumption of public water.

Reasons

- The intuitive reason behind this is that, as income increases, the purchasing power to buy bottled and distilled water increases. There is a common perception amongst HIG that public water quality is not better than bottled water.
- Another justification is that, as the area surveyed is from a metropolitan city, more and more literate women go out for work. Since public water supply is irregular, it needs to be stored. There is an opportunity cost involved in collecting water. If the number of working women is higher, it means that

less time they have for water collection and storage. Thus it is convenient to go in for bottled water and private supply. The number of literate women and number of working women are highly correlated. Thus, as women become more literate they prefer going out for work and have less number of hours left for water collection.

Results at Dissaggregated Income Levels

It is important to analyse the factors that significantly affect the consumption of public water for domestic use at disaggregated income levels, i.e. low-income group, high income group and middle-income group. In the following least squares models, regression equations for the three income groups have been estimated and the significance of the factors was tested. At the final stage, only significant factors were retained for inferences.

HIG

Model	R	R Square	Adjusted R Square	Std. Error of the Estimate
1	.993(a)	.985	.984	5.252

Model	Unstandardized Coefficients		Standardized Coefficients	t	Sig.
	B	Std. Error	Beta		
Constant	–.841	2.267		–.371	.715
Cost Incurred	.188	.006	.993	33.593	.000

MIG

Model	R	R Square	Adjusted R Square	Std. Error of the Estimate
1	.984(a)	.968	.966	1.612

Model	Unstandardized Coefficients		Standardized Coefficients	t	Sig.
	B	Std. Error	Beta		
Constant	5.156	.941		5.480	.000
Family size	1.188	.273	.164	4.357	.000
Cost incurred	.133	.006	.877	23.252	.000

LIG

Model	R	R Square	Adjusted R Square	Std. Error of the Estimate
1	.898(a)	.807	.795	2.823

Model	Unstandardized Coefficients		Standardized Coefficients	t	Sig.
	B	Std. Error	Beta		
Constant	5.290	1.665		3.177	.006
Family size	3.406	.416	.898	8.182	.000

Observations

- As the foregoing regression clearly show, in HIG cost incurred on water is significant at 0 per cent level of significance. Inclusion of other variables distorts the results. The R square value comes out to be small as well as the p value is large. Thus only cost incurred figures are treated for final analysis.
- For MIG, family size and expenditure done on water significantly explain the variation at 0 per cent level. That means they are highly significant.
- For LIG, only family size is significant in explaining the variation in water consumed. It is significant at 0 per cent level.

Inferences

- In the case of HIG, only cost incurred explains significant variation in the quantity of public water consumed, i.e. as the cost incurred on water supply increases, their consumption per say goes up. For HIG, the variation is only cost, incurred which means since this is a homogenous group with almost the same level of incomes, literacy level and family size only the expenditure done on water supply defines the variability in water consumption.

- Similarly, for MIG, family size and cost incurred are the only two factors that are variable enough in explaining the variation in water consumed. In the case of LIG, family size is the only independent variable that explains total variability.

Implications for Delhi Jal Board

Major concerns

1. Disparity in availability of water across different sections of the society.
2. Quality of water.
3. Water supply to LIG and groundwater degradation.

Disparity in water supply over time and space:

- In my sample, 16 per cent of households did not have public water connections at all and LIG accounted for most of it. 67 per cent of households in LIG did not have access to public water supply.
- 67 per cent from the LIG did not have any tap connections. They resorted to leakages from the DJB pipelines and tubewells etc.
- On the regularity front, 57.3 per cent of the respondents rated it as irregular.
- Among the LIG, more households rated water supply as irregular as compared to the HIG and MIG groups, clearly showing a bias against the unplanned and low income colonies. In HIG, a fairly large proportion marked it irregular that acts as a major

force for them to shift to private and bottled water supply.

- If we analyse the perceived quality of water through the ratings of the different households, then the following observations are recorded:
  - If we see the ratings in terms of good water quality, then there is a pattern that reflects that as we move from MIG to HIG, less number of households rate water qualities as good. This means that households in the HIG perceive public water as inferior. In case of LIG, 44.4 per cent rate quality as good since they do not have many options since private supply is expensive.
- Another aspect that is important in the case of water quality is the incidence of water-related diseases. In HIG and LIG, a negligible percentage of respondents said that they suffered regularly from water-related diseases. In LIG it was seen, only 16.7 per cent said that they never suffered from such diseases whereas a whopping 61 per cent rated it as often. This shows that quality of water going to LIG is not good in the sample considered.

Public water supply and groundwater degradation!

- Excess demand for water, Inadequate public water supply (force people to dig tubewells) and low costs of setting tubewells – are a few causes of watertable depletion.
- In the present study, LIG accounted for most tubewells (25 per cent rely on groundwater).
- Tube wells are low cost (average fixed cost of Rs. 3,800 for LIG).
- Private supply is not an alternative for LIG since its expensive.

- As there is a move from LIG to HIG, reliance on groundwater declines. (Groundwater qualitatively is not suitable for drinking and washing purposes).
- MIG (16 per cent - tubewells) and HIG (6 per cent - tube wells) depend more on private water source (as they have the purchasing power) like bottled water and tanks because of quality considerations, thus indicating a close linkage between inadequate public water supply and groundwater degradation in LIG areas.

Conclusions and Suggestions

The present study attempted to examine the significant factors that affect the consumption of public water for domestic use, at both aggregate and disaggregated income levels. Supply of public water in the selected colonies was assessed and comparisons were made in terms of quantity and quality of water supplied. Some of the major conclusions of the study are:

- Factors like household income, literate females, expenditure on public water and family size were the most significant factors explaining variations in consumption of public water for domestic use at the aggregate level. However, at the disaggregated levels, only some of the above-mentioned variables could explain variation in consumption of water.
- There is a huge disparity in water supply across different areas (planned vs. unauthorised) i.e. public water supply is tilted towards planned and high-income groups. Water being a public good, should be available to all, irrespective of income or standards of living. Since it enters the productivity function of all human beings directly, for overall development it should be the priority of the state to provide it equally and in right quantity to all its

citizens. Thus, DJB has to make available water to all the sections of the society. Any other problem area was timely availability. Tap water in a lot of areas is available only twice a day. The frequency should increase. There is a lot of scope and need to improve the quality of water, keeping in view availability of working population in their homes.

- International norms of rust-free pipelines, water-purifying technologies need to be adopted. Water contents need to be scientifically tested and recounted.
- Water availability in LIG should be assessed as it has large social and environmental impacts in terms of groundwater degradation and health hazards.

Bibliography

Ministry of Water Resources (MWR), 2006, "Annual Reports" available at *http://wrmin.nic.in/*

Batchelor R.A. (1975), Household Technology and the Domestic Demand for Water, Land Economics-51 (Aug.).

Briscoe John (2005)-'Executive summary on India's water economy', World Bank Report on India's Water Economy, 5 October.

Chaudhary Amita, Sahu Chandra Nirmal and Nanda Subasini (2004), 'Demand for Water in Behrampur: Issues and tools for Management'. *Journal of Environment and Ecological Management*, July edition.

Daga Shivani (2004) 'Private Supply of Water in Delhi'. Centre for Civil Soceity, New Delhi.

Doria, M. N. (2006) 'Bottled Water Versus Tap Water', *Journal of Water and Health*. February 4.

Ford, Tim (2006) 'Emerging Issues in Water and Health Research', *Journal of Water and Health.*

Iyer Ramaswamy R. (2001) 'Water: Charting a course for future', *Economic and Political Weekly*, March 31.

Kaur Jasleen (2005)-"Water Poverty Index; an Interstate Comparison of India". Department of Economics, Jamia Millia Islamia, Delhi., Mimeographed.

Krishna Rekha, Bhadwal Suruchi, Javed Akram, Singhal Shaleen & S. Sreekesh (2003)- "Water Stress in Indian Villages". EPW, September 13

Lee Elle J & Schwab Kellogg J (2005) – "Deficiencies in Drinking Water Distribution Systems in Developing Countries". Journal of Water and Health 3.2.

Lakshamanan P, Pouchepparadjou A & Sendhil R (2006) - "Impact of Watershed Development on Crop Productivity and Income Distribution – An Economic Analysis". Ashwatta – volume 6, issue 1.

Mazzanti Massimiliano and Montini Anna (2005) 'The Determinants of Residential Water Demand Empirical Evidence for a Panel of Italian Municipalities', *http;/ideas.repec.org/p/femwpg/2005.27.html*

Maria Augustin (2004) 'Standing on Blue Gold: Alternative Scenarios of Integrated Water Resource Management in Delhi', Communication at the India International Centre, Delhi, February 6.

Maria Saleth (1996) 'Water Institutions in India: Economics, Law and Policy, Critical Gaps in Water Economy', IEG, Delhi.

Meinzen Ruth and Mendoza Meyra (1996) 'Alternative Water Allocation Mechanisms, Indian and International experiences', EPW March 30.

Pimentel David et al (1997)-"Water Resources: Agriculture, the Environment and Society, an Assessment of Status of Water Resources". American Institute of Biological Sciences.

Raghuram T. L. (2005) 'Water Quality: Governance and Strategy Options for India', IIM Kolkata.

Renwick, Mary, Green Richard and Chester McCorkle (1998) 'Measuring the Price Responsiveness of Residential Water Demand in California's Urban Areas', Journal of Environmental Economics and Management 40, 37–55.

Renwick Mary, Archibald Sandra O (1998) 'Demand Side Management Policies for Residential Water Use: Who Bears the Conservation Burden', Land Economics, volume 74.

Shah Amita (2001) 'Water Scarcity Induced Migration: Can Watershed Projects Help', EPW, March 30.

Shah Tushaar and Ballabh Vishwa (1997)-"Water markets in North Bihar, Six village study in Muzzaffarpur District". EPW December 27.

Singh, Anil Kumar (2002) 'Emerging Trends and Issues in Water Management for Agricultural Production'. Water Technology Centre, India Agriculture Research Institute, New Delhi.

Smith, Paul J. and Gross Charles S. (1999), 'Water and Conflicts in Asia', Asia–Pacific Centre for Security Studies, Hawaii, September 17.

National Commission for Integrated Water Resources Development Plan (1999)-"Report of the Working Group on the Perspective of Water Requirements". Ministry of Water Resources, Government of India, Delhi.

Planning Commission Report (2003) 'Future Water Crisis in India'. *http://planningcommission.nic.in/*

UNDP (2000) available at *http://www.un.org/millennium/declaration/ares5520.htm.*

Winpenny, James (1994) 'Managing Water as an Economic Resource', The Water Problem, Routledge, London

Wong S T (1972) 'A Model on Municipal Water Demand: A Case Study of North Eastern Illinois', Land Economics, volume – 48.

World Bank and WHO (1993), available at *http://www.who.int/mediacentre/multimedia/2002/ind_sanitation.*

CORPWATCH – *www.corpwatchindia.org,* June 10, 2003.

UNESCO (2000), World day report on "Water Use in the World: Present Situations and Future Needs". *www.unesco.org*

Millennium Development Goals, United Nations Development Programs. *www.un.org/milleniumgoals*

Delhi Jal Board, (DJB 2006) *www.delhijalboard.nic.in*

Central Ground Water Board (CGWB), Ministry of Water Resources, Government of India. *http://cgwb.gov.in/*

Central Pollution Control Board (CPCB), Ministry of Environment & Forests, Government of India. *http://www.cpcb.nic.in/*

Chapter 22

A Black Spot on God's Own Country: Hazards of Sand Mining and Lessons from the State's Environmental Movements

Suhaib Ismail Mohammed

Introduction

Sand is one of the most important items needed in any kind of construction work. River sand is an essential raw material for building construction and an alternative is yet to be found out. Owing to the exorbitant utilisation of the river sand for construction activities, its availability has dwindled, leading to its shortage. Above all, its over-exploitation continues to destroy the environment of the river basins. Severe shortage of sand compels the dealers to embark on any method, and environmental consequences are ignored. Deterioration of environmental factors combined with the shortage of river sand have compelled miners to look for alternative sources.

Rock-crushed sand has been in use in many other Indian states. The physiographical and geological formations in such areas, particularly in the Deccan plateau region, provide plenty of rock raw materials for the sand production. The environment of Deccan regions appears relatively less affected in the short run due to the vastness of the region and the peculiar nature of the availability of rock formations. However, geological formations and the physiographic features in Kerala are not conducive to copy the same methodology here. The main rock formations in Kerala are primarily in the Western Ghats on the east and in a few traverses. These hill ranges are one of the determining factors of regions environment. The present trend in exploiting the hill ranges for sand production is definitely going to have serious consequences. Those who advocate rock sand as the alternative source for construction work have never pointed out the source of rock. If they are bent upon going ahead with the idea, definitely our hills will disappear in near future, and will affect the regular monsoons Jose and Gopal, 2003).

Ecological Features

Out of the 300 'Rare, Endangered and Threatened' species in the Western Ghats, 159 are in Kerala; of which 70 are herbs, 23 climbers, 8 epiphytes, 15 shrubs and 43 trees. The state is endowed with diverse types of ecosystems, each supporting unique assemblage of biological communities, with an impressive array of species and genetic diversity. The state has 95 per cent of the flowering plants and 90 per cent of the vertebrate fauna of the Western Ghats. Catchment areas and coastal areas are ecologically sensitive areas in Kerala. (Department of Forest & Wildlife, Government of Kerala).

Sand mining changes the physical characteristics of the river basin, disturbs the closely linked flora and fauna and alters the hydrology and soil structure. It is very acute in the Bharathpuzha, which was one of the longest rivers of the state. Soil erosion accounts for 87 per cent of the total degraded land in India. Non-adoption of proper crop rotation and extension of cultivation to lands of highly ecologically sensitive areas are some of the important

Table 22.1: Details of River Sand Mining from Various River Basins of Greater Kochi Region

River Basin	River Length (Km)	Drainage Area (km^2) reserve	Estimated Sand mining	Panchayats involved in ($10^6 m^3$/year)	Volume of sand extracted (M^3/year)	sustainable yield ($10^6 m^3$)	No.of labourers
1	2	3	4	5	6	7	8
Periyar	244	5398	26.7	15	3.11	50708	7000
Pamba	176	2235	10.62	14	0.40	17883	995
Muvattupuzha	121	1554	7.56	12	0.33	41827	1147
Manimala	90	847	3.18	10	0.47	14200	1300
Chalakkudi	130	1704	4.91	5	0.15	7810	1200
Meenachil	78	1272	3.12	-	-	-	-
Achankoil	128	1484	5.8	-	-	-	-

Source: Quoted by NEERI, 2002.

reasons contributing to soil erosion. Various mining/ quarrying activities include sand mining from the river channels and over bank areas, soil quarrying, hard rock mining, laterite sand mining, brick clay mining, etc. Around an 10 lakh hectare area is prone to moderate to severe soil erosion in the state (Kerala State Pollution Control Board).The process of environmental deterioration begins with the resource extraction activity and results in land and water degradation, expressed in terms of increased soil erosion, changes in landscapes, hydrological imbalances, water pollution, etc. Unlike the rivers of our neighbouring states, the rivers of Kerala are too small in size and resource capability. Due to the non-availability of sand in the river channels, now attention has been shifted to the river banks and flood plains. This results in slumping/caving and widening of river banks, decline in the watertable and saline water intrusion into the wells of adjacent areas. Flood plain mining in the paddy fields after removing the top clay layer has a direct bearing on the local hydrological regime and groundwater movements.

A study by the Centre for Earth Science Studies (CESS),Trivandrum, is an institution of the Kerala State Council for Science Technology & Environment, which also carry out studies in river basin evaluation, groundwater management, coastal erosion and other special problems. Their study reveals that about 168 MT of sand is being mined per day from 8 panchayats in the Neyyar river basin. Neyyar river basin is at the foot of western Ghats. There are about 320 sand-mining locations in the Periyar river basin in the central Kerala, with a total of 39 local bodies involved in sand mining, to mine about 8,372 MT (2093 truck loads)in a day. At this present rate of mining, the entire sand resource in the Periyar River would be exhausted within a decade, apart from other environmental implications. Many of the fish fauna as well as other aquatic organisms in rivers of the southern part of the state are under threat due to loss of habitat, feeding and breeding grounds, decline of food, aggravated salt water ingression, etc. resulting out of indiscriminate sand mining. The soil quarrying, mainly done for construction sector and paddy land reclamation, especially at the Neyyar dam basin, has degraded the soil with loss of nutrients, micro organisms.

Clay mining in paddy fields is taking place at an alarming rate, mainly for brick making. Apart from the land degradation problems, clay mining also leads to lowering of groundwater level. Hard rock mining is also another environmental issue in many parts of the state. River Chaliyar flows offer luxurious greenery and fertility to the valleys of Nilgiris, Wynad and Nilambur. After traveling 169 km, through the districts of Malappuram and Kozikkode, it finally joins the Arabian Sea at Chaliam (Beypore) near Feroke. The total catchment area of this river is 2,923 sq. km. Of this, 388 sq. km. are in Tamil Nadu. Hundreds of small water currents, streams and rivulets join the river at various points and it serves as a as a life stream for the Malappuram and Kozikkode districts of Kerala. The river occupies the fourth position among the 44 rivers of the state as far as water resources are concerned. River Chaliyar is an example of the eternal relationship of human civilisation and rivers. Today, five lakhs people living in 18 panchayats in Mallapuram district and an equal number from the Kozikkode district depend upon Chaliyar and its tributaries for their domestic, agricultural and industrial needs. Hundreds of small and large water projects representing various Panchayats, Municipalities and Kozikkode Corporation directly depend upon Chaliyar for their need of water. The Kavanakkallu Regulator-cum-Bridge, is a new mega-project, hinde the natural flow of the river. The natural flow of the river acts as a natural barrier and protect the river water from mixing with salt/ saline water. Besides, the project is likely to breed environmental imbalance. The sand mafia is busy in selling even the last grain of sand taken from the river. The authorised and unauthorised ways of sand mining are active throughout Chaliyar. Due to this, the water level in the ponds and wells is coming down. Thus, sand mining is directly affecting the people who depend upon Chaliyar their drinking water. The Chaliyar river banks have gone in to private hands. They have degraded the banks of the river under the pretext of cultivation and construction of granite walls. The same is the fate of its tributaries. The authorities are conniving with encroachers by allowing them to make farms and even playgrounds. Mining of sand became a topic of interest recently among the planners and entrepreneurs of our country because of the increasing demand and spiralling costs of river sand for construction purposes. Although the land of Kerala is endowed with many rivers and splendid rainfall, river channels once occupied by a sandy bed for its major course have been turned into dry, rock-bedded channels due to extensive illegal mining activity. Now-a-days, sand borrowing from most of our rivers has been temporarily stopped which created and impasse and panic on the construction front. This problem has now been solved to some extent by substituting river sand with crusher sand. Further, adulterated sand, which is a mixture of sands from estuary and coastal land, also is freely used for the construction.

Sand mining is illegal. In Kerala, not only has it been placed under interim stay by the state High Court, it also goes against the provisions of the 1991 Coastal Regulation Zone (CRZ) notification, the only specialised attempt to lay down norms for coastal development in India. Frequent offenders include seaside hotels, factories, housing projects and export-oriented shrimp farms. But, given that the government itself is a violator of coastal norms, for example, the original CRZ notification was tweaked to allow a nuclear power plant to come up on Kalpakam beach—it is little surprise that to previous page offences go unchecked.

Areas under the 1991 CRZ notification include coastal lands influenced by tidal action up to 500 m from the high-tide line, as well as land between low-tide and high-tide lines. To regulate permissible activities, the CRZ was in turn divided into four zones of varying ecological sensitivity, with provisions for stringent protections in the core zone. Powerful vested interests have lobbied over the years to water it down so that today the protective spirit of the original notification is all but lost. Approved state-level coastal zone implementation plans do not exist. For the first time, the traditional rights of India's six million-strong fishing population also received some measure of legal recognition.

The Coastal Regulation Zone (CRZ) notification has, however, been flouted countless times. To date, it has undergone at least 21 amendments. As per CRZ, fifty per cent of the royalty will be handed over by the district administration to the concerned local body; the rest will be deposited in the District Collector's river management fund. Meanwhile, transporters have been asked to carry permits signed by the local body secretary and countersigned by the block officer in efforts to stem corruption. As much of the mining activities are done for the public, a clash of interest arises from the general public when the question of public intervention comes in. Protests and agitation on the one hand, and using mined sand for the construction of one's own house. Sadly, even the Left front organisations follow the same suit. But sincere protests/vigil and practical solutions are practised by organisations like Solidarity Youth Movement and a few NGOs.

An expert committee constituted by the state government in 1997 to study problems in Bharatapuzha, recommended that inter-state transport of sand be banned, and that studies be conducted on the sand levels at each site, and permission for mining given only if sufficient sand was available. Check- posts should be established at each *kadavu* (area where sand is mined) and a watchman appointed. The technology available to minimise the use of sand for construction activities needs to be availed of. The government should implement the regulations efficiently, and the use of river sand for purposes (such as land filling) other than construction activities be prohibited.

Often proposed as an alternative to beach mining, is offshore sand mining. Extensive (and expensive) studies must be conducted before any offshore mining can be attempted. Offshore sand banks, coral reefs and sea-grass beds diffuse the energy of storm waves; if large quantities of sand are removed from offshore sand banks in locations where replenishment would not occur, serious coastal damage would result in the event of a major storm. A complex relationship exists amongst sand banks, coral reefs, marine biota, current circulation, waves and swells patterns. Offshore mining does present a possibility with an initially high price tag. As a as Kerala is concerned, the question of offshore sand mining has to be addressed impartially. A Few companies have approached the Government for getting a licence for mining of offshore sands from certain locations which are identified as potential zones. The idea has been temporarily suspended due to resistance from environmentalists.

Uncontrolled quarrying may end up in irreversible changes in the environment. This has imposed mining of sand several folds higher than the natural replenishments With demarcation of high-tide and low-tide lines still incomplete, implementation is only a farce. Considering the extent of damage to the Bharathapuzha, and the high demand for sand, it is difficult to imagine that there is a solution to the problem. Experts who have studied the river say that the damage is reversible, but restoration could take years, perhaps decades. Various studies recommended a people's movement to protect the 'natural heritage', adding that the river management fund needed to be put to better use.

A Glimmer of Hope

Despite the government's lopsided approach to sand mining and its insensitivity to its environmental consequences, Kerala's history of environmental movements provides a glimmer of hope. Organisations like Solidarity Youth Movement should learn and draw lessons from the most successful green movements of the state. It will be in the fitness of the theme of the present paper to present a resume of the two most successful green protest movements of the state, i.e. Silent Valley Struggle and Plachimada anti-cola struggle.

The Silent Valley Struggle For Environment

The Silent Valley is a shola forest (shola is the thick vegetation found at the base of the valleys in the western hills of south India). It is a storehouse of rare and valuable plants and animals. Scientists have found several varieties of wild pepper here. Cardamom grows wild, as do black gram, rice and bean. Several plants have medicinal value.The evergreen forest tree—*Hydnocarpus*, whose seeds contain the oil used to treat leprosy, and the herb-like shrub *Rauvolfia serpentina* used for treating high blood pressure, are two examples. Rare fauna include the Lion-tailed

Macaque, Great Indian Hornbill, and Nilgiri Tahr. This remote valley triggered off one of the fiercest environmental disputes the country has known. It all began with an innocent enough proposal put forward by the Kerala State Electricity Board (KSEB) to build a dam across the Kuntipuzha river to create a reservoir in Silent Valley, and then use the impounded water to generate electricity. The dam, proposed to be 130 metres high, was to be built between two hillocks in a gorge through which the river runs. The then Prime Minister Indira Gandhi, who had shown more interest in environmental matters than other political leaders before her, appointed a Committee in 1980 to look into whether the Western Ghats as a whole were in danger of damage. The Committee pointed out that Silent Valley was the last remaining example of flora and fauna that had evolved to the fullest possible extent in a tropical rainforest, and was an ecosystem undisturbed by human interference. Were the dam to be built, this unique ecosystem might be irretrievably lost. The Committee argued that the 120 megawatts of power which the dam would help generate were not so important for Kerala. Around this time, a group of schoolteachers and others, who constituted the Kerala Sastra Sahithya Parishat (KSSP), became involved in the Silent Valley. For many years, this NGO had been writing and publishing science texts in the local language, Malayalam, so that they could be read by a wider group of people, and science could become a tool for social revolution. Gradually, KSSP began to get involved in environmental issues, of which Silent Valley was the most controversial.

KSSP's arguments, based on academic knowledge and reasoning, did not appeal to everybody. The idea of conserving a virgin forest for its flora and fauna seemed irrelevant to the people living near the proposed dam and the inhabitants of the northern districts of Kerala, who were suffering due to an acute shortage of electricity, and unemployment. KSSP conducted its own studies, which showed why the Valley should not be destroyed and how the same benefits could be obtained in other ways. The support of Dr. Salim Ali, eminent ornithologist and personal acquaintance of Mrs. Indira Gandhi, was particularly valuable. KSSP was able to rouse public opinion on the need to save Silent Valley. It had science groups in many villages and its journals and newsletters reached out to a large number of people. It had an impressive membership of 7,000 thinking people. KSSP collected signatures of around 600 teachers, prominent citizens and students, and sent a memorandum to the Kerala Government. It organised street plays, exhibitions, public debates, and a unique *jatha*—a marathon march—covering 300 to 400 villages along a 6,000-km route. Leading intellectuals of Kerala, who were members of KSSP, wrote letters and articles in the press and participated in the public debates. This was the first time in India because of KSSP's involvement with students, that teenagers came out on the streets to protest against the destruction of the environment and across Kerala students proclaimed their opposition to the project. In place of numerous committees that had mushroomed in the state and in cities (Chennai and Mumbai), a local committee was formed in Mannarkad (the region to which Silent Valley belongs) to save the project. Their argument focused on the high unemployment rate and the absence of industries in the state, leading to the high rates of out-migration. Industrialisation, for which electric power was a must, was seen as the only solution.

Finally, on the basis of an examination of the costs and benefits, as well as the Prime Minister's support, the scales tilted against the project. The Government of India advised

Kerala state to abandon the project. Silent Valley was declared a National Park in 1985—which meant that no project could come up in the area. But no battle in the field of conservation is ever final, and there is no guarantee that the Valley will remain silent for all time to come. There have been faint rumblings in recent years to revive the Silent Valley project.

Plachimada Anti-Cola Struggle

The struggle against the Plachimada Plant of Coca Cola was launched on 22 April 2002 with continuous picketing/ dharna, mainly by the Adivasis. The Hindustan Coca-Cola Beverages Pvt. Ltd established this unit in 1998-99 in a 40-acre plot (previously multi-cropped paddy lands) at Plachimada of Perumatty Panchayat in Chittur Taluk of Palakkad District, Kerala, about 5 km west of the Tamil Nadu border of Pollachi in Coimbatore District and 30 km east of Palakkad on the Palakkad – Meenakshipuram – Pollachi Road. About 70 permanent workers and 150–250 casual labourers are employed in the factory .

From this plant, every day 85 lorry/truckloads of beverage products, each load containing 550–600 cases and each case containing 24 bottles with 300 ml capacity leave the factory premises. In its 17-hectare land, more than 65 borewells were sunk to extract the groundwater for the production of Coke and Maza. Every day, 15 million litres of ground water is extracted by Coca Cola free of cost. Bottle washing involves the using of chemicals and the effluents are released without treatment contaminating the groundwater, creating a water crisis for the tribals, Dalits, and the farmers. The police is giving protection to Coca Cola white the people waging the struggle, especially women and children, are behind bars. Farmers, tribals the media, NGOs and a few other voluntary organizations are leading the struggle in Plachimada.

Coca Cola had approached the High Court, seeking protection to their property. Ever since the launch of the struggle, there has been a huge police contingent stationed at the Coca Cola plant to protect it. The struggle committee filed a counter appeal and the High Court ordered the police to protect the protesters. Chief Minister V.S Achuthananthan imposed a ban on Coca Cola soon after he assumed power in 2006. This was indeed a bold step. But the poor defence of the ban in the court helped Coca Cola to remain in the state.

Following the 2004 meeting of the World Social Forum (WSF) in Mumbai, the small village of Plachimada in Palakkad district of Kerala witnessed a rare union of radical unionists, 'Green' politicians, environmentalists and social activists from around the world, expressing solidarity with villagers who have been hopelessly engaged in a battle for their right to water with the multinationals for over two years. The struggle has also exposed all the political parties, from the right to the left, that behind the veil of pious rhetoric, have all turned to be petty brokers for the Coca Cola Company. But then, this is not an isolated phenomenon but a widespread phenomenon all over the country with globalisation and opening up of predatory capital, that destroys people and livelihood resources.

References

Biswas, Nilanjana (2007): "Discontent At Rising Tide", Tehelka, 15th September, New Delhi.

Jose, Geo and Gopal, Sanjay Mangala (2003):"NAPM joins Protest against Sea-sand Mining in Kerala: Organizations Call For People's Right Over Resources". Press Release, National Alliance For People's Movements, 29, January

Krishnakumar, V (31st, January, 2004) "Fighting The Cola Giants In Kerala", Frontline Magazine.

Warhate, S.R, M.K.N. Yenkie, M.D. Chaudhari and W.K. Pokale "Impacts of Mining Activities on Water and Soil", Journal of Environmental Science & Engineering. vol. 48, No. 2, P. 81-90, April 2006

ELECTRONIC SOURCES

Governemnt of Kerala Forests & Wildlife Department Website

Kerala State Pollution Control Board, Environment (Protection) Rules, 1986.

Stuligross, D. (1999). "The Political Economy of Enviornmental Regulation in India. *Pacific Affairs"*, 72 (3), 392

Towards Sustainability: Learning from the Past, Innovating for the Future, A Publication for World Summit on Sustainable Development, Ministry of Environment and Forests, GOI and Centre for Environment Education, 2002

Chandran, M.P, *www.riverindia.in/index/chaliyar.*